Technologies To Maintain Biological Diversity

Technologies To Maintain Biological Diversity

OFFICE OF TECHNOLOGY ASSESSMENT TASK FORCE

Susan Shen, Project Director and Analyst
Edward F. MacDonald, Analyst
Michael S. Strauss, Analyst
James L. Chamberlain
David Notter
Robert Prescott-Allen
Brucc Ross-Sheriff
Linda Starke
Lisa Olson

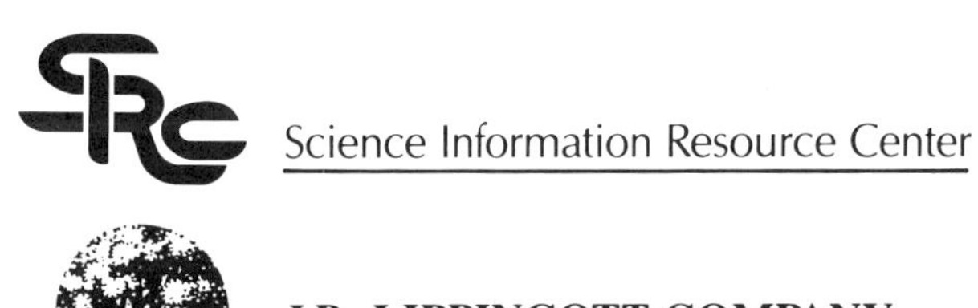

London Mexico City New York St. Louis São Paulo Sydney

Authorized Hardbound Edition, 1988

Science Information Resource Center
East Washington Square
Philadelphia, Pennsylvania 19105

Publisher's Note:

This permanent edition includes the complete text of the Office of Technology Assessment Special Report entitled *Technologies to Maintain Biological Diversity,* which represents only part of a major ongoing research effort in biological technology. Under the direction of the OTA Project Staff, and with the help of a distinguished Advisory Panel and several Workshop Groups, the Authors prepared many new research reports and reviewed and distilled the mass of background and workshop papers and reports.

Science Information Resource Center is a cooperative venture of J.B. Lippincott Company and Hemisphere Publishing Corporation, subsidiaries of Harper & Row, Publishers, Inc., New York.

Library of Congress Cataloging-in-Publication Data
Technologies to maintain biological diversity.
Includes bibliographies and index.
1. Biological diversity conservation.
2. Biotechnology. I. United States. Congress.
Office of Technology Assessment.
[QH75.T366 1988] 333.95'16 87-26419
ISBN 0-397-53005-6

Foreword

The reduction of the Earth's biological diversity has emerged as a public policy issue in the last several years. Growing awareness of this planetary problem has prompted increased study of the subject and has led to calls to increase public and private initiatives to address the problem. This interest in maintaining biological diversity has created a common ground for a variety of groups concerned with implications of a reduction or ultimate loss of the planet's genetic, species, or ecosystem diversity.

One major concern is that loss of plant, animal, and microbial resources may impair future options to develop new important products and processes in agriculture, medicine, and industry. Concerns also exist that loss of diversity undermines the potential of populations and species to respond or adapt to changing environmental conditions. Because humans ultimately depend on environmental support functions, special caution should be taken to ensure that diversity losses do not disrupt these functions. Finally, esthetic and ethical motivation to avoid the irreversible loss of unique life forms has played an increasingly major role in promoting public and private programs to conserve particular species or habitats.

The broad implications of loss of biological diversity are also reflected in the different concerns and jurisdictions of congressional committees that requested or supported this study. Requestors include the House Committee on Science, Space, and Technology; Senate Committee on Foreign Relations; and Senate Committee on Agriculture, Nutrition, and Forestry. The House Committee on Foreign Affairs; House Committee on Agriculture; and House Committee on Merchant Marine and Fisheries endorsed the requested study.

The task presented to OTA by these committees was to clarify for Congress the nature of the problems of reduction of the Earth's biological diversity and to set forth a range of policy options available to Congress to respond to various concerns. The principal aim of this report is to identify and assess the technological and institutional opportunities and constraints to maintaining biological diversity in the United States and worldwide. Two background papers (*Grassroots Conservation of Biological Diversity in the United States* and *Maintaining Biological Diversity in the United States: Data Considerations*) and a staff paper (*The Role of U.S. Development Assistance in Maintaining Biological Diversity in Developing Countries*) were also prepared in conjunction with this study.

OTA is grateful for the valuable assistance of the study's advisory panel, workgroups, workshop participants, authors of background papers, and the many other reviewers from the public and private sectors who provided advice and information throughout the course of this assessment. As with all OTA studies, the content of this report is the sole responsibility of OTA.

John H. Gibbons

JOHN H. GIBBONS
Director

Advisory Panel
Technologies To Maintain Biological Diversity

Kenneth Dahlberg, *Chair*
Department of Political Science
Western Michigan University

Stephen Brush
International Agricultural Development
University of California, Davis

Peter Carlson
Director
Crop Genetics International

Rita Colwell
Office of the Vice President for Academic Affairs
University of Maryland, Adelphi

Raymond Dasmann
Department of Environmental Studies
University of California, Santa Cruz

Clarence Dias
President
International Center for Law in Development

Donald Duvick
Senior Vice President of Research
Pioneer Hi-Bred International

David Ehrenfeld
Cook College
Rutgers University

Major Goodman
Department of Crop Science
North Carolina State University

Grenville Lucas
The Herbarium
Kew Royal Botanic Gardens

Richard Norgaard
Department of Agricultural and Resource Economics
University of California, Berkeley

Robert Prescott-Allen
Partner
PADATA, Inc.

Paul Risser
Vice President for Research
University of New Mexico

Oliver Ryder
Research Department
San Diego Zoo

Michael Soulé
Adjunct Professor
School of Natural Resources
University of Michigan

John Sullivan
Vice President of Production
American Breeders Service

NOTE: OTA appreciates and is grateful for the valuable assistance and thoughtful critiques provided by the advisory panel members. The panel does not, however, necessarily approve, disapprove, or endorse this report. OTA assumes full responsibility for the report and the accuracy of its contents.

OTA Project Staff
Technologies To Maintain Biological Diversity

Roger C. Herdman, *Assistant Director, OTA*
Health and Life Sciences Division

Walter E. Parham, *Food and Renewable Resources Program Manager*

Analytical Staff

Susan Shen, *Project Director and Analyst*

Edward F. MacDonald, *Analyst*

Michael S. Strauss, *Analyst*

Catherine Carlson,[1] *Research Assistant*

Robert Grossmann,[2] *Analyst*

Allen Ruby,[3] *Research Assistant*

Contractors

James L. Chamberlain

David Notter

Robert Prescott-Allen

Bruce Ross-Sheriff

Linda Starke[4] and Lisa Olson,[5] *Editors*

Administrative Staff

Patricia Durana,[6] Beckie Erickson,[7] and Sally Shaforth,[8] *Administrative Assistants*

Nellie Hammond, *Secretary*

Carolyn Swann, *Secretary*

---

[1]Through January 1986.
[2]Through August 1985.
[3]Summer 1985.
[4]Through August 1986.
[5]After August 1986.
[6]Through July 1985.
[7]Through October 1986.
[8]From Dec. 15, 1986.

CONTENTS

Technologies To Maintain Biological Diversity

Chapter 1
Summary and Options for Congress

CONTENTS

Tables

Box

Chapter 1
Summary and Options for Congress

Most biological diversity survives without human interventions to maintain it. But as natural areas become progressively modified by human activities, maintaining a diversity of ecosytems, species, and genes will increasingly depend on intervention by applying specific technologies. A spectrum of technologies are available to support maintenance of biological diversity (defined in box 1-A).

Box 1-A.—What Is Biological Diversity?

Biological diversity refers to the variety and variability among living organisms and the ecological complexes in which they occur. Diversity can be defined as the number of different items and their relative frequency. For biological diversity, these items are organized at many levels, ranging from complete ecosystems to the chemical structures that are the molecular basis of heredity. Thus, the term encompasses different ecosystems, species, genes, and their relative abundance.

How does diversity vary within ecosystem, species, and genetic levels? For example,

- **Ecosystem diversity**: A landscape interspersed with croplands, grasslands, and woodlands has more diversity than a landscape with most of the woodlands converted to grasslands and croplands.
- **Species diversity**: A rangeland with 100 species of annual and perennial grasses and shrubs has more diversity than the same rangeland after heavy grazing has eliminated or greatly reduced the frequency of the perennial grass species.
- **Genetic diversity**: Economically useful crops are developed from wild plants by selecting valuable inheritable characteristics. Thus, many wild ancestor plants contain genes not found in today's crop plants. An environment that includes both the domestic varieties of a crop (such as corn) and the crop's wild ancestors has more diversity than an environment with wild ancestors eliminated to make way for domestic crops.

Concerns over the loss of biological diversity to date have been defined almost exclusively in terms of species extinction. Although extinction is perhaps the most dramatic aspect of the problem, it is by no means the whole problem. The consequence is a distorted definition of the problem, which fails to account for many of the interests concerned and may misdirect how concerns should be addressed.

THE PROBLEM

The Earth's biological diversity is being reduced at a rate that is likely to increase over the next several decades. This loss of diversity —measured at the ecosystem, species, and genetic levels—is occurring in most regions of the world, although it is most pronounced in particular areas, most notably in the tropics. The principal cause is the increasing conversion of natural ecosystems to human-modified landscapes. Such alterations can provide considerable benefits when the land's capability to sustain development is preserved, but compelling evidence indicates that rapid and unintended reductions in biological diversity are undermining society's capability to respond to future opportunities and needs. Most scientists and conservationists working in this area believe that the problem has reached crisis proportions, although a few remain skeptical and maintain that this level of concern is based on exaggerated or insufficient data.

The abundance and complexity of ecosystems, species, and genetic types have defied complete inventory and thus the direct assessment of changes. As a result, **an accurate estimate of the rate of loss is not currently possible.** Determining the number of species that exist,[1] for example, is a major obstacle in assessing the rate of species extinction. But use of biological principles and data on land use conversions have allowed **biologists to deduce that the rate of loss is greater than the rate at which new species evolve.**

Reduced diversity may have serious consequences for civilization. It may eliminate options to use untapped resources for agricultural, industrial, and medicinal development. Crop genetic resources have accounted for about 50 percent of productivity increases and for annual contributions of about $1 billion to U.S. agriculture. For instance, two species of wild green tomatoes discovered in an isolated area of the Peruvian highlands in the early 1960s have contributed genes for marked increase in fruit pigmentation and soluble-solids content currently worth nearly $5 million per year to the tomato-processing industry. Future gains will depend on use of genetic diversity.

Loss of plant species could mean loss of billions of dollars in potential plant-derived pharmaceutical products. About 25 percent of the number of prescription drugs in the United States are derived from plants. In 1980, their total market value was $8 billion. Loss of tropical rain forests, which harbor an extraordinary diversity of species, and loss of deserts, which harbor genetically diverse vegetation, are of particular concern. Consequences to humans of loss of potential medicines have impacts that go beyond economic benefits. Alkaloids from the rosy periwinkle flower (*Catharantus roseus*), a tropical plant, for example, are used in the successful treatment of several forms of cancer, including Hodgkin's disease and childhood leukemia.

Photo credit: H. Iltis

A foggy, moss- and epiphyte-enshrouded tropical forest in Ecuador is about to be cleared for local agriculture, a main cause of loss of diversity.

Although research in biotechnology suggests exciting prospects, scientists will continue to rely on genetic resources crafted by nature. For example, new methods of manipulating genetic material enable the isolation and extraction of a desired gene from one plant or organism and its insertion into another. Nature provides the basic materials; science enables the merging of desired properties into new forms or combinations. Loss of diversity, therefore, may undermine societies' realization of the technology's potential.

Another threatening aspect of diversity loss is the disruption of environmental regulatory

[1]Approximately 1.7 million species have been identified. Millions more, however, have yet to be discovered. Recent research indicates that species of tropical insects alone could number 30 million.

Photo credit: H. Iltis

A dense stand of *Zea diploperennis* in Sierra de Manantalan, Jalisco, Mexico. This ancient wild relative of corn could be worth billions of dollars to corn growers around the world because of its resistance to seven major diseases plaguing domesticated corn.

functions that depend on the complex interactions of ecosystems and the species that support them.

Diverse wetlands provide productive and protective processes of economic benefit. Millions of waterfowl and other birds of economic value depend on North American wetlands for breeding, feeding, migrating, and overwintering. About two-thirds of the major U.S. commercial fish, crustacean, and mollusk species depend on estuaries and salt marshes for spawning and nursery habitat. Wetlands temporarily store flood waters, reducing flow rates and protecting people and property downstream from flood and storm damage. One U.S. Army Corps of Engineers' estimate places the present value of the Charles River wetlands (in Massachusetts) for its role in controlling floods at $17 million per year. Although placing dollar values on such ecosystem services is problematic and reflects rough approximations, the magnitude of the economic benefit stresses the importance of these often overlooked values.

Humans also value diversity for reasons other than the utility it provides. Esthetic motivations have played important parts in promoting initiatives to maintain diversity. Cultural factors, as reflected in the way Americans identify with the bald eagle or the American bison or how plants and animals form a fundamental aspect of human artistic expression, illustrate these values.

Forces that contribute to the worldwide loss of diversity are varied and complex. Historically, concern for diversity loss focused on commercial exploitation of threatened or endangered species. Increasingly, however, attention has been focused more on indirect threats that are nonselective and more fundamental and sweeping in scope.

Most losses of diversity are unintended consequences of human activity. Air and water pollution, for example, can cause diversity loss far from the pollution's source. The decline of several fish species in Scandinavia and the near extinction of a salmon species in Canada have been attributed to acidification of lakes due to acid rain. Population growth in itself may not be intrinsically threatening to biological diversity. A populous country like Japan is an example of how a high standard of living, appropriate government policies, and a predominantly urbanized population can limit the rate of ecosystem disruption. However, when population growth is compounded by poverty, a negative impact is characteristic. In many tropical developing countries, high population growth and the practice of shifting agriculture employed by peasant farmers are considered the greatest threats to diversity.

This report assesses the potential of diversity-maintenance technologies and the institutions developing and applying these technologies. But **maintaining biological diversity will depend on more than applying technologies.** Technologies do not exist to re-create the vast majority of ecosystems, species, and genes that are being lost, and there is little hope that such technologies will be developed in the foreseeable future. Therefore, efforts to maintain diversity must also address the socioeconomic, political, and cultural factors involved.

INTERVENTIONS TO MAINTAIN BIOLOGICAL DIVERSITY

There are two general approaches to maintaining biological diversity. It may be maintained where it is found naturally (onsite), or it may be removed from the site and kept elsewhere (offsite). Onsite maintenance can focus on a particular species or population or, alternatively, on an entire ecosystem. Offsite maintenance can focus on organisms preserved as germplasm or on organisms preserved as living collections. Table 1-1 lists examples of management systems. These management systems have somewhat different objectives, but all four are necessary components of an overall strategy to conserve diversity. Conservation objectives can be enhanced by investing in any combination of the four systems and by improving links to take advantage of their potential complementariness. The objectives of the management systems are summarized in table 1-2.

Maintaining plants, animals, and microbes onsite—in their natural environments—is the most effective way to conserve a broad range of diversity. Onsite technologies primarily focus on establishing an area to protect ecosystems or species and on regulating species harvest. To date, the guidelines for optimal design of protected areas are limited, however.

Offsite maintenance technologies are applied to conserving a small but often critical part of

Table 1-1.—Examples of Management Systems To Maintain Biological Diversity

Onsite		Offsite	
Ecosystem maintenance	Species management	Living collections	Germplasm storage
National parks	Agroecosystems	Zoological parks	Seed and pollen banks
Research natural areas	Wildlife refuges	Botanic gardens	Semen, ova, and embryo banks
Marine sanctuaries	*In-situ* genebanks	Field collections	Microbial culture collections
Resource development planning	Game parks and reserves	Captive breeding programs	Tissue culture collections

Increasing human intervention ⟶

⟵ Increasing emphasis on natural processes

SOURCE: Office of Technology Assessment, 1986.

Table 1-2.—Management Systems and Conservation Objectives

Onsite		Offsite	
Ecosystem maintenance	Species maintenance	Living collections	Germplasm storage
Maintain:	Maintain:	Maintain:	Maintain:
• a reservoir or "library" of genetic resources	• genetic interaction between semidomesticated species and wild relatives	• breeding material that cannot be stored in genebanks	• convenient source of germplasm for breeding programs
• evolutionary potential	• wild populations for sustainable exploitation	• field research and development on new varieties and breeds	• collections of germplasm from uncertain or threatened sources
• functioning of various ecological processes	• viable populations of threatened species	• offsite cultivation and propagation	• reference or type collections as standard for research and patenting purposes
• vast majority of known and unknown species	• species that provide important indirect benefits (for pollination or pest control)	• captive breeding stock of populations threatened in the wild	• access to germplasm from wide geographic areas
• representatives of unique natural ecosystems	• "keystone" species with important ecosystem support or regulating function	• ready access to wild species for research, education, and display	• genetic materials from critically endangered species

SOURCE: Office of Technology Assessment, 1986.

the total diversity. Technologies for plants include seed storage, *in vitro* culture, and living collections. Most animals are commonly maintained offsite as captive populations. Cryogenic storage of seeds, *in vitro* cultures, semen, or embryos can improve the efficiency of offsite maintenance and reduce costs.

Microbial diversity is important for both its beneficial and its harmful effects. That is, microbes (e.g., bacteria and viruses) can present serious threats to human health. By the same token, these organisms are used in a range of beneficial activities, such as for developing vaccines or for treating wastes.

Scientists are hampered in their storage, use, and study of microbial diversity by their inability to isolate most micro-organisms. For those micro-organisms that have been isolated and identified, offsite maintenance is the most cost-effective technique.

Links between onsite and offsite management systems are important to increasing the efficiency and effectiveness of efforts to maintain diversity. Some technologies developed for domesticated species, for instance, can be adapted to wild species. Embryo transfer technologies developed for livestock are now being adapted for endangered wild animals.

Photo credit: B. Dresser

Staff of the Cincinnati Wildlife Research Federation working on an anesthetized white rhinoceros in an effort to develop embryo transfer techniques. Proper equipment must be developed for collection of embryos from the more common white rhino before it is tested on the endangered black rhino. The white rhino would then be used as a surrogate for embryos from black rhinos.

Determining the efficacy and appropriateness of technologies depends on biological, sociopolitical, and economic factors. Taken together, these factors influence decisionmaking and must be considered in defining objectives for maintaining diversity and for identifying strategies to meet these objectives.

Biological considerations are central to the objectives and choice of systems. Only some diversity is threatened; therefore, the task of maintaining it can focus on elements that need special attention. A biologically unique species (one that is the only representative of an entire genus or family) or a species with high esthetic appeal may be the focus of intensive conservation management.

Political factors also influence conservation objectives and management systems. Commitments of government resources, policies, and programs determine the focus of attention, and to a large extent, such commitments reflect public interests and support. For example, a disproportionate share of U.S. resources is devoted to programs for a few of the many endangered species. Substantial sums have been spent in 11th-hour efforts to save the California condor and the black-footed ferret, while other endangered organisms such as invertebrate species receive little attention.

The applicability of management systems also depends on economic factors. Costs of alternative management systems and the value of resources to be conserved may be relatively clear in the case of genetic resources. For example, the benefits of plant breeding programs compared with the cost of seed maintenance justify germplasm storage technologies. However, cost-benefit analysis is more difficult when benefits are diffuse and accrue over a long period. And onsite maintenance programs compete with other interests for land, personnel, and funds.

Success in maintaining biological diversity depends largely on institutions that develop and apply the various technologies. Within the United States, a variety of laws in addition to public and private programs address various aspects of diversity conservation. But while some aspects of diversity are covered, other aspects are ignored. Table 1-3 lists major Federal mandates pertinent to diversity maintenance.

Because U.S. interest in biological diversity extends beyond its borders, the United States subscribes to a number of international conservation laws and supports programs through bilateral and multilateral assistance channels. However, many of these programs have too little support to be effective in resolving internationally important problems.

Both domestic and international institutions deal with aspects of diversity. Some focus attention exclusively on maintaining certain agricultural crops, such as wheat, and others focus on certain wild species, such as whales and migratory waterfowl. A shift has occurred in recent years from the traditional species protection approach to a more encompassing ecosystem maintenance approach.

Much of the work important to diversity maintenance is done in isolation and is too disjunct to address the full range of concerns. And some concerns receive little or no attention. For example, the objectives of the USDA's National Plant Germplasm System (NPGS) place primary emphasis on economic plants and little emphasis on non-crop species. Similarly, programs to protect endangered wild species direct attention away from species that are threatened but not listed as endangered. The lack of connections between programs is another institutional constraint. Linkages help define common interests and areas of potential cooperation—important steps in defining areas of redundancy, neglect, and opportunity.

THE ROLE OF CONGRESS

Given the implications and irreversible nature of biological extinction, policymakers must continue to address the problem of diminishing biological diversity. A significant increase in attention and funding in this area seems consistent with U.S. interests, in view of the benefits the United States currently derives from biological diversity and the advances that biotechnology might achieve given a diversity of genetic resources. In addition, enough information exists to define priorities for diversity maintenance and to provide a rationale for taking initiatives now, although further research and critical review of the nature and extent of diversity loss are also warranted.

OTA has identified options available to Congress. These options are discussed under five major issues:

1. strengthening the national commitment,
2. increasing the Nation's ability to maintain biological diversity,
3. enhancing the knowledge base,
4. supporting international initiatives, and
5. addressing loss of biological diversity in developing countries.

For each issue, alternative or complementary options are presented. These range from legislative initiatives to programmatic changes within Federal agencies. Options also define opportunities to cultivate or support private sector initiatives. In a number of areas, however, success will depend on increased or redirected commitments of resources. Table 1-4 provides a summary of policy issues and options.

Strengthen the National Commitment To Maintain Biological Diversity

The national commitment to maintain biological diversity could be strengthened. Despite society's reliance on biological resources for sustenance and economic development, loss of diversity has yet to emerge as a major concern among decisionmakers. About 2 percent of the national budget is spent on natural resources-related programs, which include diversity-conservation programs as one subset.

A number of government and private programs address maintenance of biological diversity, but most programs have objectives too nar-

Table 1-3.—Federal Laws Relating to Biological Diversity Maintenance

Common name	Resource affected	U.S. Code
Onsite diversity mandates:		
Lacey Act of 1900	wild animals	16 U.S.C. 667, 701
Migratory Bird Treaty Act of 1918	wild birds	16 U.S.C. 703 et seq.
Migratory Bird Conservation Act of 1929	wild birds	16 U.S.C. 715 et seq.
Wildlife Restoration Act of 1937 (Pittman-Robertson Act)	wild animals	16 U.S.C. 669 et seq.
Bald Eagle Protection Act of 1940	wild birds	16 U.S.C. 668 et seq.
Whaling Convention Act of 1949	wild animals	16 U.S.C. 916 et seq.
Fish Restoration and Management Act of 1950 (Dingell-Johnson Act)	fisheries	16 U.S.C. 777 et seq.
Anadromous Fish Conservation Act of 1965 (Public Law 89-304)	fisheries	16 U.S.C. 757a-f
Fur Seal Act of 1966 (Public Law 89-702)	wild animals	16 U.S.C. 1151 et seq.
Marine Mammal Protection Act of 1972	wild animals	16 U.S.C. 1361 et seq.
Endangered Species Act of 1973 (Public Law 93-205)	wild plants and animals	7 U.S.C. 136 16 U.S.C. 460, 668, 715, 1362, 1371, 1372, 1402, 1531 et seq.
Magnuson Fishery Conservation and Management Act of 1977 (Public Law 94-532)	fisheries	16 U.S.C. 971, 1362, 1801 et seq.
Whale Conservation and Protection Study Act of 1976 (Public Law 94-532)	wild animals	16 U.S.C. 915 et seq.
Fish and Wildlife Conservation Act of 1980 (Public Law 96-366)	wild animals	16 U.S.C. 2901 et seq.
Salmon and Steelhead Conservation and Enhancement Act of 1980 (Public Law 96-561)	fisheries	16 U.S.C. 1823 et seq.
Fish and Wildlife Coordination Act of 1934	terrestrial/aquatic habitats	16 U.S.C. 694
Fish and Game Sanctuary Act of 1934	sanctuaries	16 U.S.C. 694
Historic Sites, Buildings, and Antiquities Act of 1935	natural landmarks	16 U.S.C. 461-467
Fish and Wildlife Act of 1956	wildlife sanctuaries	15 U.S.C. 713 et seq. 16 U.S.C. 742 et seq.
Wilderness Act of 1964 (Public Law 88-577)	wilderness areas	16 U.S.C. 1131 et seq.
National Wildlife Refuge System Administration Act of 1966 (Public Law 91-135)	refuges	16 U.S.C. 668dd et seq.
Wild and Scenic Rivers Act of 1968 (Public Law 90-542)	river segments	16 U.S.C. 1271-1287
Marine Protection, Research and Sanctuaries Act of 1972 (Public Law 92-532)	coastal areas	16 U.S.C. 1431-1434 33 U.S.C. 1401, 1402, 1411-1421, 1441-1444
Federal Land Policy and Management Act of 1976 (Public Law 94-579)	public domain lands	7 U.S.C. 1010-1012 16 U.S.C. 5, 79, 420, 460, 478, 522, 523, 551, 1339 30 U.S.C. 50, 51, 191 40 U.S.C. 319 43 U.S.C. 315, 661, 664, 665, 687, 869, 931, 934-939, 942-944, 946-959, 961-970, 1701, 1702, 1711- 1722, 1731-1748, 1753, 1761-1771, 1781, 1782
National Forest Management Act of 1976 (Public Law 94-588)	national forest lands	16 U.S.C. 472, 500, 513, 515, 516, 518, 521, 576, 581, 1600, 1601-1614
Public Rangelands Improvement Act of 1978 (Public Law 95-514)	public domain lands	16 U.S.C. 1332, 1333 43 U.S.C. 1739, 1751- 1753, 1901-1908
Offsite diversity mandates:		
Agricultural Marketing Act of 1946 (Research and Marketing Act)	agricultural plants and animals	5 U.S.C. 5315 7 U.S.C. 1006, 1010, 1011, 1924-1927, 1929, 1939-1933, 1941-1943, 1947, 1981, 1983, 1985, 1991, 1992, 2201, 2204, 2212, 2651-2654, 2661-2668 16 U.S.C. 590, 1001-1005 42 U.S.C. 3122
Endangered Species Act of 1973 (Public Law 93-205)	wild plants and animals	7 U.S.C. 136 16 U.S.C. 460, 668, 715, 1362, 1371, 1372, 1402, 1531 et seq.
Forest and Rangeland Renewable Resources Research Act of 1978 (Public Law 95-307)	tree germplasm	16 U.S.C. 1641-1647

NOTE: Laws enacted prior to 1957 are cited by Chapter and not Public Law number.

SOURCE: Office of Technology Assessment, 1986.

Table 1-4.—Summary of Policy Issues for Congressional Action Related to Biological Diversity Maintenance

Issue	Finding	Options
Strengthen national commitment	*Adopt a comprehensive approach to maintaining biological diversity*	Establish a national biological diversity act Prepare a national conservation strategy Amend appropriate legislation of Federal agencies
	Increase public awareness of biological diversity issues	Establish a national conservation education act Amend the International Security and Development Cooperation Act
Increase ability to maintain biological diversity	*Improve research, technology development and application*	Direct National Science Foundation to establish a conservation biology program Establish a national endowment for biological diversity
	Fill gaps and inadequacies in existing programs	Provide sufficient funding for existing maintenance programs Improve link between onsite and offsite programs Establish new programs to fill specific gaps in current efforts
Enhance knowledge base	*Improve data collection, maintenance, and use*	Establish a clearinghouse for biological data Enhance existing natural heritage network of conservation data centers
Support international initiatives	*Provide greater leadership in the international arena*	Increase support of existing international programs Continue oversight hearings of multilateral development banks' activities
	Promote the exchange of genetic resources	Examine U.S. options on international exchange of germplasm Amend the Export Administration Act to affirm U.S. commitment to free exchange of germplasm
Address loss in developing countries	*Amend Foreign Assistance Act*	Adopt broader definition of biological diversity in Foreign Assistance Act
	Enhance capability of the Agency for International Development	Direct AID to adopt strategic approach to diversity conservation Increase AID staffing of personnel with environmental training
	Establish alternative funding sources for biological diversity projects	Create special account for natural resources and the environment Apply more Public Law 480 funds to effort

SOURCE: Office of Technology Assessment, 1987.

rowly defined to address the broad scope of biological diversity concerns. Nor do the ad hoc programs use coordination and cooperation to build a systematic approach to tackle the issue. State and private efforts fill some gaps in Federal programs, but they do not provide a comprehensive national commitment and thus leave many aspects of the problem uncovered.

Federal agencies, for example, coordinate the onsite conservation activities mentioned specifically in Federal species protection laws, such as those under the authority of the Endangered Species Act of 1973 (Public Law 93-205), but no formal institutional mechanism exists for the thousands of plant, animal, and microbial species not listed as threatened or endangered. Mandates for offsite conservation are equally vague about which species they are to consider. For example, the Research and Marketing Act of 1946 is intended to "promote the efficient production and utilization of products of the soil" (7 U.S.C.A. 427), but it is interpreted narrowly by the Agricultural Research Service (ARS) to mean economic plant species and varieties. Thus, little government attention has been given to conserving the multitude of wild plant species offsite. Even less attention is given to offsite conservation of domesticated and wild animals.

FINDING 1: A comprehensive approach is needed to arrest the loss of biological diversity. Significant gaps in existing programs could be identified with such an approach, and the resources of organizations concerned with the issue could be better allocated. Improved coordination could create opportunities to enhance effectiveness and efficiency of Federal, State, and private programs without interfering with achievement of the programs' goals.

The broad scale of the problem of diversity loss necessitates innovative solutions. Various laws and programs of Federal, State, and private organizations already provide the framework for a concerted comprehensive approach. At this time, however, few of these programs state maintenance of biological diversity as an explicit objective. As a result, diversity is given cursory attention in most conservation and resource management programs. Some of them, such as the Endangered Species Program, address diversity more directly but are concerned with only one facet of the problem. Duplication of efforts, conflicts in goals, and gaps in geographic and taxonomic coverage are consequences.

To resolve this institutional problem, a comprehensive approach to maintaining biological diversity is needed. The implication is not that all programs should address the full range of approaches; rather, organizations should view their own programs within the broader context of maintaining diversity and should coordinate their programs with those of other organizations. Programs and organizations would thereby benefit from one another. Gaps could be identified and eventually filled, and duplicate efforts could be reduced. And organizations could improve efficiency by taking the responsibilities for which they are best suited. Moreover, financial support for diversity maintenance could be more effectively distributed. A step in this direction has been taken in recent initiatives, but congressional commitment to such an endeavor is necessary to ensure that efforts will be made to achieve a comprehensive approach to maintaining biological diversity.

Option 1.1: Enact legislation that recognizes the importance of maintaining biological diversity as a national objective.

Current legislation addressing the loss of biological diversity in the United States is largely piecemeal. Although many Federal laws affect conservation of diversity, few refer to it specifically. The National Forest Management Act of 1976 is the only legislation that mandates the conservation of a "diversity of plant and animal communities," but it offers no explicit direction on the meaning and scope of diversity maintenance.

Consequently, existing Federal programs focus on sustaining specific ecosystems, species, or gene pools, or on protecting endangered wildlife. Species protection laws authorize Federal agencies to manage specific animal populations and their habitats. Habitat protection laws authorize the acquisition or designation of habitats under Federal stewardship. Federal laws for offsite maintenance of plants authorize the collection and genetic development of plant species that demonstrate potential economic value.

The Endangered Species Act authorizes protection of species considered threatened or endangered in the United States. However, listing endangered species does not eliminate the problem; efforts are hampered by slow listing procedures, by emphasis on vertebrate animals at the expense of plants and invertebrates, and by concerns about conflicts that endangered status might create.

Congress could pass a National Biological Diversity Act to endorse the importance of the issue and to provide guidance for a comprehensive approach. Such an act could explicitly state maintenance of diversity as a national goal, establish mechanisms for coordinating activities, and set priorities for diversity conservation. A national policy could bring about cooperation among Federal, State, and private efforts, help reduce conflicting activities, and improve efficiency and cost-effectiveness of programs.

To be effective, a new act would require a succinct definition of biological diversity and explicit goals for its maintenance. Otherwise, ambiguities would lead to misinterpretation and confusion. Diversity, for example, could be interpreted broadly when authorities and funding are being sought and narrowly when responsibilities are assigned. Identifying goals is likely to be a long and politically sensitive process. Decisionmakers and the public will have to determine if conserving maximum diversity is the desirable goal. Finally, to be effective, the law must have both public support and adequate resources, or it would simply provide a false reassurance that something is being done.

Option 1.2: Develop a National Conservation Strategy for U.S. biological resources.

Another means of comprehensively addressing diversity maintenance is to develop a National Conservation Strategy (NCS). This strategy could be developed in conjunction with, or in lieu of, a mandate as suggested in the preceding option. The process would initiate coordination of Federal programs. Program administrators could identify measures to reduce overlap and duplication, to minimize jurisdictional problems, and to develop new initiatives.

A national strategy could minimize potential competition, conflict, and duplication among programs in the private and public sectors. In addition, preparation of an NCS would strengthen efforts to promote NCSs in other countries. Some 30 countries (mostly developing countries, but also including Canada and the United Kingdom) have initiated concrete steps to prepare an NCS. U.S. action might reinforce the momentum for NCSs in other countries.

Congress could establish an independent commission to prepare the NCS. Members of the commission could serve part-time and be provided a budget for meetings and administrative support. The commission could include representatives from government, academia, and the private sector. The Public Land Law Review Commission and the National Water Commission are potential models.

In developing a national strategy, such a commission could do the following:

- assess the adequacy of existing programs to conserve biological diversity;
- formulate a national policy on maintenance of biological diversity;
- identify measures required to implement the policy, any obstacles to such measures, and the means to overcome those obstacles;
- determine how biological diversity maintenance relates to other conservation and development interests; and
- include a public consultation and information program to build a consensus on the content of the national conservation strategy.

Another way to prepare a strategy is to tap the resources of an established government agency. An appropriate body could be the Council for Environmental Quality (CEQ), which is part of the Office of the President. Created by the National Environmental Policy Act of 1969, CEQ already prepares annual reports for the President on the state of the environment. In doing so, it uses the services of public and private agencies, organizations, and individuals and hence has the experience and authority to bring together various interest groups and expertise. On the other hand, CEQ, though fully staffed in the 1970s with a range of environmental experts, now has only a small staff of administrators. Coordinating and guiding the substantive development of an NCS is thus beyond the council's current capacity except through use of consultants.

Because the success of an NCS depends on participation of a broad spectrum of interest groups, its preparation could be a daunting prospect. The number, size, and nature of U.S. Government agencies and the different sectors involved could make preparation and implementation of a strategy difficult.

Option 1.3: Amend the legislation of Federal agencies to make maintenance of biological diversity an explicit consideration in their activities.

Yet another means for Congress to encourage a comprehensive approach is to make maintenance of biological diversity an explicit consideration of Federal agencies' activities. A number of Federal programs affecting biological diversity are scattered throughout different agencies, but the lack of coordination results in inefficient and inadequate coverage of the problem.

These amendments could involve the creation of new programs, or they could lead to modified objectives for existing programs. In either case, the amendments should redirect certain policies, consolidate conservation efforts, and provide criteria for settling conflicts. An amendment for Federal land managing agencies, for example, could require that these agencies make diversity conservation a priority in decisions relating to land acquisition, disposal, and exchange.

Such amendments would probably be resisted by individual Federal agencies, which could argue that they are already maintaining diversity and do not need more explicit direction from Congress. In addition, agencies could argue that they could not increase their activities without new appropriations; otherwise, the quality of existing work could be compromised.

Before such amendments are written, a systematic review of all Federal resource legislation will be needed to determine how existing statutory mandates and programs affect the conservation of diversity and how they complement or contradict one another, and to designate which programs are most in need of revision. Such a complex review will take time and money and is likely to be opposed by agencies.

FINDING 2: Because maintenance of biological diversity is a long-term problem, policy changes and management programs must be long lasting to be effective. But, such policies and programs must be understood and accepted by the public, or they will be replaced or overshadowed by shorter term concerns. Conveying the importance of biological diversity requires formulating the issue in terms that are technically correct yet understandable and convincing to the general public. To undertake the initiative will require not only biologists but also social scientists and educators working together.

Diversity loss has not captured public attention for three reasons. First, it is a complex concept to grasp. Rather than attempt to improve understanding of the broad issue, organizations soliciting support have made emotional appeals to save particular appealing species or spectacular habitats. This approach is effective in the short-term, but it keeps the constituency and the scope of the problem narrow. Second, the more pervasive threats to diversity, such as loss of habitat or diminished genetic bases for agricultural crops, are gradual processes rather than dramatic events. Third, most benefits of maintaining diversity are often diffuse, unpriced, and reaped over the long-term, resulting in relatively low economic values being assigned to the goods and services provided. The benefits of diversity, therefore, are not presented concretely and competitively with other issues. Consequently, the public and policymakers generally lack an appreciation of possible consequences of diversity loss.

Notwithstanding these difficulties, environmental quality has been a major public policy concern since the 1970s, and it remains firmly entrenched in the consciousness of the American public. A 1985 Harris poll, for example, indicated that 63 percent of Americans place greater priority on environmental clean-up than on economic growth. And because stewardship of the environment includes maintaining diversity, this predisposition of Americans could be built on to develop support for diversity maintenance programs.

Biological diversity benefits a variety of special interest groups; its potential constituency is enormous but fragmented. It includes, for example, the timber and fishing industries as well as farmers, gardeners, plant breeders, animal breeders, recreational hunters, indigenous peoples, wilderness enthusiasts, tourists, and all those who enjoy nature. The combined interests of all these groups could cultivate a national commitment to maintaining biological diversity, if properly orchestrated.

Option 2.1: Promote public education about biological diversity by establishing a National Conservation Education Act.

Just as sustaining support to enhance environmental quality required public education programs, so too will a concerted national effort to conserve biological diversity require a strong public education effort. A National Conservation Education Act could be patterned after the Environmental Education Act of 1971 (Public Law 91-516), which authorized the U.S. Commissioner of Education to establish education programs that would encourage understanding of environmental policies.[2]

A new act could support programs and curricula that promote, inter alia, the importance of biological diversity to human welfare. A small grants program could support research and pilot public education projects. Funds could be made available to evaluate methods for curricula development, dissemination of curricula, teacher training, ecological study center design, community education, and materials for mass media programs. The act could support interaction among existing State environmental education programs, such as those in Wisconsin and Minnesota, and encourage the establishment of new programs in other States. The Department of Education could provide consulting services to school districts to develop education programs.

An attempt to establish additional environmental education legislation might be opposed because of the trend to reduce the Federal Government's role in education and to rely more on State and private sector initiatives. Therefore, it could be argued that private organizations, such as the Center for Environmental Education, are the appropriate agents to increase public awareness. It could also be argued that Federal agencies are already educating the public about environmental issues and could easily include biological diversity in their programs without new legislation. Besides, new legislation would require additional appropriations, and in a time of budgetary constraints, funding requests for conservation education programs would probably be opposed.

Option 2.2: Amend the International Security and Development Act of 1980 to increase the awareness of the American public about international diversity conservation issues that affect the United States.

Even more difficult than increasing the public's awareness of domestic issues in biological diversity is increasing their awareness of the relevance of diversity loss in other countries. In addition to humanitarian and ethical reasons, maintaining diversity in other countries benefits the United States by sustaining biological resources needed for American agriculture, pharmacology, and biotechnology industries, and by sustaining natural resources necessary for commerce and economic development.

Maintaining biological diversity for security and quality of life enhancement, and the wisdom of incorporating such issues into U.S. foreign assistance efforts, are justification for Congress to promote public awareness of the global nature of the problem.

Mechanisms for educating the public about such international issues are already in place. Specifically, several nongovernmental organizations (NGOs) have international conservation operations. A coalition of these groups actively participated in the U.S. Interagency Task Force on biological diversity that formulated the *U.S. Strategy on the Conservation of Biological Diversity in Developing Countries.* As a group, they have identified public education as a major role for NGOs.

The grassroots approach of NGOs is conducive to heightening public awareness, as illustrated by the support for programs to alleviate famine in Africa. Recognizing the potential of NGOs to stimulate public awareness and discussion of the political, economic, technical, and social factors relating to world hunger and poverty, Congress amended the International Security and Development Cooperation Act of

[2]This act was repealed by Public Law 97-35 in 1981, and the Department of Education has requested no funds for environmental education in its 1987 budget.

1980 with Title III, Section 316, to further the goals of Section 103.[3]

This amendment provides NGOs with Biden-Pell matching grants to support programs that educate U.S. citizens about the links between American progress and progress in developing countries. The Agency for International Development (AID) has used these grants mainly to promote American understanding of the problems faced by farmers in developing countries and how resolution of those problems benefits Americans. Recently, use of the grants has been broadened to include public education on international environmental issues. *Congress could encourage this action by expressing its approval during oversight hearings or by further amending the International Security and Development Cooperation Act specifically to authorize support for education programs on environmental issues, especially on biological diversity.*

Increase the Nation's Ability To Maintain Biological Diversity

The ability to maintain biological diversity depends on the availability of applicable technologies that are useful and affordable and on programs designed to apply these technologies to clearly identified needs. Thus, increasing the Nation's ability to maintain diversity will require an improved system for identifying needs and for developing or adapting technologies and programs to address these needs.

At present, technologies and programs are not sufficient to prevent further erosion of biological resources. The problem of diversity loss has been recognized relatively recently, and scientists have just begun to focus attention on it. Progress is slow partly because basic research is poorly funded, and institutions are not organized to follow-up basic research with synthesis of results, technology development, and technology transfer. The last reason implies a need for goal-oriented research.

All too often, the Nation's current research programs related to biological diversity do not have a goal-oriented approach. Institutional reward systems and prestige factors deter many scientists from engaging in work that translates basic science into practical tools. Several Federal agencies support basic biology and ecology research, but too little support exists for synthesis of the research into technologies.

Improved links between research and management systems, that is, technology transfer, can increase efficiency, effectiveness, and ability for maintaining diversity. For example, understanding how to maintain and propagate wild endangered species has been preceded by efforts to maintain domestic species. Perhaps the most dramatic linkage is embryo transfer technology developed for livestock now being adapted for endangered wildlife. Similarly, plant storage technologies developed for agricultural varieties, such as cryogenics and tissue culture, may be valuable tools for maintaining rare or threatened wild plant species, even if only as backup collections.

FINDING 3: Current technologies are insufficient to prevent further erosion of biological resources. Thus, increasing the Nation's ability to maintain biological diversity will require acceleration of basic research as well as research in development and implementation of resource management technologies.

Most resource management technologies were developed to meet narrow needs. Onsite technologies are generally directed toward a particular population or species, and offsite technologies are generally directed toward organisms of economic importance. This restricted focus of basic research and technology development is not sufficient to meet the broad goal of maintaining diversity, given the number of species involved and the time and funds available.

[3]Sec. 103, entitled "Agriculture, Rural Development and Nutrition," recognizes that the majority of people in developing countries live in rural areas and close to subsistence. It authorizes the President to furnish assistance to alleviate hunger and malnutrition, enhance the capacity of rural people, and to help create productive on- and off-farm employment. Sec. 316 encourages private and voluntary organizations to facilitate widespread public discussion, analysis, and review of the issues of world hunger. It especially calls for increased public awareness of the political, economic, technical, and social factors affecting hunger and poverty.

To accelerate research and application of diversity-conserving technologies, a shift of emphasis is necessary in research funding. Agencies that fund or conduct research (e.g., the National Science Foundation (NSF) and the Agricultural Research Service of the USDA) generally do not focus on applying research to technology development; they usually are oriented toward supporting basic research. For example, research funds are available for descriptive studies of population genetics but not for studies on applications of genetic theory to onsite population management. Scientists are rewarded for research that tests hypotheses relatively quickly and for publication of research results in academic journals. These incentives discourage broad, long-term studies and neglect analyzing research results to develop technology systems.

Another avenue to increasing the ability to maintain diversity is to encourage development and implementation of programs by private organizations. Although many private efforts are not defined in terms of diversity conservation per se, activities to conserve aspects of diversity (i.e., ecosystems, wild species, agricultural crops, and livestock) have had significant impact. These efforts are not likely to replace public or national programs, but they could be an integral part of the Nation's attempt to maintain its biological heritage.

Option 3.1: Direct the National Science Foundation to establish a program for conservation biology.

The field of conservation biology seeks to develop scientific principles and then apply those principles to developing technologies for diversity maintenance. Recently, the development of this discipline has gained momentum through the establishment of study programs at some universities and the formation of a Society of Conservation Biology, with its own professional journal. Nevertheless, conservation biology is only beginning to be recognized by the academic community as a legitimate discipline. No research funds support it explicitly. Therefore, few scientists can afford to conduct innovative conservation biology research.

Current funding for research and technology development in conservation biology is negligible, in large part because NSF considers it to be too applied, while other government agencies consider it to be too theoretical. *Congress could encourage scientists to specialize in conservation biology by establishing within NSF a separate conservation biology research program that would support the broad spectrum of basic and applied research directed at developing and applying science and technology to biological diversity conservation.*

To enhance interprogram links, this program could fund studies that integrate onsite and offsite methods—at the ecosystem, species, and genetic levels. Such a program would also bring much needed national recognition, research funding, and scientific expertise to the field of conservation biology. This support would accelerate its acceptance and growth within the scientific community and the development of new principles and technology.

Current statutory authority of NSF would cover such a program. NSF programs are supposed to support both basic and applied scientific research relevant to national problems involving public interest; the maintenance of biological diversity is such a problem.

NSF might resist establishing such a program, because NSF views conservation biology as a mission-oriented activity. Since conservation biology includes technology development, NSF might view a diversity program as a potentially dangerous precedent to its role as the Nation's major supporter of basic research. Furthermore, NSF might argue that a new research program is not needed because its Division of Biotic Systems and Resources already supports about 60 basic research projects that address biological diversity issues. These projects, however, largely ignore the social, economic, political, and management aspects of biological diversity, and conservation is usually of secondary importance to the projects.

An alternative to establishing an NSF program could be to enhance or redirect existing programs in other agencies to promote research in diversity maintenance. The Institute of

Museum Services (IMS), a federally sponsored program, already provides a small amount of funding for research on both onsite and offsite diversity maintenance. IMS supports activities from ecosystem surveys to captive breeding. However, the principal focus of IMS is public education, and its small budget is spread over a wide range of programs (e.g., art museums and historic collections), many of which are unrelated to biological research. Thus, IMS would be unable, with its current funding, to take greater responsibility for technology development; new appropriations would be necessary.

Development and application of diversity-conserving technologies could also be funded through other Federal agencies' research programs. *Congress could encourage appropriate agencies to increase emphasis on development of diversity technology.* One source of funding is through the USDA Competitive Research Grants Office (CRGO). At present, the only research related to genetic resources funded by USDA-CRGO is in the area of molecular genetics. As a result, little funding is available for scientists seeking to conduct research in germplasm preservation, maintenance, evaluation, and use.

Option 3.2: Establish a National Endowment for Biological Diversity.

Congress could establish a National Endowment for Biological Diversity to fund private organizations in research, education, training, and maintenance programs that support the conservation of biological diversity. Currently, no central institution funds such efforts.

Efforts, however piecemeal, of private organizations and individuals are currently making significant contributions to the maintenance of the Nation's diversity. Frequently, they undertake activities that Federal and State agencies cannot or do not address. Through their special interests, these groups as a whole also play a major role in raising public awareness and concern about the loss of diversity. In this way, they increase the constituency backing government programs that maintain natural areas as well as those that collect and safeguard genetic resources.[4] Funding, however, is a major constraint for nearly all these private activities. A program of small grants with a ceiling of perhaps $25,000 per grant (similar to the grants awarded by IMS) could make a substantial contribution to the shoestring budgets of these small organizations and thus enhance national efforts to maintain biological diversity at relatively little cost.

A National Endowment for Biological Diversity could provide funds to private organizations to carry out the following:

- support research and application of methods to conserve biological diversity,
- award fellowships and grants for training,
- foster and support education programs to increase public understanding and appreciation of biological diversity, and
- buy necessary equipment such as small computers.

This national endowment could be created by amending the act that authorizes other national endowment (of arts and humanities) programs. The National Foundation on Arts and Humanities Act of 1965 (Public Law 89-209) declares that the encouragement and support of national progress is of Federal concern and supports scholarships, research, the improvement of education facilities, and encouragement of greater public awareness.

A major constraint to establishing an endowment is the availability of funds during this period of severe budget cutbacks. However, even a small program could significantly encourage private sector initiatives in diversity maintenance. Thus, the total amount needed for such an endowment could be modest, and it might be feasible to use only startup funds and a partial contribution from the Federal Government and raise the remainder of the endowment from private sector contributions.

[4]For further discussion, see U.S. Congress, Office of Technology Assessment, *Grassroots Conservation of Biological Diversity in the United States*, Background Paper #1, OTA-BP-F-38 (Washington, DC: U.S. Government Printing Office, February 1986).

FINDING 4: Many Federal agencies sponsor diversity maintenance programs that are well designed but not fully effective in achieving their objectives because of inadequate funding and personnel, lack of links to other programs, or lack of complementary programs in related fields.

Much is already being done to maintain certain aspects of diversity in the United States, but efforts are constrained by shrinking budgets and personnel. And as noted earlier, the programs addressing biological diversity are piecemeal rather than comprehensive or strategic. Whether or not Congress chooses to promote a comprehensive strategy for diversity maintenance, specific attention is needed to remedy the major gaps and inadequacies in existing programs.

Option 4.1: Provide increased funding to existing programs for maintenance of diversity.

A number of governmental programs for diversity maintenance already exist, some because of congressional mandates. Yet the full potential of some of those programs has not been realized because funding is insufficient. Two such programs are the National Plant Germplasm System (NPGS) and the Endangered Species Program, though others would also benefit from higher levels of funding.

The NPGS of the Agricultural Research Service has functioned for years on severely limited funds and, consequently, is in danger of losing some of the storehouse of plant germplasm. This desperate situation is best illustrated by the National Seed Storage Laboratory (NSSL), which is expected to exceed its storage capacity in 2 years. At the same time, NSSL is being pressured to increase collection and maintenance of wild plant germplasm. NPGS is attempting to respond to various criticisms about its effectiveness, but progress has been slow because of lack of funds and personnel. The 1986 appropriation for germplasm work is approximately $16 million, but to support current programs adequately would cost about $40 million (1981 dollars) annually.

Similarly underfunded and understaffed is the Endangered Species Program of the Fish and Wildlife Service. A review of this program shows a substantial and growing backlog of important work. The rate of proposing species for the threatened and endangered list is so slow that a few candidates (e.g., Texas Henslow's sparrow) may have become extinct while awaiting listing. Critical habitat has been determined for only one-fourth of the listed species, and recovery plans have been approved for only some of the listed species.

Congress could provide adequate funding for these and other programs to achieve their goals in maintaining diversity. NPGS could, as a result, increase the viability of stored germplasm through more frequent testing and regeneration of accessions. NSSL could increase its efficiency by expanding storage capacity and adopting new technologies. For example, cryogenic storage could be used to reduce maintenance cost and space, thereby enabling a larger collection of germplasm. Likewise, the Endangered Species Program would be able to assess candidate species faster and to develop and implement recovery plans for those already listed species.

Option 4.2: Amend appropriate legislation to improve the link between onsite and offsite maintenance programs.

Coordination between onsite and offsite programs is inadequate. *By amending appropriate legislation, Congress could encourage the complementary use of onsite and offsite technologies.* For example, the Endangered Species Act could be amended to encourage use of captive breeding and propagation techniques. Such methods have been used with some endangered species, such as the red wolf, whooping crane, and grizzly bear. But for other species, such as the California condor, black-footed ferret, and dusky seaside sparrow, recovery plans do not exist or were too long delayed. Recovery plans for endangered species seldom include the use of offsite techniques, partly because captive breeding and propagation are outside the scope of natural resource management agencies; rather, they are in the province of zoos, botanic gardens, arboretums, and agricultural research stations.

By mandating that recovery plans give specific consideration to captive breeding and propagation, Congress could encourage links between separate programs. The approach could be broadened to encourage cooperative efforts between public and private organizations working offsite and onsite to conserve ecosystem and genetic diversity. A model for such efforts exists in the emerging cooperation between the Center for Plant Conservation (network of regional botanic institutions) and NSSL.

Option 4.3: Establish programs to fill gaps in current efforts to maintain biological diversity.

One of the most obvious gaps in domestic programs is the lack of a formal national program to maintain domestic animal genetic resources. *Congress could establish a program to coordinate activities for animal germplasm conservation, thereby reducing duplication and encouraging complementary actions.* Such a program could be established through clarification of the Agricultural Research Service mandate. An animal program could parallel the National Plant Germplasm System, but other structures should be explored as well. Alternatively, a separate program established to be semi-independent from government agencies might serve a greater variety of interests. The best structure for such a program is at present unclear.

A congressional hearing could be held to identify the main issues in establishing an animal germplasm program and to discuss alternative structures and scope of such a program.

Coordination of international efforts is also needed to preserve the diversity of agriculturally important animals. Some efforts have already been made, and the concept of an international program is gaining support. *Congress could encourage the establishment of an International Board for Animal Genetic Resources (IBAGR).* This program could parallel the International Board for Plant Genetic Resources (IBPGR). An IBAGR could set standards and coordinate the exchange and storage of germplasm between countries and address related issues such as quarantine regulations. It could foster onsite management of genetic resources for both minor and major breeds.

Another major gap is protection of U.S. ecosystem diversity. Numerous types of ecosystems, such as tall grass prairie, are not included in the Federal public lands system. *Congress could direct Federal land-managing agencies to include representative areas of major ecosystems in protected areas.*

One vehicle for this is the Research Natural Area (RNA) system. Since 1927, the RNA system, with the cooperation of multiple Federal agencies and private groups, has developed the most comprehensive coverage of natural ecosystem types in the United States. RNAs, however, are small scale and are mainly established on land already in public ownership. Therefore, the RNA system, may not be able to cover the major ecosystems without some additional mechanism to acquire land not already in the Federal domain, possibly through land exchanges. Nevertheless, *Congress could recognize the RNA system as a mechanism and direct agencies to work toward filling the program gaps.*

Enhance the Knowledge Base

Developing effective strategies to maintain diversity depends on knowing the components of biological systems and how they interact. Information on the status and trends in biological systems is also needed for public policy. The first step in developing such information is fundamental descriptions of the various components—species, communities, and ecosystems. Data can then be analyzed to determine how best to maintain biological diversity. More specifically, baseline data are needed for the following activities:

- assessing the abundance, condition, and distribution of species, communities, and ecosystems;
- disclosing changes that may be taking place;

- monitoring the effectiveness of resource management plans once they are implemented; and
- determining priorities for areas that merit special efforts to manage natural diversity that would benefit from protection, and that deserve particular attention to avoid biological disruption or to initiate mitigative actions.

To be effective and efficient, the acquisition, dissemination, and use of data must proceed within the context of defined objectives. For the most part, biological data used in diversity maintenance programs has been acquired without the direction of a coordinating goal. Not surprisingly, these data are widely scattered and generally incompatible. Geographical and taxonomical data gaps exist. Some taxonomic groups are ignored in field inventories, while others, particularly plants and animals with economic or recreational value, are monitored extensively. Finally, there is little data on the social, economic, and institutional pressures on biological diversity. Consequently, available data cannot be used easily in decisionmaking directed at maintaining biological diversity.

FINDING 5: Congress and other policymakers need improved information on biological diversity. Such information cannot be supplied without improvements in data collection, maintenance, and synthesis.

Policymakers need comprehensive information on the ramifications and scope of diversity loss. Information provided by the scientific community should be a basis for resource policy and management decisions. To serve in the context of public policy, data should satisfy four criteria:

1. The data must be of *high quality*; that is, it must meet accepted standards of objectivity, completeness, reproducibility, and accuracy.
2. The data must have *value*; that is, it must address a worthwhile problem.
3. The data must be *applicable*; that is, it must be useful to decisionmakers responsible for making policy.
4. The data must be *legitimate*; that is, it must carry a widely accepted presumption of accuracy and authority.

Much information is already available but not in an assimilated form useful to decisionmakers. Data on the status and trends of biological diversity are scattered among Federal, State, and foreign agencies and private organizations. Consolidation of these data is necessary to identify gaps, to provide a comprehensive understanding of the status of the Earth's biota, and especially to define priorities for action.

Option 5.1: Establish a small clearinghouse for data on biological diversity.

The purpose of a clearinghouse would be to coordinate data collection, synthesis, and dissemination efforts. It could serve government agencies, private organizations, corporations, and individuals. The clearinghouse could perform the following functions:

- survey and catalog existing Federal, State, private, and international databases on biological resources;
- evaluate the quality of databases;
- provide small grants and personnel support services to strengthen existing databases; and
- publish annual reports on the status and needs of the biological data system.

Success in these endeavors would accelerate progress toward several objectives:

1. setting of priorities for conservation action;
2. monitoring trends;
3. developing an alert system for adverse trends;
4. identifying gaps and reviewing needs to fill them;
5. facilitating development of environmental impact assessments; and
6. evaluating options, actions, and successes and failures.

As a data-coordinating body, the clearinghouse could guide efforts to collect data on biological diversity, which will provide a comprehensive perspective that Federal agencies cannot supply because of their varied mandates. Access to previously inaccessible data would be facilitated, which should reduce duplication of efforts. By evaluating the quality of information, the clearinghouse could help eliminate a general distrust among users of other databases. Access to a diversity of databases means that no standardized system is forced on data users, which has been a formidable obstacle to database integration and use.

The clearinghouse *would not necessarily* maintain its own primary database. Commercial databases in the public domain could be included in the system, and proprietary and other limited-access databases could be reviewed regularly, with permission. Database enhancements to cover gaps could be funded by small grants. The clearinghouse's information systems could be made available through a library service and special searches. It could charge appropriate fees for all its services.

The same clearinghouse could assess information on biological diversity in international databases. It could provide a small amount of financial and personnel aid to help international organizations improve their databases. In addition, it could work with development assistance agencies to support the participation of other countries' national databases in such international and regional networks as the International Union for the Conservation of Nature and Natural Resources Conservation Monitoring Center, the United Nations Educational, Scientific, and Cultural Organization's (UNESCO) Man and the Biosphere Program (MAB), and The Nature Conservancy International.

Possible objections to such a clearinghouse include the following: 1) that lack of a uniform system of data collection for the United States would hinder national data analysis and use, and 2) that evaluating the quality of other agencies' databases would be politically sensitive. Questions such as the size, administrative structure, and cost of a clearinghouse program must be answered as well. Because it would not maintain its own primary database, however, such a clearinghouse would not need to be a large-scale operation.

Option 5.2: Provide funding to enhance the existing network of natural heritage conservation data centers.

A number of State governments, aided by The Nature Conservancy (TNC), have already established a network of Natural Heritage Data Centers in many States and in some foreign countries. These centers collect and organize biological data specifically for diversity conservation. All centers use a standardized format to collect and synthesize data. The result has been a vehicle to exchange and to aggregate information about what is happening to biological resources at State and local levels and, more recently, around the Nation and across the Western Hemisphere.

Funding for these data centers comes from a combination of Federal, State, and private (including corporate) sources. Progress has been limited, however, by the amount of available funds. *Congress could enhance these efforts by providing a consistent source of additional funding.* By increasing support for the Federal-State-private partnership, the action by Congress could reinforce the application of standard methods, enhance interagency compatibility, improve the efficiency of biological data collection and management, and facilitate the free exchange of useful information. Moreover, the partnership could accelerate the rate at which data centers spread to the remaining States and nations.

An appropriation of $10 million per year, for example, could be divided among several data center functions: supporting central office activities in research, development, documentation, and training; conducting taxonomic work; and matching grants from States and other participants. One source of funding could be the Land and Water Conservation Fund. Although this fund is used mainly for land acquisition, it could also support preacquisition activities such as identification of lands to be acquired. Data centers are key to such activities.

This option does not necessarily replace the need for an information clearinghouse because diverse databases and information systems will continue to operate. The two options could be complementary. Some clearinghouse functions might be handled by TNC, but others, such as facilitating improvement of and access to data sources, could be best handled by a separate entity that functions much like a library.

Support International Initiatives To Maintain Biological Diversity

Most biological resources belong to individual nations. However, many benefits from diversity accrue internationally. American agriculture, for example, depends on foreign sources for genetic diversity to keep ahead of constantly evolving pests and pathogens. And many bird populations important to controlling pests in the United States overwinter in the forests of Latin America.

Solutions to problems that cause diversity loss must be implemented locally, but many of these will be effective only if supported by international political and technical cooperation. Examples of such problems include the international trade in rare wildlife, the greenhouse effect of carbon dioxide on the atmosphere, the effects of acid rain on freshwater lakes and forests, and damage to oceans by pollution and overfishing. The United States has the political prestige needed to initiate international cooperation, and it leads the world in much of the technical expertise needed, such as fundamental biology and information processing. Thus, the United States has both motive and ability to participate and to provide leadership in international conservation efforts.

The United States has historically played a leading role in promoting international conservation initiatives, and precedence exists for extending this leadership to an international or global approach for conserving biological diversity. A variety of international conventions and multilateral programs already specify biological diversity as an aspect of broader conservation objectives (e.g., biosphere reserve program). Such internationally recognized obligations can be important policy tools in concert with technical, administrative, and financial measures to encourage programs for conserving diversity. Obligations confirmed by international conventions provide conservation authorities with the justification frequently needed to strengthen their national programs.

FINDING 6: The United States has begun to abdicate leadership in international conservation efforts, with the result that international initiatives are weakened or stalled in the tropical regions where diversity losses are most severe. Renewed U.S. commitment could accelerate the pace of international achievements in conservation.

The United States has been a model and an active leader in international conservation activity. The movement toward establishment of national parks worldwide grew out of the United States. In the early 1970s, the United States was a leader in international environmental and resource deliberations, notably in the 1972 UN-sponsored Stockholm Conference on the Human Environment. U.S. leadership, for example, played an important role in establishing the United Nations Environment Programme (UNEP), and in securing the Convention on International Trade in Endangered Species of Wild Fauna and Flora (CITES) and the World Heritage Convention, all important foundations of current international efforts to support maintenance of biological diversity.

However, U.S. support for these kinds of initiatives has declined. The retrenchment in support reflects austerity measures as well as dissatisfaction with the performance of specific international organizations. Effective international projects, such as UNESCO's Man and the Biosphere Program, have suffered by association.

U.S. support of international conservation efforts is pivotal in that the United States has greater resources and stronger technical abilities than most other countries to address the complex issue of diversity loss. Without greater initiative and access to resources, many countries will be unable to arrest loss of diversity within their borders. Under existing conditions, countries that harbor the greatest diversity are expected to devote a large part of their national resources to address the problem, even though benefits commonly extend beyond their countries. It would seem equitable for those countries that benefit, including the United States, to share more fully in efforts to conserve diversity in countries otherwise unable to do so.

Option 6.1: Sustain or increase support of international organizations and conventions.

International conservation initiatives are important tools for long-term conservation of biological diversity. Yet, existing international agreements are often poorly implemented because of lack of adequate administrative machinery (e.g., adequately funded and staffed secretariats), lack of financial support for on-the-ground programs (e.g., equipment, training, and staff), and lack of reciprocal obligations that could serve as incentives to comply.

An exception is CITES, which has mechanisms to facilitate reciprocal trade controls and a technical secretariat. The existence of this machinery in large part accounts for the relative success of this convention. The United States has been globally influential in supporting CITES and has reinforced it through national legislation that prohibits import into the United States of wildlife taken or exported in violation of another country's laws. The amendment to the Lacey Act of 1900 (Public Law 97-79) in 1981 backs efforts of other nations seeking to conserve their wildlife resources. This law has been a powerful tool for wildlife conservation throughout the world because the United States is a major importer of wildlife specimens and products.

U.S. contributions to international conservation programs have been diminishing recently. The appropriation cycle for funding such programs has been an annual tug-of-war between Congress and the Administration. The budget of the World Heritage Convention in 1985 was $824,000. The United States, one of the major forces behind the Convention's founding, usually contributes at least one-fourth of the budget. In the fiscal years of 1979 to 1982, U.S. contributions averaged $300,000. But from fiscal year 1982 to 1984, the United States made no contributions. But in fiscal year 1985, $238,903 was contributed. In fiscal year 1986, $250,000 had been appropriated, but the amount was cut to $239,000 under Gramm-Rudman-Hollings Balanced Budget and Emergency Deficit Control Act.

Congress could maintain or increase U.S. support of international organizations and programs in several ways. *Congress could ensure that these organizations receive adequate annual appropriations and could conduct oversight hearings to encourage the Administration to carry out the intent of Congress.*

One possible drawback associated with contributions to international intergovernmental organizations is their lack of accountability. Relative to bilateral assistance channels, the United States has little control over how or to whom intergovernmental organizations direct their resources. The consequence is that U.S. funds go to countries that are unfriendly or even adversarial to the United States and its policies.

It should be recognized, however, that many international activities specific to maintenance of biological diversity, especially activities of UNEP, UNESCO-MAB, and IBPGR, operate largely within scientific channels, which tends to reduce the political overtones inherent in intergovernmental organizations. Also, objectivity can be enhanced in programs willing to establish protocols. For example, establishing criteria to determine which areas qualify for biosphere reserve status or which unique areas warrant (natural) World Heritage status provides objectivity in directing resources.

Congress could also encourage or direct Federal agencies to assign technical personnel to international organizations or to the secretariats of the various conventions. This option could be difficult to implement without legislating special allowances for agency personnel ceilings and budgets. Otherwise, agencies will be reluctant to assign personnel overseas in light of a shrinking Federal work force and budget.

Option 6.2: Continue to direct U.S. directors of multilateral development banks (MDBs) to do the following: 1) press for more specific and

systematic MDB efforts to promote sound environmental and resource policies akin to the World Bank's wildland policy, 2) work to make projects consistent with international and recipient country environmental policies and regulations, and 3) seek to involve recipient country environmental officials and nongovernmental organizations in project formulation processes.

A significant part of all international development assistance efforts are funded by the World Bank and regional MDBs. Thus, these organizations are uniquely situated to influence environmental aspects of development, including the maintenance of biological diversity. In fact, the MDBs' priorities and policies can be the single most important influence on the development model adopted by developing countries. MDB agricultural, rural development, and energy programs all have profound effects on biological resources in developing countries.

In 1986, the World Bank promulgated a new policy on the treatment of wildlands in development projects. The bank recognizes that although further conversion of some natural land and water areas to more intensive uses will be necessary to meet development objectives, other pristine areas may yield benefits to present and future generations if maintained in their natural state. These are areas that, for example, may provide important environmental services or essential habitats to endangered species. To prevent the loss of these wildland values, the policy specifies that the Bank will normally decline to finance projects in these areas and instead prefer projects on already converted lands. Conversion of less important wildlands must be justified and compensated by financing the preservation of an ecologically similar area in a national park or nature reserve, or by some other mitigative measures. The policy provides systematic guidance and criteria for deciding which wildlands are in need of protection, which projects may need wildland measures, and what types of wildland measures should be provided.

In 1980, the World Bank, Inter-American Development Bank, Asian Development Bank, and six other multilaterals signed a "Declaration of Environmental Policies and Procedures Relating to Economic Development," and formed the Committee on International Development Institutions on the Environment (CIDIE), under the auspices of the United Nations Environment Programme. The agencies agreed to systematic environmental analysis of activities funded for environmental programs and projects. However, a subsequent study found that these policy statements by the MDBs were not effectively translated into action. Criticisms of how well MDBs implement environmental policies remain strong. And it is too soon to determine the effectiveness of the World Bank's wildland policy.

The United States is limited in its ability to effect change at MDBs because the banks are international institutions run collectively by member nations. Since the United States is a large contributor, however, it does have considerable influence on bank policies, which are determined by boards of directors.

The primary way Congress affects policies of these banks is by requesting that the U.S. executive directors—who are responsible to the Secretary of the Treasury—carry out congressionally approved policies. These requests may be made at oversight hearings or in the language of appropriation legislation. For instance, the 1986 House Committee on Appropriations Report stated guidelines for the U.S. executive directors (Sec. 539), which included the addition of relevant staff, development of management plans, and commitment to increase the proportion of programs supporting environmentally beneficial projects. To continue this guidance, *Congress could require the U.S. executive directors of MDBs to encourage the adoption of a policy similar to the World Bank's wildlands policy statement.*

FINDING 7: Constraints on international exchange of genetic resources could jeopardize future agricultural production and progress in biotechnologies. Such constraints are becoming more likely because developing countries with sovereignty over most such resources believe that the industrial nations have benefited at their expense. Debates on

the issue could benefit from a more informed and less impassioned approach.

All countries benefit from the exchange of genetic resources. Many of the major crops currently grown in various countries have originated elsewhere. Coffee, for example, is native to the highlands of Ethiopia. Yet, today, it represents an important source of income for farmers in other parts of Africa, Asia, and Latin America. Maize, originally from Central America, is grown as a staple crop in North America and Africa. Countries continue to depend on access to germplasm from outside their borders to maintain or enhance agricultural productivity. Political and economic considerations, however, are now prompting national governments to restrict access to their germplasm. Behind these efforts is an implicit desire by some countries to obtain greater compensation for the genetic resources that are currently made freely available.

The International Board for Plant Genetic Resources (IBPGR) is the main international institution dealing with the offsite conservation of plant genetic diversity. Established in 1974, it promotes the establishment of national programs and regional centers for the conservation of plant germplasm. It has provided training facilities, carried out research in techniques of plant germplasm conservation, supported numerous collection missions, and provided limited financial assistance for conservation facilities. However, it does not operate any germplasm storage facilities itself.

Due in part to the success of IBPGR in focusing attention on the need to conserve genetic diversity, the issue of germplasm exchange has become embroiled in political controversy. Some critics regard the IBPGR as implicitly working for agribusiness interests of industrial nations. Central to the issue is a perception on the part of many developing countries that they have been freely giving genetic resources to industrial nations which, in turn, have profited at their expense.

This controversy led the United Nations Food and Agricultural Organization (FAO) to sponsor an International Undertaking on Plant Genetic Resources. The undertaking proposed an international germplasm conservation network under the auspices of FAO. It declared that each nation has a duty to make all plant genetic materials—including advanced breeding materials—freely available. IBPGR was to continue its current work, but it would be monitored by FAO.

FAO then established the Commission on Plant Genetic Resources to review progress in germplasm conservation. The commission held its first meeting in March 1985, with the United States present only as an observer. Much of the discussion focused on the concerns expressed in the undertaking and on onsite conservation.

The continuing controversy includes charges that the current international system enables countries to restrict access to germplasm in international collections for political and economic reasons. Also of concern to some parties is the impact of plant patenting legislation.

Current charges and arguments in the FAO forum tend to oversimplify the complexity of how germplasm is incorporated into plant varieties and to distort the actual nature of genetic exchange between and among industrial and developing countries. Restrictions on export of germplasm, for example, appear to be more common for developing countries. Nevertheless, the perception of inequity in the current situation is real, and it could result in increasing national restrictions on access to and export of germplasm. Further, the issue of control over genetic resources could become a significant stumbling block to establishing international commitment and cooperation in the maintenance of overall biological diversity.

Option 7.1: Closely examine the actions available to the United States regarding the issue of international exchange of genetic resources.

Efforts to address the conservation and exchange of plant genetic resources in the FAO forum have been controversial. It is not yet apparent how the United States should act in this regard. Congress could give increased attention to determining what options are available.

One possible action is for Congress to request that an independent organization, such as the National Academy of Sciences, study this issue. In fact, NAS has already indicated interest in investigating this as a part of its current 3-year study of global genetic resources . Such a study could draw on other agencies and individuals with interest and expertise in this area to define several general actions the United States might take in regard to international exchange of genetic resources and the consequences associated with it.

Another option is to favor the status quo, ignoring the criticisms and avoiding the risk that new political actions might disrupt effective scientific working arrangements. A practical international flow of germplasm is likely to continue in the future, with or without the formal international arrangements envisioned by the FAO undertaking. In time, the political issues may be resolved equitably without pushing nations into conflicts over breeders' rights or access to genetic materials.

Another possibility would be for the United States to associate with the FAO Commission on Plant Genetic Resources. U.S. influence might strengthen the international commitment to free flow of germplasm and reduce the risk that germplasm will increasingly be withheld for political or economic reasons.

Unless Congress chooses to restrict plant breeders' rights in the United States, the U.S. Government will be unable to join the undertaking without major reservations. Such a change in domestic law seems politically unlikely, given domestic benefits provided by plant breeders' rights and the effective lobbying efforts of the seed industry. However, the United States could consider renegotiating the FAO undertaking to require a commitment to grant global access to genetic resources—with appropriate exceptions for certain privately held materials—within the context of an internationally supported commitment to help countries conserve and develop their genetic resources. Parallel agreements also might be developed for domestic animal, marine, and microbial resources. Such agreements could also define national and international obligations to collect and conserve the germplasm that is being displaced by new varieties or by changing patterns of agricultural developments.

Finally, U.S. representatives could consider promoting a discussion of genetic resource exchanges outside formal channels in an effort to separate the technical issues from emotional ones. The Keystone Center, an environmental mediation organization, is exploring the possibility of conducting a policy dialog on this topic in the near future.

Option 7.2: Affirm the U.S. commitment to the free flow of germplasm through an amendment to the Export Administration Act.

Specific allegations have been made that the United States has restricted the access to germplasm in national collections (at the National Plant Germplasm System) for political reasons. The government, however, maintains that it adheres to the principles of free exchange.

To reinforce recent executive affirmations of the free flow of germplasm, *Congress could exempt the export of germplasm contained in national collections from Export Administration Act restrictions or political embargoes imposed for other reasons.* Comparable provisions are already included in this act with respect to medicine and medical supplies (50 U.S.C. app. sec. 2405 (g), as amended by Public Law 99-64, July 12, 1985). Because this germplasm is already accessible through existing mechanisms, such a provision would only reaffirm the U.S. position and remove from the current debate the allegations of U.S. restrictions of access to germplasm.

On the other hand, the process of amending the act may generate support for restricting germplasm—by excluding certain countries from such an exemption. Restricting access in such a manner would likely lead to an international situation counter to U.S. interests. In such a case, no action would be preferable to an amendment.

Address Loss of Biological Diversity in Developing Countries

The United States has a stake in promoting the maintenance of biological diversity in developing countries. Many of these nations are in regions where biological systems are highly diverse, where pressures that degrade diversity are generally most pronounced, and where the capacity to forestall a reduction in diversity is least well-developed. The rationale for assisting developing countries rests on: 1) recognition of the substantial existing and potential benefits of maintaining a diversity of plants, animals, and microbes; 2) evidence that degradation of specific ecosystems is undermining the potential for economic development in a number of regions; and 3) esthetic and ethical motivations to avoid irreversible loss of unique life forms.

The U.S. Congress, recognizing these interests, passed Section 119 of the Foreign Assistance Act of 1983, specifying conservation of biological diversity as a specific objective of U.S. development assistance. The U.S. Agency for International Development (AID), as the principal agency providing development assistance, was given a mandate to implement this policy, which reads in part:

> In order to preserve biological diversity, the President is authorized to furnish assistance to countries in protecting and maintaining wildlife habitats and in developing sound wildlife management and plant conservation programs. Special effort should be taken to establish and maintain wildlife sanctuaries, reserves, and parks; to enact and enforce anti-poaching measures; and to identify, study, and catalog animal and plant species, especially in tropical environments.

A review of AID initiatives since 1983 suggests that despite the formulation of a number of policy documents, the agency lacks a strong commitment to implementing the specific types of projects identified in Section 119. This lack of commitment is due to several factors, including: 1) a belief that the agency is already addressing biological diversity to the extent it should, 2) reduced levels of budgets and staff to initiate projects, and 3) an inadequate number of trained personnel to address conservation concerns generally.

Several questions arise in relation to the capacity and the appropriateness of U.S. commitments to support diversity conservation efforts through bilateral development assistance. First, it is unclear whether Section 119, as the principal legislation dealing with concerns over diversity loss outside the United States, defines U.S. interests too narrowly. Second, it is uncertain how Section 119 relates to the principal goals of foreign assistance, as specified in section 101. Finally, questions remain concerning the commitment of resources and personnel to address U.S. interests in maintaining diversity in developing countries.

FINDING 8: Existing legislation may be inadequate and inappropriate to address U.S. interests in maintaining biological diversity in developing countries.

Maintaining diversity will depend primarily on onsite maintenance. The "special effort" initiatives identified in Section 119 are important components of a comprehensive program. What is not clear, however, is whether the emphasis is appropriate within the context of U.S. bilateral development assistance. That is, establishing protected areas and supporting anti-poaching measures can have adverse impacts on populations that derive benefits from exploiting resources within a designated area. These populations are characteristically among the "poorest majority" intended to be the principal beneficiaries of U.S. development assistance (Sec. 101). However, demands of local populations (e.g., for fuelwood or agricultural land) may threaten diversity and even the sustainability of the resource base on which they depend. It does, however, raise questions on the appropriateness of supporting activities that could place increased stress on these populations.

Second, existing legislation identifies concern over diversity loss separately from conversion of tropical forests and degradation of environment and natural resources (Sec. 118

and 117, respectively). Clearly, these concerns are interrelated, although not synonymous. It is questionable whether such a distinction is appropriate within the context of development assistance legislation. An argument can be made that U.S. development assistance should approach diversity maintenance within the context of conservation—that is, as a wise use of natural resources, as elaborated in the World Conservation Strategy. In doing so, the objectives of diversity maintenance and development interests could be made more compatible.

Finally, although Section 119 speaks of biological diversity, the thrust of the legislation addresses a narrower set of concerns—that of species extinction. While certainly a prominent concern, and perhaps even the central motivation behind the legislation, it fails to address the broader set of U.S. concerns over diversity loss in developing countries. As noted earlier, a focus on unique populations would be a more appropriate, though more problematic, approach. This is particularly important with regard to preserving genetic resources of potential benefit to agriculture or industry, which is the most strongly argued rationale for conserving biological diversity. Existing legislation does not specifically identify these interests.

Option 8.1: Restructure existing sections of the Foreign Assistance Act to reflect the full scope of U.S. interests in maintaining biological diversity in developing countries.

The U.S. Foreign Assistance Act (FAA) comes up for reauthorization in 1987. Major restructuring of the act is already being considered. Revamping could provide an opportunity to recast certain provisions of the legislation to better account for U.S. interests in maintaining diversity in developing countries.

Providing for conservation of natural resources and the environment in general, and of biological diversity and tropical forests in particular, are important considerations in a restructuring of FAA. Less clear, however, is whether the language and disaggregation of these interests is appropriate in the context of bilateral development assistance.

One specific consideration could be to resolve potential conflicts of interests that exist in the language of Section 119—that of emphasizing the need to establish protected areas and poaching controls without specific reference to impacts on indigenous populations. *Congress could correct this potential conflict by adding language to Section 119 such as, "Support for biological diversity projects should be consistent with the interests, particular needs, and participation of local populations."* It is widely recognized that the viability of protected areas is largely contingent on these provisions. Adding such language would thus provide greater consistency within the objectives of FAA as well as specify criteria that heighten chances of project success.

In addition, *Congress could recast the language of existing legislation to provide a fuller accounting of U.S. interests in maintaining diversity in developing countries.* Such changes could expand from a focus on endangered species to the loss of biological systems, including ecosystems and genetic resources. Such an effort might also emphasize practical aspects of conservation initiatives of particular interest to developing countries and stress the goal of promoting ability and initiatives of the countries themselves.

Finally, *Congress could combine those sections of FAA that deal with natural resources and environmental issues to reflect the interrelatedness of these amendments.* Provisions could be made to account for specific concerns over species extinctions currently emphasized in Section 119. But approaches and concerns reflected in these amendments are probably best considered together. Provision of funding within such a restructuring would also be important.

FINDING 9: AID could benefit from additional strategic planning and conservation expertise in promoting biological diversity projects.

Congress has already taken steps to earmark funds for biological diversity projects within AID's budget. The existing mechanisms within the agency to identify and promote diversity projects are not well established, however. Because funding is minimal, it is all the more important to devise a strategy that allows priority initiatives to be defined.

Environmental expertise within AID is slim. In recent years, in-house expertise in this area has declined, and that which does exist has been severely overextended. Addressing biological diversity will, therefore, require both increasing the number of AID staff with environmental training and an increased reliance on expertise outside AID, in other government agencies and in the private sector. AID has already taken steps to cultivate this environmental expertise, but further actions could be taken.

Option 9.1: Direct AID to adopt a more strategic approach in promoting initiatives for maintenance of biological diversity.

The *U.S. Strategy on the Conservation of Biological Diversity: An Interagency Task Force Report to Congress* was delivered to Congress in February 1985, in response to provisions in Section 119. A general criticism of the document was that although it contained 67 recommendations, it lacked any sense of priority or indication of funding sources to undertake these recommendations. In an attempt to apply the recommendations to specific agency programs, AID drafted an *Action Plan on Conserving Biological Diversity in Developing Countries* (January 1986). Comments received from AID overseas suggest that problems exist in translating the general principles and recommendations of an agency plan into specific initiatives at the country level.

A more refined approach to addressing diversity interests within the agency may be required. Such an approach would seek to incorporate biological diversity concerns into AID development activities at different levels of the agency, ranging from general policy documents at the agency level to more strategic efforts at the regional bureau and mission levels.

At least two efforts could be considered at the agency level. First, *Congress could direct AID to prepare a policy determination (PD) on biological diversity*. A PD would serve as a general statement that maintaining diversity is an explicit objective of the agency. In developing a PD, AID should review provisions contained in the recent World Bank wildlands policy statement.

Existence of a PD could mean that consideration of diversity concerns would, where appropriate, become an integral part of sectoral programming and project design. Further, it would require that projects be reviewed and evaluated by the Bureau of Program and Policy Coordination for consistency with the objectives of the PD. Because of the increase in bureaucratic provisions this would create, the formulation of a PD on diversity probably would not be well received within AID.

A second effort is to establish a centrally funded project within AID's Bureau of Science and Technology. AID has already developed a concept paper along these lines as a prelude to a more concrete project identification document. As conceived, the concept paper examines the possibility of establishing a biological diversity project. One major benefit of such a project would be the establishment of a focal point for coordinating funding and technical assistance on biological diversity. The Science and Technology Bureau's emphasis on technical assistance, research, training, and institutional development would make it the appropriate bureau for such a program. A constraint to this approach is that biological diversity projects may continue to be separate rather than an integral part of development programs.

The three regional bureaus of AID (i.e., Africa, Asia and Near East, and Latin America and the Caribbean) could also prepare documents that identify important biological diversity initiatives in their regions. The Asia and Near East Bureau, in fact, has already prepared such a document that could be used in highlighting regional priorities. A reluctance to direct scarce funds to diversity projects, at the expense of more traditional development projects, has limited the utility of the document to date. Nevertheless, the development of such reports for each regional bureau is considered an effective way to identify priorities for existing diversity projects, especially given the earmarking of funds.

The most important focus of biological diversity strategies is at the mission level, where projects are implemented. Congress has already mandated that Country Development Strategy Statements and other country-level documents prepared by AID address diversity concerns. Most missions, however, lack the expertise or adequate access to expertise needed to address this provision of Section 119 as amended.

Option 9.2: Direct AID to acquire increased conservation expertise in support of biological diversity initiatives.

The ability of AID to promote biological diversity in developing countries is seriously undermined by its lack of personnel trained in environmental sciences. While true at the agency headquarters, the problem is particularly acute in its overseas missions. Although AID designates an environmental officer at each mission, the person usually has little professional experience or training in the area. Often environmental duties are combined with numerous other duties; few AID personnel are full-time environmental officers. Under these circumstances, it is difficult to envision how AID can effectively promote biological diversity maintenance.

Congress could direct AID to recruit and hire additional personnel with environmental science backgrounds or at a minimum provide increased training for existing staff. The near-term prospects for AID, however, point to a reduction in an already overworked staff. It seems unlikely, therefore, that significant in-house conservation expertise will be developed. Consequently, addressing biological diversity within AID will depend on providing access to conservation expertise within other government agencies and in the private sector. Even drawing on outside expertise, AID will need some increase in environmental officers to manage and coordinate projects.

AID already draws on other government agencies to participate in projects supporting biological diversity maintenance. Mechanisms such as Participating Agency Service Agreements (PASA) and Resource Services Support Agreements (RSSA) allow interagency exchanges of experts and services. AID currently has a RSSA with Fish and Wildlife Service for the services of a technical advisor to handle biological diversity issues. These mechanisms could be used to facilitate further access to conservation experts in other government agencies.

A biological diversity program could be established within the existing Forestry Support Program, for example. The Forestry Support Program is an RSSA between AID and the U.S. Department of Agriculture (USDA) to provide technical assistance to AID in the area of forestry and natural resources. A diversity program would likely be an RSSA between AID, the Department of the Interior, and USDA. Such a program would provide AID missions with access to conservation expertise within the Department of the Interior, the USDA, and through a roster of consultants.

A constraint to the RSSA and PASA is agency personnel ceilings and the limited number of personnel with international experience. In light of a reduction of the Federal work force, agencies may be reluctant to devote their staff to nonagency projects. Although some Federal programs have been successfully used in supporting AID projects, expertise within the private sector will also be needed to address AID's requirements.

The Peace Corps is also seen as having special potential to support biological diversity projects. Cooperative agreements with the National Park Service, Fish and Wildlife Service, The Man in the Biosphere Program, and World Wildlife Fund/U.S. have increased the Peace Corps' capacity and access to talent and training in this area. Another area of potential collaboration is between the Peace Corps and the Smithsonian Institution, especially given the Smithsonian's newly established Biological Diversity Program. Precedence exists for such a cooperative relationship, in the form of the Smithsonian-Peace Corps Environmental Program, which was terminated in the late 1970s. With the emergence of special interests in diversity maintenance, *Congress could direct both agencies to investigate re-establishing a similar initiative focused on biological diversity projects.*

Section 119 of FAA states:

> . . . whenever feasible, the objectives of this section shall be accomplished through projects managed by appropriate private and voluntary organizations, or international, regional, or national nongovernmental organizations which are active in the region or country where the project is located.

A number of nongovernmental organizations (NGOs) are already working with AID in developing capacity to maintain diversity in developing countries. These include important initiatives in the areas of conservation data centers, of supporting development of national conservation strategies, and of implementing field projects. AID is also using a private NGO to maintain a listing of environmental management experts. Such partnership could continue to be encouraged by Congress through oversight hearings, for instance. Encouraging joint public-private initiatives through matching grants should also be stressed.

FINDING 10: A major constraint to developing and implementing diversity-conserving projects in developing countries is the shortage of funds. Present funding levels are insufficient to address the scope of the problem adequately.

Recently passed legislation earmarked $2.5 million of AID's 1987 funds for biological diversity projects. Given that this amount is intended to be used to address diversity loss over three continents and is guaranteed for only 1 year, its adequacy can be questioned. Faced with prospects of further cuts in an already reduced foreign assistance budget and a shift in the composition of this budget to proportionally less development and food aid in favor of military aid and economic support funds, it is difficult to see where further funding for diversity maintenance could be derived.

Option 10.1: Establish a new account within the AID budget to support biological diversity initiatives identified in the Foreign Assistance Act.

Sections 117, 118, and 119 of FAA all define congressional interest in conservation as an integral aspect of development. With the exception of the 1987 earmarking of funds for biological diversity, no formal funding source has been attached to these sections. The result is that support for conservation initiatives generally has been weak. Support has been further eroded recently because those functional accounts used for conservation projects—Agriculture, Rural Development, and Nutrition; and Energy, Private Voluntary Organization, and Selected Development Activities—have received disproportionate funding cuts.

Congress could define its support for the importance of conservation to development by establishing a separate fund, perhaps called an Environment and Natural Resources Account, that could be used by AID to support diversity maintenance activities. Concerns exist that functional accounts generally tend to reduce AID's flexibility, and consideration has even been given to eliminating them entirely. If established, however, an Environment and Natural Resources account could be used to define congressional concerns in this area. Specific earmarking for biological diversity could be considered within this new functional account.

Option 10.2: Amend the Agricultural Trade Development and Assistance Act of 1954 specifying that funds from the Food for Peace Program (Public Law 480) could be used for projects that directly promote the conservation of biological diversity.

An existing source of funds for biological diversity projects is Public Law 480 Food for Peace program. Titles I and III make commodities available at concessional rates with long-term, low-interest financing for debts incurred. Recipient countries resell the U.S. commodities and are required by contract to apply part of the currency to self-help projects agreed on between the country and the AID mission. The country can eventually cancel some of its debt by applying equivalent funds to long-term development projects. Title II provides U.S. commodities to developing countries in cases of emergency or for nutrition and development programs. This Food for Work program has conducted reforestation and resource management projects in which laborers are paid with food and with wages generated from the resale of U.S. commodities. Hence, Public Law 480 funds are already being used to finance projects that promote diversity maintenance. More could be done if *Congress amends Public Law 480 specifying that funds could be used for diversity conservation projects.*

Other existing funding mechanisms could be redirected to include funding of diversity projects. In response to funding cuts at AID, conservation groups have proposed certain ways to provide money for biological diversity projects. One such mechanism is the use of economic support funds for additional development assistance programs. Though primarily used for other purposes, economic support funds are the most flexible of AID's funds, with the fewest restrictions on their use. Therefore, *Congress could direct the General Accounting Office to examine such funding mechanisms and assess their feasibility as funding sources for maintenance of biological diversity.*

Part I
Introduction and Background

Chapter 2
Importance of Biological Diversity

CONTENTS

Table

Figure

Boxes

Chapter 2

Importance of Biological Diversity

HIGHLIGHTS

- Biological diversity benefits human welfare directly, as various organisms are used to satisfy basic human needs, and indirectly, as diversity supports many processes essential to human survival and progress.
- The constituency for maintaining biological diversity is large but fragmented because many groups focus on various aspects of biological resources rather than on diversity per se. Constituents who are politically articulate in support of diversity are usually motivated by its intrinsic values rather than its substantial economic values.

Human welfare is inextricably linked to, and dependent on, biological diversity. Diversity is necessary for several reasons: 1) to sustain and improve agriculture, 2) to provide opportunities for medical discoveries and industrial innovations, and 3) to preserve choices for addressing unpredictable problems and opportunities of future generations. Actual and potential economic uses range from subsistence foraging to genetic engineering. The essential services of ecosystems, such as moderating climate; concentrating, fixing, and recycling nutrients; producing and preserving soils; and controlling pests and diseases are also dependent on biological diversity. Finally, diversity has esthetic and ethical values.

DEFINITION

Biological diversity refers to the variety and variability among living organisms and the ecological complexes in which they occur. Diversity can be defined as the number of different kinds of items and their relative frequency in a set (97). Items are organized at many levels, ranging from complete ecosystems to the chemical structures that are the molecular basis of heredity. Thus, the term encompasses the numbers and relative abundance of different ecosystems, species, and genes. (Box 2-A describes major components of biological diversity.)

Species diversity, for example, decreases when the number of species in an area is reduced or when the same number exists but a few become more abundant while others become scarce. When a species no longer exists in an area, it is said to be locally eliminated. The extreme effect of species diversity loss is extinction—when a species no longer exists anywhere.

Biological diversity is the basis of adaptation and evolution and is basic to all ecological processes. It contributes to research and education, cultural heritage, recreation and tourism, the development of new and existing plant and animal domesticates, and the supply of harvested resources (table 2-1). The intrinsic importance of biological diversity lies in the uniqueness of all forms of life: each individual is different, as is each population, each species, and each association of species. Major functional and utilitarian benefits of ecosystem, species, and genetic diversity are described in the next five sections; evaluation of diversity and the constituencies of diversity are discussed in the final sections.

Box 2-A.—Components of Biological Diversity

- **Genetic** characteristics are the features of plants and animals that are passed from one generation to the next through the mechanism of genes. Genes control the chemistry and structure of whole organisms; therefore, the greater the diversity of genes, the greater the chance of having some individuals that can survive gradual environmental changes and a greater chance of having potential direct utility for people.
- **Species** is a taxonomic category ranking immediately below genus and including closely related, morphologically similar individuals that actually or potentially interbreed.
- **Habitat** is the place where a species finds the required combination of food, cover, water, and other resources to meet its biological needs. Each species is adapted generally to a specific arrangement and amount of essential resources.
- A **population** is a subset of all the individuals of one species. It consists of individuals that are found in a distinct portion of the species range that interbreed with some regularity. Thus, the members of a population share a common set of genetic characteristics. Populations can have some characteristics not found in other populations of the same species, in which case such terms as subspecies, variety, or breed are used to describe them.
- A **community** is a collection of species present in one place at one time. Most communities contain groups of species that interact in a variety of ways. There are usually two consequences of these interactions: 1) certain species cannot be maintained without other interacting species; and 2) some species so strongly affect other species that the community changes significantly if that species is removed.
- Communities are part of a higher level of organization called **ecosystems**. An ecosystem is a dynamic complex of plant and animal communities, which along with their abiotic environment, constitutes one functioning whole. An understanding of ecosystem functioning can greatly influence the kinds of conservation and management applied.

Table 2-1.—Examples of Benefits From Ecosystem, Species, and Genetic Diversity

Ecological processes	Research	Cultural heritage	Recreation and tourism	Agriculture and harvested resources
Ecosystem diversity				
Maintenance of productivity; buffering environmental changes; watershed and coastal protection	Natural research areas; sites for baseline monitoring (e.g., Serengeti National Park, Zambesi Teak Forest)	Sacred mountains and groves; historic landmarks and landscapes (e.g., Mount Fuji; Voyageurs Park, Minnesota)	700 to 800 million visitors per year to U.S. State and national parks; 250,000 to 500,000 visitors per year to mangrove forests in Venezuela	Rangelands for livestock production (e.g., 34 in the U.S.); habitats for wild pollinators and pest enemies (e.g., saving $40 to $60 per acre for grape growers)
Species diversity				
Role of plants and animals in forest regeneration, grassland production, and marine nutrient cycling; mobile links; natural fuel stations	Models for research on human diseases and drug synthesis (e.g., bristlecone pine, desert pupfish, medicinal leeches)	National symbols (bald eagles); totems; objects of civic pride (e.g., port orford cedar, bowhead whale, *Ficus religiosa*)	95 million people feed, observe, and/or photograph wildlife each year; 54 million fish; 19 million hunt	Commercial logging, fishing, and other harvesting industries ($27 billion/year in U.S.); new crops (e.g., kiwi fruit, red deer, catfish, and loblolly pine)
Genetic diversity				
Raw material of evolution required for survival and adaptation of species and populations	Fruit flies in genetics, corn in inheritance, and *Nicotiana* in virus studies	Breeds and cultivars of ceremonial, historic, esthetic, or culinary value (e.g., Texas longhorn cattle, rice festivals (Nepal))	100,000 visitors per year to Rare Breeds Survival Trust in the United Kingdom	Required to avoid negative selection and enhancement programs; pest and disease resistance alleles

SOURCE: Office of Technology Assessment, 1986.

BENEFITS TO ECOLOGICAL PROCESSES

Ecological processes include—

- **regulation**: monitoring the chemistry and climate of the planet so it remains habitable;
- **production**: conversion of solar energy and nutrients into plant matter;
- **consumption**: conversion of plant matter into animal matter;
- **decomposition**: breakdown of organic wastes and recycling of nutrients;
- **protection processes**: protection of soil by grasslands and forests and protection of coastlines by coral reefs and mangroves, for example; and
- **continuation of life**: processes of feeding, breeding, and migrating.

Knowledge of the relationship between diversity and ecological processes is fragmentary, but it is clear that diversity is crucial to the functioning of all major life processes, for diversity helps maintain productivity and buffers ecosystems against environmental change. Diversity within ecosystems is essential for protective, productive, and economic benefits. Species diversity is necessary for a stable food web. And diversity of genetic material allows species to adapt to changing environmental conditions.

Ecosystem Diversity

Ecosystems are systems of plants, animals, and micro-organisms, together with the non-living components of their environment (45). It can be recognized on many scales, from biome—the largest ecological unit—to microhabitat (box 2-B). Ecosystem diversity refers to the variety that occurs within a larger landscape. Loss of ecosystem diversity can result in both the loss of species and genetic resources and in the impairment of ecological processes.

In eastern and southern Africa, for instance, the mosaic of ephemeral ponds, flood plains, and riparian woodlands enable antelope, elephant, and zebra to survive long cycles of wet and dry years (16,23). On the American continent, many animal species cope with oscillations in weather and climate by migrating between biomes—spending the rainy season in the tropical dry forest and the dry season in the rain forest, that is, summer in temperate forest and winter in tropical forest. Others use different habitats within the same biome; for example, leaf-eating primates and flower-pollinating bats move from dry sites in the rainy season to evergreen riparian trees in the dry season (32,48).

Several types of ecosystems are closely associated with protective and productive processes of direct economic benefit. Cloud forests, for

Box 2-B.—Scales of Ecosystem Diversity

Several ways exist to classify the many scales of ecosystem diversity. An example using the Pacific Northwest to illustrate four levels of ecosystems is shown below. Animal species characteristic of each level are noted.

1. **Biome**: temperate coniferous forest
   —Rufous hummingbird
   —Mountain beaver
2. **Zone**: western hemlock
   —Coho salmon
   —Oregon slender salamander
3. **Habitat**: old growth forest
   —Vaux's swift
   —Spotted owl
4. **Microhabitat**: fallen tree
   —Clouded salamander
   —California red-backed vole

The fallen tree component of old growth and mature forests illustrates the contribution of ecosystem diversity to ecological processes. Fallen trees provide a rooting medium for western hemlock and other plants that is moist enough for growth to continue during the summer drought, a reserve of nitrogen and other nutrients, and a source of food and shelter for animals and micro-organisms that play key roles in redistributing and returning the nutrients to the regenerating forest. For example, the rotten wood provides habitat for truffles, and the truffles are eaten by the California red-backed vole, which spreads the truffle spores, so helping the growth of Douglas fir trees, which require mycorrhizal fungi (such as truffles) for uptake of nutrients (56).

example, increase precipitation, often substantially (38). Watershed forests generally reduce soil erosion and thereby help protect downstream reservoirs, irrigation systems, harbors, and waterways from siltation (45). Coral reefs are productive oases in otherwise unproductive tropical waters. Algae living inside coral polyps enable the corals to build the reefs (8,49). The reefs, in turn, support local fisheries and protect coastlines.

Wetlands are another example of an ecosystem with protective processes linked to economic output. Millions of waterfowl and other birds of great economic value depend on the diverse North American wetlands—coastal tundra wetlands, inland freshwater marshes, prairie potholes, coastal saltwater marshes, and mangrove swamps—for breeding, feeding, migrating, and overwintering.

These wetlands also support most commercial and recreational fisheries in the United States. About two-thirds of the major U.S. commercial fish, crustacean, and mollusk species depend on estuaries and salt marshes for spawning and nursery habitat (88,90). Other wetland services include water purification (by removing nutrients, processing organic wastes, and reducing sediment loads), riverbank and shoreline protection, and flood assimilation. Wetlands temporarily store flood waters, reducing flow rates and protecting people and property downstream from flood and storm damage.

For example, the U.S. Army Corps of Engineers chose protection of 8,500 acres of wetlands over construction of a reservoir or extensive walls and dikes as the least-cost solution to flooding problems in the Charles River basin in Massachusetts. It was estimated that loss

Photo credit: U.S. Fish and Wildlife Service—L. Childers

Ecosystems such as these wetlands have protective and productive functions linked to economic output.

of the Charles River wetlands would have resulted in an average of $17 million per year in flood damage (80,88,90). (Data on wetlands ecosystem losses are given in chapter 3.)

Species Diversity

Some species play such an important role in particular ecosystems that the ecosystems are named after them. Zambezi Teak Forest and Longleaf-Slash Pine Forest are examples. But the ecological processes that maintain dominant species often depend on other species. For example, elephants and buffaloes make a crucial contribution to regeneration of Zambezi teak by burying seeds, providing manure, and destroying competing thicket species (72).

Depletion of species can have a devastating impact higher up the food chain. For example, catches of common carp in the Illinois River are one-tenth of what they were in the early 1950s. This decrease appears to be the result of pollution-caused die-off in the 1950s of fingernail clams, mayfly larvae, and other river-bottom macro-invertebrates. These macro-invertebrates are still scarce, for river-bottom sediment is slow to recover from pollution, much slower than water quality, for example (44).

Certain species have a greater effect on productive processes than is indicated by their position in a food web (figure 2-1). Earthworms, for instance, improve the mixing of soil, increase the amount of mineralized nitrogen available for plant growth, aerate the soil, and improve its water-holding capacity (98). Ants also contribute to soil formation in temperate regions and the tropics. They contribute to the aeration, drainage, humidification, and enrichment of both forest and grassland soils (99).

In East Africa, species diversity increases the productivity of grasslands. For example, grazing by wildebeest promotes the lush regrowth eaten by gazelles (59,60). Similar interactions have been observed in North American grasslands between prairie dogs and bison. Although the standing crop of grass in prairie dog towns

Photo credit: National Park Service, Department of the Interior

Wind Cave National Park, South Dakota, contains a variety of wildlife including bison, elk, prairie dogs, pronghorn, and deer. Interactions between species such as prairie dogs and bison increase the productivity of grasslands.

Figure 2-1.—Krill: The Linchpin to the Antarctic Foodweb

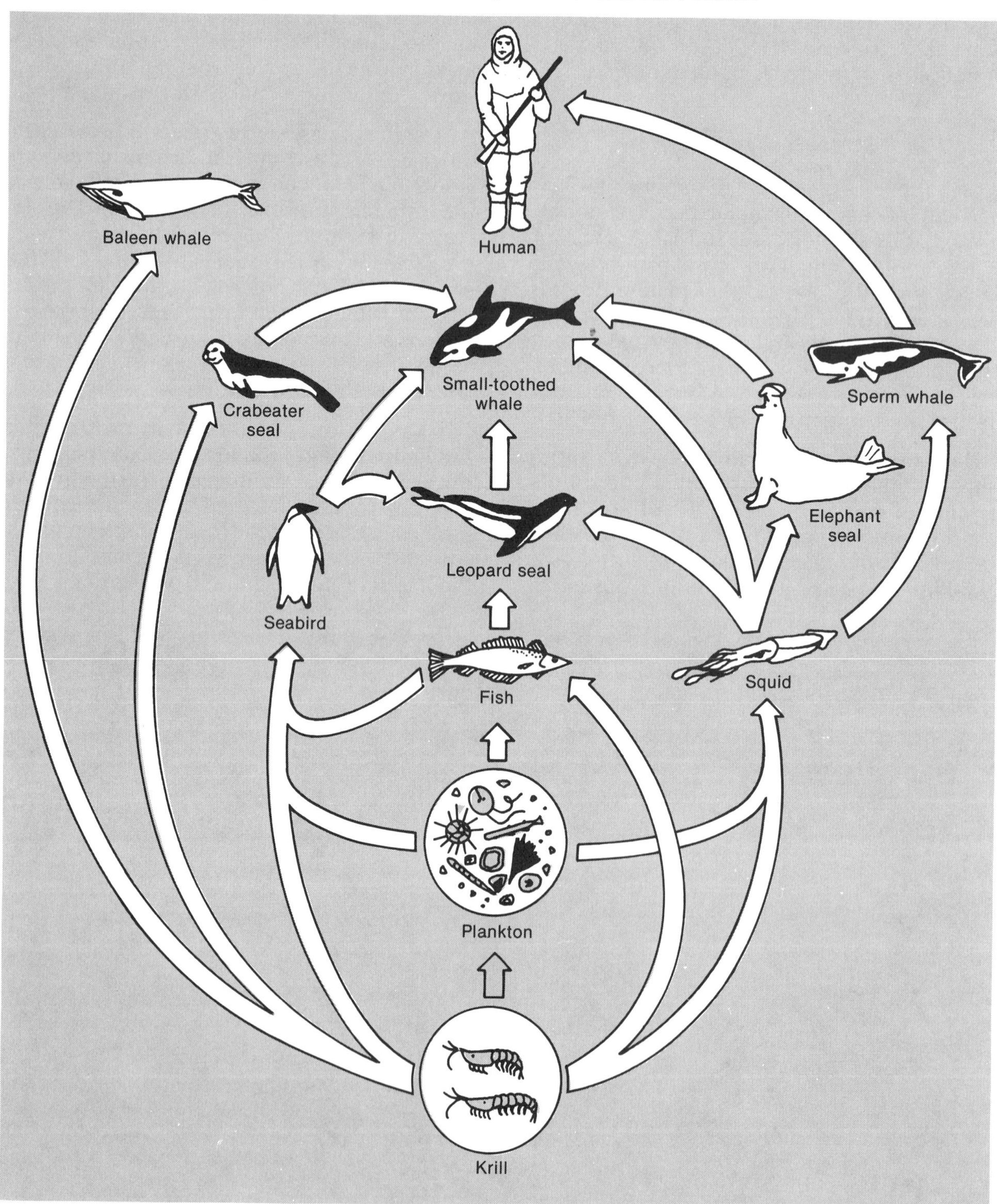

Antarctic waters are among the most productive in the world. The main link in this foodweb is the small krill, shrimp-like creatures that feed on plankton. Krill, in turn, support seabirds, fish, and squid, which are the mainstay of seals and whales.

is half that of grass outside, protein levels and digestibility are significantly higher. In Wind Cave National Park, prairie dog towns occupy less than 5 percent of the area, but bison spend 65 percent of their time per unit area in the towns, mostly feeding (28).

Some species have an unusually prominent position in food webs, being major predators of species on lower levels of the food chain and major prey of species on higher levels. Arctic cod, for example, feed on herbivorous and carnivorous zooplankton (amphipods, copepods, and decapods). Cod, in turn, is an important food of many bird and marine mammal species including gulls, narwhals, belugas, and harp seals (25).

Genetic Diversity

Intraspecific genetic diversity allows species to adapt to changing conditions, thus sustaining ecosystem and species diversity; it also helps produce plants and animals that will support more productive agriculture and forestry. Genetic diversity is distributed unevenly among and within species. Some groups of species appear to be more variable than others: reptile, bird, and mammal species have less than half the genetic variation found in invertebrate species and less than a quarter of that found in many insects and marine invertebrates (34).

The greater the amount of genetic variation in a population, the faster its potential rate of evolution (7). Certain genes are directly important for survival (e.g., genes conferring disease resistance). In addition, genetic diversity enables species to adapt to a wide range of physical, climatic, and soil conditions and to changes in those conditions. Genetic diversity is positively correlated with fitness, vigor, and reproductive success (7,85).

Among marine animals, and probably among terrestrial animals as well, high genetic variability is associated with high species diversity, which in turn is associated with a number of spatially different microhabitats (e.g., tropical and deep sea environments). It seems likely that the high genetic variability provides the flexibility to make finely tuned adjustments to microhabitats.

BENEFITS TO RESEARCH

Research may hold answers to many of the questions facing this complex world. The results of research on the patterns and processes of temperate forests have provided methods for sustainable management of those ecosystems. Knowledge of tropical rain forests will result in similar strategies. Without diversity of species, researchers would not have the needed plant material to develop many vaccines, intravenous fluid, or other medicines. The potential for further advancement has not been fully realized, yet a loss of species diversity will adversely affect future research. Protection of genetic diversity is equally essential, because materials from plants and animals have provided valuable knowledge on viruses, immunology, and disease resistance.

Ecosystem Diversity

Many contributions of ecosystem diversity to global ecological processes, e.g., the role of wetlands in the Earth's oxygen balance, have yet to be demonstrated quantitatively. But the research required to develop and test these hypotheses depends on the full range of diversity. By studying natural ecosystems, scientists are better able to understand how the Earth works.

Knowledge of the role of ecosystem diversity in ecological processes is substantial and growing, largely because of the availability of natural research areas such as the Olympic National Park and the H.J. Andrews Experimental Ecological Reserve in Willamette National Forest (42,81). Relatively undisturbed grasslands in the

Serengeti National Park (Tanzania) and Wind Cave National Park (South Dakota) provide research significant for range management. Research includes, for example, studies of the extent to which grazing intensity increases primary production and the protein content and digestibility of grasses (28). Research on species and natural gene pools also requires ecosystem maintenance.

Representative examples of major ecosystems are used as reference sites for baseline monitoring on productivity, regeneration, and adaptation to environmental change. In addition, evaluation of development projects to ensure they are both economical and sustainable calls for assessment of, among other things, their environmental effects measured against unaltered sites with similar vegetation, soils, and climate.

The Zambezi Teak Forest ecosystem, for example, which yields Zambia's most valuable timber, is declining rapidly, due to excessive logging, fire, and shifting cultivation. If present trends continue, this forest would effectively disappear in 50 years. Attempts at artificial regeneration have met with little success. To improve understanding of natural regeneration, an undisturbed tract of the forest in Kafue National Park is being studied. Continued monitoring of the Kafue tract will provide data needed for assessing costs and benefits of any silviculture system for the Zambezi Teak Forest (72,74).

Ecosystems are also living classrooms. The University of California's Natural Land and Water Reserves System includes 26 reserves representing 106 of the 178 habitat types identified for the State. The reserves are used for instruction and research in botany, geology, ecology, archeology, ethology, paleontology, wildlife management, genetics, zoology, population biology, and entomology (52). Enabling children and adults to experience different ecosystems is an effective way to teach ecological processes, genetic variation, community composition and dynamics, and human relations with the natural world.

Species Diversity

Species diversity is the basis for many fields of scientific research and education. The array of invertebrates used in research illustrates the importance of diversity to the advancement of science. The 100 or so species of Hawaiian picture-winged fruit flies are the organisms of choice for basic research on genetics, evolutionary biology, and medicine. Tree snails of Hawaii and the Society Islands provide ideal material for research on evolution and genetic variation and differentiation (57).

Bristlecone pines, the oldest known living organisms and found only in the U.S. Southwest, are used to calibrate radiocarbon dates and hence, are important for archeology, prehistory, and climatology (62). Contributions of plant and animal species to biomedical research and drug synthesis abound (63,71). Examples include:

- Desert pupfishes, found only in the Southwest, tolerate salinity twice that of saltwater and are valuable models for research on human kidney disease (63).
- Sea urchin eggs are used extensively in experimental embryology, in studies of cell structure and fertilization, and in tests on the teratological effects of drugs (98).
- Medicinal leeches are important in neurophysiology and research on blood clotting (98).
- An extract of horseshoe crabs provides the quickest and most sensitive test of vaccines and intravenous fluids for contamination with bacterial endotoxins (98).
- Butterfly species are used in research on cancers, anemias, and viral diseases (82).
- The study of sponges is making substantial contributions to structural chemistry, pharmaceutical chemistry, and developmental biology and has also resulted in the discovery of novel chemical compounds and activities. D-arabinosyl cytosine, an important synthetic antiviral agent, owes its development to the discovery of spongouridine, which was isolated from a Jamaican

Photo credit: J. Cohen, National Zoo

The armadillo is one of only two animal species known to contract leprosy. These animals now serve as research models to find a cure.

sponge. Three derivatives of this compound have been patented as antiviral and anticancer drugs (10).

Genetic Diversity

Genetic variability is one of the characteristics of fruit flies, tree snails, and butterflies that makes them so useful for research. The unusual range of diversity among the races, varieties, and lines of corn contributes to its enormous value for basic biological research. One example is the discovery and analysis of regulatory systems that control gene expression, which added a new dimension to the study of inheritance (21).

The genus *Nicotiana* has also been used widely in genetic and botanical research largely because of the great variation among its species (84). The varied reactions to specific viruses characteristic of many *Nicotiana* species provide a potential tool for separating and identifying viruses. *Nicotiana* species have been involved in numerous discoveries of virus research (e.g., virus transmissibility, purification, and mutability) (35).

Special genetic stocks are essential research tools. For example, inbred lines of chickens developed at the University of California at Davis are used worldwide for research on immunology and disease resistance of chickens. Mutant stocks of chickens also serve as genetic models for scoliosis (lateral curvature of the spine) and muscular dystrophy in humans (58).

BENEFITS TO CULTURAL HERITAGE

Throughout history, societies have put great value on physical features of their environment. In developed and developing countries, a diversity of ecosystems is a source of esthetic, historic, religious, and ritualistic values. Species diversity assures people of national and state symbols, and many such symbols are protected. Genetic diversity continues in part because of the cultural value of plants and animals. Gardeners around the world share seed material ensuring genetic survival.

Ecosystem Diversity

Natural ecosystems have great cultural (including religious, esthetic, and historic) importance for many people. Mountains are the focus of religious celebrations and rituals throughout the world: Mount Kenya, Mount Everest, Mount Fuji, Mount Taishan in China, and Black Mesa in Arizona. Forests also have great spiritual value: probably the only surviving examples of primary forest in southwestern India are sacred groves—ancient natural sanctuaries where all living creatures are protected by the deity to which the grove is dedicated. Removing even a twig from the grove is taboo (36).

People who lead subsistence-based lives identify closely with the ecosystems on which they depend. Two examples are the Guarao people in the mangrove swamps and savannas of Venezuela's Orinoco Delta (39) and the Inuit people in the tundra of the North American Arctic (9,24). The economic, social, and spiritual elements of the relationship between such peoples and the ecosystems that support them are inseparable.

Ecosystems define and symbolize relationships between human beings and the natural world and express cultural and national identity. In the United States, the landscapes protected in wilderness areas, national parks, monuments, and preserves are full of historical meaning and show the close ties between America the nation and America the land. Examples of these are pre-Columbian Indian habitations at Mesa Verde in Colorado; symbols of the opening of the Midwest and West at Voyageurs Park in Minnesota; and combinations of wilderness preservation and human occupation including current subsistence-use at Kobuk Valley in Alaska (66,94).

Species Diversity

Whereas the Continental Congress in 1782 adopted the bald eagle as a national symbol; and

Whereas the bald eagle thus became the symbolic representation of a new nation under a new government in a new world; and

Whereas by that act of Congress and by tradition and custom during the life of this Nation, the bald eagle is no longer a mere bird of biological interest but a symbol of the American ideals of freedom . . .

—Bald and Golden Eagle Protection Act of 1940

Photo credit: National Wildlife Federation

Cultural value of species is exemplified by the bald eagle, adopted by the Continental Congress as a symbol of the United States.

When Congress adopted the bald eagle as a national symbol, it was responding to an ancient human need to identify with other species. All over the world and throughout history, people have adopted animals and plants as emblems, icons, symbols, and totems and invested them with ideals and values, adopted them as representations of particular characteristics of their culture and society, sought the power and authority they stand for, or venerated them as embodiments of fruitfulness and life itself.

The endangered bowhead whale plays a pivotal cultural role in several Yupik and Inupiat Eskimo villages in northern Alaska. Bowhead whale hunting is the first and most important activity in the subsistence cycle. It is a major social unifier, providing community identity and continuity with the past. The division, distribution, and sharing of bowhead whale meat and skin involve the entire community, strengthening kinship and communal bonds. Important ceremonies, celebrations, and feasts accompany the harvest of a bowhead whale and the distribution and sharing of its meat (4,5).

Port Orford Cedar (*Chamaecyparis lawsoniance*), prized for its cultural and economic values, has become the focus of a recent controversy. It grows only in a small area of southern Oregon and northern California, where it produces some of the area's highest priced timber. Top quality may cost as much as $3,000 per 1,000 board-feet. This price reflects demand from Japan, where it is used in homes and temples as a substitute for the no longer available Japanese Hinoki cypress. It also has great cultural importance for Native Americans of the Hupa, Yurok, and Karok tribes in northwestern California, who regard it as sacred and use the wood in homes and religious ceremonies. Management of remaining stands of the cedar has become controversial, because mature trees are in short supply and threatened by a tree-killing root-rot disease, spread partly by logging operations (22).

Native Americans seek to reserve all the Port Orford Cedar growing on formal tribal land—now administered by the U.S. Forest Service—for ceremonial purposes. Other citizens' groups seek a management plan that would control logging operations and restrict loggers' access to some areas to reduce the spread of the fungus. Scientists at the Forest Service and Oregon State University are exploring the genetic diversity of the species in an effort to develop strains resistant to the fungus (22).

In South and Southeast Asia, trees, Asian elephants, monkeys, cobras, and birds figure prominently in tribal religions and have been taken into the pantheons of Hinduism and Buddhism. Certain tree species, such as *Ficus religiosa*, are sacrosanct and may not be cut down (2,20); political authorities often invoke the sanction of animals to win popular support (61). Interspecific loyalties persist; the hornbill, central figure of the Gawai Kenya-lang or Hornbill Festival of the Iban people in Sarawak, Malaysia, is also the official emblem of the state (50).

In urban North America, species also express community identity. Inwood, Manitoba, proclaims itself the garter snake capital of the world (after the mass matings of red-sided garter snakes that occur nearby) (67), and Pacific Grove, California, dubs itself Butterfly Town, USA (after the spectacular colonies of Monarch butterflies that overwinter there) (98). These actions are partly commercial acumen—the phenomena are tourist attractions—but they also reflect civic pride and perhaps something deeper as well.

Genetic Diversity

Many crop varieties and livestock breeds persist because they are culturally valuable to different societies. This group includes plants and animals with religious and ceremonial significance—such as the festival rices of Nepal and Mithan cattle in northern Burma and northeastern India (40)—as well as varieties valued for their contribution to the traditional diet. Farmers in the Peruvian Andes commonly plant their potato fields with many varieties (often 30 or more), producing a mixture of colors, shapes, textures, and flavors to enhance the diet (14). In northwestern Spain, a mosaic

Photo credit: G. Nabhan

Hopi Indian garden of mixed crops illustrates ancient horticultural traditions that persist on this continent.

of local varieties of beans and other legumes is grown, each variety intended for a particular dish in the traditional cuisine (3).

A growing number of Americans value traditional cultivars and breeds for their history and for their esthetic and culinary qualities. Native Americans, helped by grassroots organizations, continue to grow traditional varieties of corn, chiles, beans, and squash (91). Hispanic-American farmers in the Southwest prefer native corn for its texture, flavor, and color, even though its yield is only one-third to one-fourth of hybrid corn (64). The cultural value of rare livestock breeds is exemplified by Texas Longhorn cattle (which have a prominent place in American history) and Navaho sheep (whose fleece is important to Navaho weaving).

Gardeners have organized national and regional networks to conserve some plant varieties because they have better taste, have links with national, local, and ethnic history; are suitable for the home garden; and because of the abundance of colors and forms found among old and local varieties of potatoes, corn, beans, and other crops (33,43,47,64,91).

BENEFITS TO RECREATION AND TOURISM

Millions of people worldwide derive benefits from recreation and tourism provided by biological diversity. Without diverse ecosystems, countries would lose tremendous amounts of foreign exchange. Without wilderness areas, national parks, or national forests, city dwellers would have no place to "escape" the daily pressures. Species diversity is essential to the millions of wildlife photographers, bird lovers, and plant and animal watchers. And without genetic diversity, horticulturists, gardeners, animal breeders, and anglers would find little enjoyment in their avocations.

Ecosystem Diversity

State and National Parks in the United States attract 700 to 800 million visitors per year (73,74), and National Forests receive some 200 million visitors per year (93). One reason for these visits—indeed, some surveys suggest the main reason—is to enjoy the variety of landscapes the parks and forests protect (83). Sightseeing accounts for more recreation-visitor days (52 million) in National Forests than any other recreation activity except camping (60 million) (93).

Ecosystem diversity is a significant recreational asset in developing countries as well. In Venezuela, the mangrove forests of Morrocoy National Park attract 250,000 to 500,000 visitors per year (39); in Nepal, mountain landscapes, rhododendron forests, and fauna bring in foreign exchange (55).

Species Diversity

About 95 million Americans a year participate in nonconsumptive recreational uses of wildlife (observing, feeding, or photographing wild plants and animals); each year 54 million Americans fish and 19 million Americans hunt for sport. In the process they spend $32.4 billion per year (95).

Surveys of American recreational uses of wildlife reveal that a number of different species interest people. Recreational hunters in North America pursue some 90 species (73,74). Millions of Americans take time to observe not only birds and mammals, but also amphibians, reptiles, butterflies, spiders, beetles, and other arthropods (95).

Little data exist on wildlife recreational use by people in developing countries, but for several nations wildlife-based tourism is big business. The spectacular wild animals of east and southern Africa are the resource base of a tourist industry that brings millions of dollars in foreign exchange. In 1985, Kenya netted about $300 million from almost 500,000 visitors, making wildlife tourism the country's biggest earner of foreign exchange (1).

Genetic Diversity

Millions of home gardeners and members of horticultural and animal breed associations derive recreational benefit from genetic diversity. So, too, do millions of anglers who take advantage of stocking and enhancement programs. Tourism associated with genetic diversity involves fewer people, although the Rare Breeds Survival Trust in the United Kingdom receives 100,000 visitors a year. In North America, at least 10 million people visit the some 200 living historical farms—open-air museums that recreate and interpret agricultural and other activities of a particular point in history (91).

BENEFITS TO AGRICULTURE AND HARVESTED RESOURCES

In agriculture, a diversity of ecosystems, species, and genetic material provides increased amounts and quality of yields. In a world where population is rapidly increasing, assuring a continued increase in harvested resources is essential. Diversity in an agroecosystem provides habitat for predators of crop pests and breeding sites for pollinators. Diversity of species can be a buffer against economic failure and can also play an important role in pest management. Further, the use of genetic materials by breeders has attributed to at least 50 percent of the increase in agriculture yields and quality.

Ecosystem Diversity

Both diversity and isolation affect the ability of pests to invade a crop. They also affect the supply of pests' enemies. Uncultivated habitats next to croplands contain wildflowers, which contain important nutrients for the adult stages of predatory and parasitic insects (37). Wildflowers also support essential alternate hosts for parasites, especially in seasons when pests they prey on are not present. In California, for instance, wild brambles (*Rubus*) provide an off-season reservoir of prey for wasps, which control a major grape pest. This arrangement saves grape growers $40 to $60 per acre in reduced pesticide costs (6,54).

A variety of wild habitats also provides food, cover, and breeding sites for pollinators. Wild pollinators (chiefly insects) make major contributions to the production of at least 34 crops grown or imported by the United States, with a combined annual average value of more than $1 billion. They are the main pollinating agents in the production of cranberry and cacao, the propagation of red clover, and the production and propagation of cashew and squash. They are also significant pollinators for such crops as coconut, apple, sunflower, and carrot. The abundance of wild pollinators is largely determined by the availability of ecosystem diversity (woods, scrub, bare ground, moist areas, patches of flowers) within flight range of the crops to be pollinated (73,74).

Permanent pastures and rangelands occupy one-fourth of the Earth's land surface (31). Because they support most of the world's 3 billion head of domesticated grazing animals (45), rangelands can be considered harvested ecosystems, where the nutrients and solar energy of marginal lands are converted into meat, milk, wood, and other goods.

In the United States, 34 rangelands are involved and include plains, prairie, mountain grassland, and Texas savanna (93). Pastoral nomadism and migrations by wild herbivores are traditional ways of using these resources. Modern ways include hauling sheep between summer and winter ranges, which may be 300 to 400 kilometers apart in the intermountain region (12).

Species Diversity

Diversity of harvestable species acts as a buffer that allows people in fluctuating environments to cope with extremes. For instance, in Botswana, five wild plant species are extensively used by pastoralists and river people, but an additional 50 or more species are resorted to in times of drought (17).

Harvested species provide much of the subsistence of indigenous peoples and rural communities throughout the world. Wild bearded pig and deer contribute about 36,000 tons of meat a year to rural diets in Sarawak, Malaysia. This amount of meat from domestic animals would cost about $138 million. (15). Per capita consumption of harvested food by Inuit in the North American Arctic averages annually from 229 kg (504 lb) to 346 kg (761 lb). The per capita cost of buying substitute food (usually of lower nutritional and cultural value) was estimated to be $2,100 per year (1981 figures) (4,101).

The commercial timber, fishery, and fur industries obtain most of their resources by harvesting wild species. Harvested resources are also major contributors to the pharmaceutical industry, and to many other industries as well. The average annual value of the wild resources produced and imported by the United States between 1976 and 1980 was about $27.4 billion, of which $23 billion was timber (73,74).

Many species are involved, but most of them are economically significant only to the tradesmen involved. Even so, the number of harvested species might run up to more than a hundred. For example, it takes on average 70 species to make up 90 percent of the annual value of U.S. commercial fishery landings (74).

In agriculture, two types of diversity are useful in pest management programs: crop diversity and pest enemy diversity. Crop diversity (multiple cropping) can promote the activity of beneficial insects. For example, to attract *Lycosa* wolf spiders, the main predators of corn borers in Indonesia, farmers interplant the corn with peanuts (46). In California, lygus bugs, one of the main pests of cotton, are controlled somewhat by strip-planting alfalfa, which the bugs prefer to cotton (11). Pest enemy diversity includes introduced as well as native enemies. The Florida citrus industry saves $35 million per year by using three parasitic insect species that were imported and established at a cost of $35,000. Some 200 foreign insect pests in the United States are controlled by introduced parasites and predators (63).

A long-standing use of wild species diversity is as a source of new domesticates. In the United States, the combined farm sales and import value of domesticated wild species is well over $1 billion per year. The domestication of two major groups of resources—timber trees and aquatic animals—has only begun and is at about the same stage that agricultural domestications were some 5,000 years ago. But agricultural and horticultural domestications are still occurring.

Among the successful new food crops developed this century are kiwifruit, highbush blueberry, and wild rice (most of the wild rice

Photo credit: M.A. Altieri

Two intercropping systems—fava beans and brussels sprouts, and wild mustard and brussels sprouts—demonstrate the benefits of diversity to agriculture. Both systems benefit the brussels sprouts crop: wild mustard acts as a trap crop of flea beetles, and fava beans fix nitrogen with possible benefits to brussels sprouts yields.

produced in the United States is domesticated). New and incipient forage crops include Bahia grass, desmodium, and several of the wheatgrasses. Red deer and aquaculture species such as catfish, hardshell clam, and the giant freshwater prawn, are among the newly domesticated livestock. Loblolly pine, slash pine, Parana pine, and balsa are some of the new timber domesticates (73,74).

Domestication of wild species increases the economic benefits of wild species by improving product quality and by raising yields. It can also make a valuable contribution to rural development in areas that are marginal for conventional crops and livestock. Nepal's Department of Medicinal Plants has organized the farming of two native species (*Rauvolfia serpentina* and *Valeriana wallichii*) for example, and it is investigating propagation of several other wild species that are sources of drugs, perfumes, and flavors for export. Scientists in Zambia and Botswana are working on the domestication of mungongo tree, whose fruits are used for food and oil and whose wood is valued for carvings (74).

Genetic Diversity

Health and long-term productivity of wild resource species—from game animals to timber trees to food and sport fish—depend on genetic diversity within and among the harvested populations. If the best individuals (biggest animals, tallest trees) are harvested before they repro-

Photo credit: M. Plotkin

Medicine from nature: *Croton* sp., known as "sangre de grado" in the Peruvian Amazon. This tree produces a sap used for a variety of medicinal purposes.

duce, then the productivity and adaptability of the population will progressively decline.

In addition, certain populations are better adapted to particular locations than others. For example, chinook, coho, and sockeye salmon from different rivers are genetically distinct; these distinctions reflect differences in the physical and chemical characteristics of the streams in which they originated (69,70). Diversity needs to be maintained so that any restocking to compensate for overharvesting or habitat degradation can use populations that are adapted to the specific environmental conditions.

In agriculture, genetic diversity in the form of readily available genes reduces a crop's vulnerability to pests and pathogens. Resistance genes can be introduced as long as a high degree of genetic diversity is maintained in offsite collections, onsite reserves, and agroecosystems. U.S. plant breeders keep a substantial supply of diversity in cultivars, parental lines, synthetic populations, and other breeders' stocks ready for use (13,26).

The genetic variation in domesticated plants and animals and in their wild relatives is the raw material with which breeders increase yields and improve the quality of crops and livestock. Use of genetic resources during this century has revolutionized agricultural productivity. In the United States from 1930 to 1980, yields per unit area of rice, barley, and soybeans doubled; wheat, cotton, and sugarcane yields more than doubled; fresh-market tomato yields tripled; corn, sorghum, and potato yields more than quadrupled; and processing-tomato yields quintupled (65,92).

At least half of these increases have been attributed to plant breeders' use of genetic diversity. The gain due to breeding is estimated to be 1 percent per year for corn, sorghum, wheat, and soybeans, due mainly to improvements in grain-to-straw ratio, standability, drought resistance, tolerance of environmental stress, and

Photo credit: United Nations—M. Tzovaras

Plant breeders' use of genetic diversity has significantly increased the productivity of crops such as wheat.

pest and disease resistance (18,27,79). Similarly, the average milk yield of cows in the United States has more than doubled during the past 30 years; about one-fourth of this increase is due to genetic improvement (89).

Developing countries have also achieved increased production of major crops. The Green Revolution that has transformed heavily populated Asian countries is founded on use of particular genes. High-yielding varieties of rice, for example, rely on a gene from a traditional variety for the "dwarf" stature that enables the plant to channel nutrients from fertilizers into grain production without getting top-heavy and falling over before harvest time. Although the dwarfing trait is effective in many locations, the high-yielding varieties need other genetic characteristics from many different varieties. The rice variety IR36, used in many countries to sustain yield gains, was derived by crossbreeding 13 parents from 6 countries (19,87).

Progress in tomato improvement in the United States has followed the use of exotic germplasm (traditional cultivars, wild forms of the domesticated species, and exclusively wild species). Fruit quality (color, sugar content, solids content); adaptations for mechanized harvesting; and resistance to 15 serious diseases have been transferred to the tomato from its wild relatives. One researcher noted:

> Resistance to some of these diseases is mandatory for economic production of the crop in California, and it is doubtful whether the State's tomato industry would exist without these and other desired traits derived from exotics (77).

Rice and tomato illustrate the importance of maintaining as much of the genetic variation remaining within the domesticates and their wild relatives as possible, because both crops have benefited from genes occurring in a single population and nowhere else. Asian rice cultivars get their resistance to grassy stunt virus, a disease that in one year destroyed 116,000 hectares (287,000 acres), from one collection of *Oryza nivara* (53). The gene for a jointless fruit-stalk (a trait that assists mechanized harvesting and is worth millions of dollars per year) in tomato is found in a single population of a wild relative (*Lycopersicon cheesmanii*) unique to the Galapagos Islands (78).

A variety of genetic resources is being used in the breeding of livestock, particularly cattle and sheep. Crossbreeding Brahman cattle with Hereford, Angus, Charolais, and Shorthorn breeds has had a major impact on commercial beef production in North America (30). A number of African cattle breeds are notable sources of disease and pest resistance (West African Shorthorn to trypanosomiasis, N'dama and Baole to dermatitis, Zebu to ticks) (34). The Finnish landrace of sheep was almost lost before its high level of reproductive efficiency was discovered. It has now been incorporated into commercial mating lines in the United Kingdom and North America (30).

Yield and quality improvements can continue to be made and defended against pests and pathogens, provided plant and animal breeding continues to be supported and the genetic diversity that breeders draw on is maintained. Indeed, there is no option but to go on improving crops and livestock if world agriculture is to respond successfully to economic and environmental changes and to the new strains of pests and diseases that evolve to overcome existing resistance.

VALUES AND EVALUATION OF BIOLOGICAL DIVERSITY

Biological diversity benefits everyone, is valued by many (in a variety of ways), but is owned by no one. Thus, its evaluation is fraught with complexity. There are two broad classes of value: economic and intrinsic.

Economic Value

Economic evaluation potentially covers all functional benefits described in this chapter, ranging from tangible benefits from harvested resources and breeding materials to spiritual and other cultural benefits. The ability to calculate these values varies, however. In the cases where markets exist, calculations are easily determined (at least $27.4 billion per year in the United States for commercially harvested wild species, as noted earlier). In other cases, values are more difficult to calculate, and "shadow prices" may be used to approximate values for such benefits as ecological processes and recreation. For cultural and esthetic values, economic valuation may be impossible.

If humans interacted in a system with limited resources, then markets would allow equilibrium prices to emerge for all commodities, services, amenities and resources. These prices would reflect the relative values (including social values) of each item. The essential premises for economic valuation are utility and scarcity (75).

But for most benefits of biological diversity, free market principles do not apply. Maintenance of biological diversity is a "nonrival" good (it benefits everybody), and it is a "nonexclusive" good (no person can be excluded from the satisfaction of knowing a species exists), as are many of its benefits (research and education, cultural heritage, nonconsumptive recreation, use of genetic resources). And it is not clear that market-oriented logic is adequate to deal with two cardinal features of biological diversity: its potential for indefinite renewability (long-time horizon) and for extinction (irreversibility) (75).

Intrinsic Value

Intrinsic evaluation acknowledges that other creatures have value independent of human recognition and estimation of their worth. The concept is both ancient and universal. A spokesperson of the San people of Botswana put it this way:

> Once upon a time, humans, animals, plants, and the wind, sun, and stars were all able to talk together. God changed this, but we are still a part of a wider community. We have the right to live, as do the plants, animals, wind, sun, and stars; but we have no right to jeopardize their existence (16).

This preceding statement might be supported by Americans who believe in "existence values" —values that are defined independently of human uses (68). This belief implies a human obligation not to eradicate species or habitats, even if doing so harms no human. A 3-year study of American attitudes toward wildlife found that the majority seemed willing to make substantial social and economic sacrifices to protect wildlife and its habitats (51). Advocates of wildlife protection maintain that "it makes me feel better to know there are bears in the area, even though I'd just as soon never run into one" (76). Proponents of biological diversity argue that even if diversity is functionally redundant or has no utilitarian worth, it should be maintained just "because it is there."

CONSTITUENCIES OF DIVERSITY

Biological diversity benefits a variety of interest groups, so its constituency is enormous but fragmented by the interests of particular groups. Each group may appear small compared with the Nation as a whole. Collectively, however, these groups and their combined concern amount to the national interest in maintaining biological diversity.

Public Awareness

A major obstacle to promoting effective and long-term maintenance of biological diversity is the lack of awareness on the part of the general public of the importance of diversity (in the broader sense). It is easy to understand why the loss of biological diversity has difficulty capturing public attention. First, the concept is complex to grasp. For this reason, efforts to solicit support have appealed to emotionalism associated with the loss of particularly appealing species or spectacular habitats (86). Although effective in many cases, this approach has the effect of limiting the constituency and the boundaries of the problem. A second reason is that the more pervasive threats to diversity, such as habitat loss or narrowing of agricultural crop genetic bases, are not dramatic events that occur quickly. The difficulty is one of responding to a potentially critical problem that, for the average person, seems to lack immediacy.

Finally, promoting the case for biological diversity maintenance is also difficult because of the proliferation of environmental problems brought to public attention in the last decade or two, including acid rain, ozone depletion, the greenhouse effect, and loss of topsoil. "All these environmental problems have the apocalyptic potential to destroy, yet in every case the cause, imminence, and scope of that power are subject to polarizing (and eventually paralyzing) interpretation" (29).

Notwithstanding these difficulties, the environmental movement of the 1970s elevated environmental quality to a major public policy concern. Although the momentum of public attention may have slowed in the 1980s, it is clear that concern for the environment remains firmly entrenched in the collective consciousness of the American public. A 1985 Harris poll, for example, indicated that 63 percent of Americans place greater priority on environmental cleanup than on economic growth (41).

Balancing Interests and Perspectives

In assessing the level of public resources to be directed toward maintaining biological diversity, it is important to maintain a frame of reference of how, when, and for whom biological diversity is important. Such a perspective should consider:

1. varying perceptions on the value of biological diversity and threats to it;
2. an awareness that only some diversity can be or probably will be saved; and
3. a recognition that resources available to address efforts are limited.

As mentioned earlier, biological diversity is not at present a pervasive concern for many people, or at least there is no consensus that as much diversity must be conserved as possible. While earlier sections of this chapter identified large constituencies that value biological diversity, some elements of society remain apathetic to the issue, and others support efforts to eliminate various components of diversity. For example, considerable resources are directed to reducing populations or even eliminating entire species of pests, pathogens, or predators that threaten agriculture and human health. In terms of public policy, such efforts imply a need to recognize that in some cases diversity maintenance and other human interests can conflict. It should be noted, however, that conflicts stem less from the existence of diversity than from the altered abundance of particular species.

CHAPTER 2 REFERENCES

1. Achiron, M., and Wilkinson, R., "Africa: The Last Safari?" *Newsweek*, Aug. 18, 1986.
2. Agrawal, S.R., "Trees, Flowers and Fruits in Indian Folk Songs, Folk Proverbs, and Folk Tales," *Glimpses of Indian Ethnobotany*, S.K. Jain (ed.) (New Delhi, Bombay, Calcutta, India: Oxford and IBH Publishing Co., 1981).
3. Alaman Castro, M.C., Casanova Pena, C., and Bueno Perez, M.A., "La Recognida de Germoplasma por el Noroeste de Espana," *Plant Genetic Resources Newsletter* 53:41-43, 1983.
4. Alaska Consultants and Stephen Braund & Associates, *Subsistence Study of Alaska Eskimo Whaling Villages* (Washington, DC: U.S. Department of the Interior, 1984).
5. Alaska Eskimo Whaling Commission, *Alaska Eskimo Whaling* (Barrow, AK: 1985).
6. AliNiazee, M.T., and Oatman, E.R., "Pest Man-

agement Programs," *Biological Control and Insect Pest Management*, D.W. Davis, et al. (eds.) (Oakland, CA: University of California, Division of Agricultural Sciences, 1979).

7. Ayala, F.J., "The Mechanisms of Evolution," *Scientific American* 239(3):56-69, 1978.
8. Basson, P.W., Burchard, Jr., J.E., Hardy, J.T., and Price, A.R.G., *Biotopes of the Western Arabian Gulf: Marine Life and Environments of Saudi Arabia*, Aramco Department of Loss Prevention and Environmental Affairs (Dhahran: 1977).
9. Berger, T.R., *Village Journey: The Report of the Alaska Native Review Commission* (New York: Hill & Wang, 1985).
10. Berquist, P.R., "Sponge Chemistry—A Review," *Colloques Internationaux du CNRS* 291:383-392, 1978.
11. Bishop, G.W., Davis, D.W., and Watson, T.F., "Cultural Practices in Pest Control," *Biological Control and Insect Pest Management*, D.W. Davis, et al. (eds.) (Oakland, CA: University of California, Division of Agricultural Sciences, 1979).
12. Box, T.W., "Potential of Arid and Semi-Arid Rangelands," *Potential of the World's Forages for Ruminant Animal Production*, 2d ed., R.D. Child and E.K. Byington (eds.) (Morrilton, AR: Winrock International Livestock Research and Training Center, 1981).
13. Brown, W.L., "Genetic Diversity and Genetic Vulnerability—An Appraisal," *Economic Botany* 37(1):4-12, 1983.
14. Brush, S.B., Carney, H.J., and Huaman, Z., "Dynamics of Andean Potato Agriculture," *Economic Botany* 35(1):70-88, 1981.
15. Caldecott, J., and Nyaoi, A., "Hunting in Sarawak," report prepared for the Ad hoc Subcommittee on Mammals and Birds of the Sarawak State Legislative Assembly Special Select Committee on Flora and Fauna, Sarawak Forest Department, Kuching, Malaysia, 1985.
16. Campbell, A.C., "Traditional Wildlife Populations and Their Utilization," *Which Way Botswana's Wildlife*? proceedings of the Symposium of the Kalahari Conservation Society, Gaborone, Botswana, Apr. 15-16, 1983.
17. Campbell, A.C., *The Use of Wild Food Plants and Drought in Botswana* (Gaborone, Botswana: National Museum, no date).
18. Castleberry, R.M., Crum, C.W., and Krull, C.F., "Genetic Yield Improvement of U.S. Maize Cultivars Under Varying Fertility and Climatic Environments," *Crop Science* 24(1):33-36, 1984.
19. Chang, T.T., Adair, C.R., and Johnston, T.H., "The Conservation and Use of Rice Genetic Resources," *Advances in Agronomy* 35:37-91, 1982.
20. Chaudhuri, R.H.N., and Pal, D.C., "Plants in Folk Religion and Mythology," *Glimpses of Indian Ethnobotany*, S.K. Jain (ed.) (New Delhi, Bombay, Calcutta, India: Oxford and IBH Publishing Co., 1981).
21. Coe, E.H., and Neuffer, M.G., "The Genetics of Corn," *Corn and Corn Improvement*, 2d ed., Monograph 18, G.F. Sprague (ed.) (Madison, WI: American Society of Agronomy, 1977).
22. Cohn, L., "The Port Orford Cedar," *American Forests*, July 1986, pp. 16-19.
23. Cooke, H.J., "The Kalahari Today: A Case of Conflict Over Resource Use," *The Geographical Journal* 151(1):75-85, 1985.
24. Damas, D. (ed.), *Arctic, Volume 5: Handbook of North American Indians* (Washington, DC: Smithsonian Institution, 1984).
25. Davis, R.A., Finley, K.J., and Richardson, W.J., *The Present Status and Future Management of Arctic Marine Mammals in Canada* (Government of Northwest Territories, Yellowknife: Science Advisory Board, 1980).
26. Duvick, D.N., "Genetic Diversity in Major Farm Crops on the Farm and in Reserve," *Economic Botany* 38(2):161-178, 1984.
27. Duvick, D.N., "Plant Breeding: Past Achievements and Expectations for the Future," *Economic Botany* 40(3):289-297, 1986.
28. Dyer, M.I., Detling, J.K., Coleman, D.C., and Hilbert, D.W., "The Role of Herbivores in Grasslands," *Grasses and Grasslands: Systematics and Ecology*, J.R. Estes, R.J. Tyrl, and J.N. Brunken (eds.) (Norman, OK: University of Oklahoma Press, 1982).
29. Ebisch, R., "A Layman's Guide to Modern Menaces," *TWA Ambassador*, 18(10):49-58, October 1985.
30. Fitzhugh, H.A., Getz, W., and Baker, F.H., "Status and Trends of Domesticated Animals," OTA commissioned paper, 1985.
31. Food and Agriculture Organization (FAO) of the United Nations, *1984 FAO Production Yearbook*, FAO Statistics Series 38 (Rome: 1985).
32. Foster, R.B., "Heterogeneity and Disturbance in Tropical Forest Vegetation," *Conservation Biolog: An Evolutionary-Ecological Perspective*, M.E. Soulé and B.A. Wilcox (eds.) (Sunderland, MA: Sinauer Associates, 1980).
33. Fowler, G., "Report on Grassroots Genetic Conservation Efforts," OTA commissioned paper, 1985.
34. Frankel, O.H., and Soulé, M.E., *Conservation*

and Evolution (Cambridge, MA: Cambridge University Press, 1981).

35. Fulton, R.W., "Nicotianas as Experimental Virus Hosts," *Nicotiana: Procedures for Experimental Use*, Technical Bulletin 1586, R.D. Durbin (ed.) (Washington, DC: U.S. Department of Agriculture, 1979).
36. Gadgil, M., and Vartak, V.D., "Sacred Groves of Maharashtra: An Inventory," *Glimpses of Indian Ethnobotany*, S.K. Jain (ed.) (New Delhi, Bombay, Calcutta, India: Oxford and IBH Publishing Co., 1981).
37. Hagen, K.S., and Bishop, G.W., "Use of Supplemental Foods and Behavioral Chemicals to Increase the Effectiveness of Natural Enemies," *Biological Control and Insect Pest Management*, D.W. Davis, et al. (eds.) (Oakland, CA: University of California, Division of Agricultural Sciences, 1979).
38. Hamilton, L.S., and King, P.N., *Tropical Forested Watersheds: Hydrologic and Soils Response to Major Uses or Conversions* (Boulder, CO: Westview Press, 1983).
39. Hamilton, L.S., and Snedaker, S.C. (eds.), *Handbook for Mangrove Area Management* (Honolulu, HI: East-West Center, 1984).
40. Harlan, J.R., *Crops and Man* (Madison, WI: American Society of Agronomy, 1975).
41. Harris, L., "Current Public Perceptions, Attitudes, and Desires on Natural Resources Management," *Transactions of the 50th North American Wildlife and Natural Resources Conference* (Washington, DC: Wildlife Management Institute, 1985), pp. 68-71.
42. Harris, L.D., *The Fragmented Forest: Island Biogeographic Theory and the Preservation of Biotic Diversity* (Chicago, IL, and London: University of Chicago Press, 1984).
43. Henson, E.L., "An Assessment of the Conservation of Animal Genetic Diversity at the Grassroots Level," OTA commissioned paper, 1985.
44. Illinois Natural History Survey, "Slow Growth and Short Life Spans of Illinois River Carp," *The Illinois Natural History Survey Reports* 257, 1986.
45. International Union for Conservation of Nature and Natural Resources (IUCN), *World Conservation Strategy: Living Resource Conservation for Sustainable Development* (Gland and Nairobi: International Union for Conservation of Nature and Natural Resources-United Nations Environmental Program-World Wildlife Fund, 1980).
46. IRRI Cropping Systems Program, *1973 Annual Report* (Los Banos, Philippines: The International Rice Research Institute, 1973).
47. Jabs, C., *The Heirloom Gardener* (San Francisco, CA: Sierra Club Books, 1984).
48. Janzen, D.H., "Guanacaste National Park: Tropical Ecological and Cultural Restoration," project report to Servicio de Parques Nacionales de Costa Rica, Fundacion de Parques Nacionales de Costa Rica, Fundacion Neotropica, Nature Conservancy International Program. Department of Biology, University of Pennsylvania, Philadelphia, 1986.
49. Kaplan, E.H., *A Field Guide to Coral Reefs of the Caribbean and Florida Including Bermuda and the Bahamas* (Boston, MA: Houghton Mifflin, 1982).
50. Kavanagh, M., "Planning Considerations for a System of National Parks and Wildlife Sanctuaries in Sarawak," *Sarawak Gazette* 61:15-29, 1985.
51. Kellert, S.R., "Americans' Attitudes and Knowledge of Animals," *Transactions of the 45th North American Wildlife and Natural Resources Conference, 1980* (Washington, DC: Wildlife Management Institute, 1980).
52. Kennedy, J.A., "Protected Areas for Teaching and Research: The University of California Experience," *National Parks, Conservation, and Development: The Role of Protected Areas in Sustaining Society*, J.A. McNeely and K.R. Miller (eds.) (Washington, DC: Smithsonian Institution Press, 1984).
53. Khush, G.S., Ling, K.C., Aquino, R.C., and Aguiero, V.M., "Breeding for Resistance to Grassy Stunt in Rice, *Plant Breeding Papers* 1(4b):3-9, 1977.
54. Kido, H., Flaherty, D.L., Bosch, D.F., and Valero, K.A., "Biological Control of Grape Leafhopper, *California Agriculture* 37:4-6, 1983.
55. Lucas, P.H.C., "How Protected Areas Can Help Meet Society's Evolving Needs," *National Parks, Conservation, and Development: The Role of Protected Areas in Sustaining Society*, J.A. McNeely and K.R. Miller (eds.) (Washington, DC: Smithsonian Institution Press, 1984).
56. Maser, C., and Trappe, J.M. (eds.), *The Seen and Unseen World of the Fallen Tree*, General Technical Report PNW-164 (Portland, OR: U.S. Department of Agriculture, Forest Service, Pacific Northwest Forest and Range Experiment Station, 1984).
57. Mayr, E., *The Growth of Biological Thought:*

Diversity, Evolution, and Inheritance (Cambridge, MA, and London: Belknap Press of Harvard University Press, 1982).

58. McGuire, P.E., and Qualset, C.O. (eds.), *Proceedings of a Symposium and Workshop on Genetic Resources Conservation for California, Napa*, Apr. 5-7, 1984.
59. McNaughton, S.J., "Serengeti Migratory Wildebeeste: Facilitation of Energy Flow by Grazing," *Science* 191:92-94, 1976.
60. McNaughton, S.J., "Grazing as an Optimization Process: Grass-Ungulate Relationships in the Serengeti," *American Naturalist* 113:691-703, 1979.
61. McNeely, J.A., and Wachtel, P.S., *Power, Politics, Religion, Nature, and Animals: How the Coalition Has Worked in Southeast Asia* (Gland, Switzerland: International Union for Conservation of Nature and Natural Resources and World Wildlife Fund, 1986).
62. Mirov, N.T., and Hasbrouck, J., *The Story of Pines* (Bloomington, IN: Indiana University Press, 1976).
63. Myers, N., *A Wealth of Wild Species: Storehouse for Human Welfare* (Boulder, CO: Westview Press, 1983).
64. Nabhan, G., and Dahl, K., "Role of Grassroots Activities in the Maintenance of Biological Diversity: Living Plant Collections of North American Genetic Resources," OTA commissioned paper, 1985.
65. National Plant Genetic Resources Board, *Plant Genetic Resources: Conservation and Use* (Washington, DC: U.S. Department of Agriculture, 1979).
66. The Nature Conservancy, *Preserving Our Natural Heritage, Volume I: Federal Activities* (Washington, DC: U.S. Department of the Interior, National Park Service, 1977).
67. Nikiforuk, A., "Secrets in a Snake Pit," *Maclean's*, May 28, 1984.
68. Norton, B., "Value and Biological Diversity," OTA commissioned paper, 1985.
69. Okazaki, T., "Genetic Study on Population Structure in Chum Salmon (*Oncorhynchus Keta*)," *Bulletin of Far Seas Fisheries Research Laboratory* 19:25-116, 1982.
70. Okazaki, T., "Genetic Structure of Chum Salmon (*Oncorhynchus Keta*) River Populations," *Bulletin of the Japanese Society of Scientific Fisheries* 49(2):189-196, 1983.
71. Oldfield, M.L., *The Value of Conserving Genetic Resources* (Washington, DC: U.S. Department of the Interior, National Park Service, 1984).
72. Piearce, G.D., "The Zambezi Teak Forests: A Case Study of the Decline and Rehabilitation of a Tropical Forest Ecosystem," prepared for the 12th Commonwealth Forestry Conference, Victoria, BC, Canada, Division of Forest Research, Kitwe, September 1985.
73. Prescott-Allen, C., and Prescott-Allen, R., *The First Resource: Wild Species in the North American Economy* (New Haven, CT, and London: Yale University Press, 1986).
74. Prescott-Allen, R., "National Conservation Strategies and Biological Diversity," a report to International Union of Conservation of Nature and Natural Resources, Conservation for Development Centre (Gland, Switzerland: September 1986).
75. Randall, A., "An Economic Perspective on the Valuation of Biological Diversity," OTA commissioned paper, 1985.
76. Randall, A., "Human Preference, Economics, and the Preservation of Species," *The Preservation of Species*, B. Norton (ed.) (Princeton, NJ: Princeton University Press, 1986).
77. Rick, C.M., "Collection, Preservation, and Use of Exotic Tomato Genetic Resources," *Proceedings of a Symposium and Workshop on Genetic Resources Conservation for California, Napa*, P.E. McGuire and C.O. Qualset (eds.), Apr. 5-7, 1984.
78. Rick, C.M., and Fobes, J.F., "Allozymes of Galapagos Tomatoes: Polymorphism, Geographic Distribution, and Affinities," *Evolution* 29:443-457, 1975.
79. Russell, W.A., "Dedication: George F. Sprague, Corn Breeder and Geneticist," *Plant Breeding Reviews. Volume 2*, J. Janick (ed.) (Westport, CT: AVI Publishing Co., 1984).
80. Saenger, P., Hegerl, E.J., and Davie, J.D.S., *Global Status of Mangrove Ecosystems*, Commission on Ecology Papers 3 (Gland, Switzerland: International Union for Conservation of Nature and Natural Resources, 1983).
81. Sedell, J.R., Bisson, P.A., June, J.A., and Speaker, R.W., "Ecology and Habitat Requirements of Fish Populations in South Fork Hoh River, Olympic National Park," *Ecological Research in National Parks of the Pacific Northwest: Proceedings, 2d Conference on Scientific Research in the National Parks, 1979*, E.E. Starkey, J.F. Franklin, and J.W. Matthews (eds.), Oregon State University, Forest Research Laboratory, Corvallis, 1982.
82. Smart, P., *The Illustrated Encyclopedia of the Butterfly World* (London: Salamander Books, 1976).

83. Smith, G.C., and Alderdice, D., "Public Responses to National Park Environmental Policy," *Environment and Behavior* 11(3)329-350, 1979.
84. Smith, H.H., "The Genus as a Genetic Resource," *Nicotiana: Procedures for Experimental Use*, Technical Bulletin 1586, R.D. Durbin (ed.) (Washington, DC: U.S. Department of Agriculture, 1979).
85. Soulé, M.E., "Thresholds for Survival: Maintaining Fitness and Evolutionary Potential," *Conservation Biolog: An Evolutionary-Ecological Perspective*, M.E. Soulé and B.A. Wilcox (eds.) (Sunderland, MA: Sinauer Associates, 1980).
86. Stanfield, R.L., "In the Same Boat," *National Journal*, Aug. 16, 1986, pp. 1992-1997.
87. Swaminathan, M.S., "Rice," *Scientific American* January 1984, pp. 81-93.
88. Tiner, R.W., *Wetlands of the United States: Current Status and Recent Trends* (Washington, DC: U.S. Department of the Interior, Fish and Wildlife Service, 1984).
89. U.S. Congress, Office of Technology Assessment, *Impacts of Applied Genetics: Micro-Organisms, Plants, and Animals*, OTA-HR-132 (Washington, DC: U.S. Government Printing Office, April 1981).
90. U.S. Congress, Office of Technology Assessment, *Wetlands: Their Use and Regulation*, OTA-O-206 (Washington, DC: U.S. Government Printing Office, March 1984).
91. U.S. Congress, Office of Technology Assessment, *Grassroots Conservation of Biological Diversity in the United States—Background Paper #1*, OTA-BP-F-38 (Washington, DC: U.S. Government Printing Office, February 1986).
92. U.S. Department of Agriculture, *Agricultural Statistics 1981* (Washington, DC: U.S. Government Printing Office, 1981).
93. U.S. Department of Agriculture, Forest Service, *An Assessment of the Forest and Range Land Situation in the United States*, Forest Resource Report 22 (Washington, DC: 1981).
94. U.S. Department of the Interior, National Park Service, *Index: National Park System and Related Areas as of June 1, 1982* (Washington, DC: 1982).
95. U.S. Department of the Interior and U.S. Department of Commerce, U.S. Fish and Wildlife Service and U.S. Bureau of the Census, *1980 National Survey of Fishing, Hunting, and Wildlife-Associated Recreation* (Washington, DC: 1982).
96. Vartak, V.D., and Gadgil, M., "Studies on Sacred Groves Along the Western Ghats From Maharashtra to Goa: Role of Beliefs and Folklores," *Glimpses of Indian Ethnobotany*, S.K. Jain (ed.) (New Delhi, Bombay, Calcutta, India: Oxford and IBH Publishing Co., 1981).
97. Webb, M., Environmental Affairs Office, World Bank, Washington, DC, personal communication, June 1986.
98. Wells, S.M., Pyle, R.M., and Collins, N.M., *The IUCN Invertebrate Red Data Book* (Gland, Switzerland: International Union for Conservation of Nature and Natural Resources, 1983).
99. Wilson, E.O., *The Insect Societies* (Cambridge, MA: Belknap Press of Harvard University Press, 1971).
100. World Resources Institute/International Institute for Environment and Development, *World Resources 1986* (New York: Basic Books, Inc., 1986).
101. Worrall, D., *A Baffin Region Economic Baseline Study: Economic Development and Tourism* (Yellowknife: Government of the Northwest Territories, Department of Economic Development and Tourism, 1984).

Chapter 3
Status of Biological Diversity

CONTENTS

Tables

Figures

Boxes

Chapter 3

Status of Biological Diversity

HIGHLIGHTS

- A general consensus exists that biological diversity is being lost or degraded in most regions of the world, but acute threats are largely localized. Despite a weak knowledge base and lack of precise measurements, enough is now known to direct activities to critical areas.
- Concern over the loss of diversity have been defined almost exclusively in terms of species extinction. Although extinction is perhaps the most dramatic aspect of the problem, it is by no means the whole problem. The consequence is a distorted definition of the problem, which fails to account for the various interests concerned and may misdirect how concerns should be addressed.
- The immediate causes of diversity loss usually relate to unsustainable resource development, but the root causes for such development are complex issues of population growth, economic and political organization, and human attitudes. The complexity of the causes implies a need for multi-faceted approaches that deal with both the immediate and the root causes of diversity loss.

INTRODUCTION

Since life began, extinction has always been a part of evolution. Mass extinctions occurred during a few periods, apparently the results of relatively abrupt geological or climatic changes. But in most periods, the rate of species formation has been greater than the rate of extinctions, and biological diversity has gradually increased. Recently in evolutionary history, the human species has derived great economic value from ecosystem, species, and genetic diversity and recognized the intrinsic values of diversity. But now that the values are being recognized, there is evidence that the world may be entering another period of massive reduction in diversity. This time, humans are the cause, and it appears that the consequence will be loss of a substantial share of the Earth's valuable resources.

Diversity is abundant at a global level. About 1.7 million species of plants and animals have been named, classified, and described (57). (Descriptions are only superficial for most of these.) The remainder are still unidentified (figure 3-1). It is estimated that the world contains 5 to 10 million species, and many of these have hundreds or even thousands of distinct genetic types. A recent inventory of insect species in the canopy of a tropical forest suggests that many more insect species may exist than previously thought, pushing the estimate for the total of all species to as high as 30 million (14).

Understanding of biological diversity issues has improved in recent years, in terms of knowing the extent of diversity and understanding the causes and consequences of changes. Enough information is available in all regions of the world to intervene in the processes that cause diversity loss.

Drastic reductions in populations of wild animals and plants are not new and have long been recognized as consequences of over-intensive hunting, fishing, and gathering. For example, great bison herds of North America were depleted in the 19th century, as were stocks of various whales and bird species (52). The now barren hills of southern China's coasts and is-

Figure 3-1.—Currently Described Species

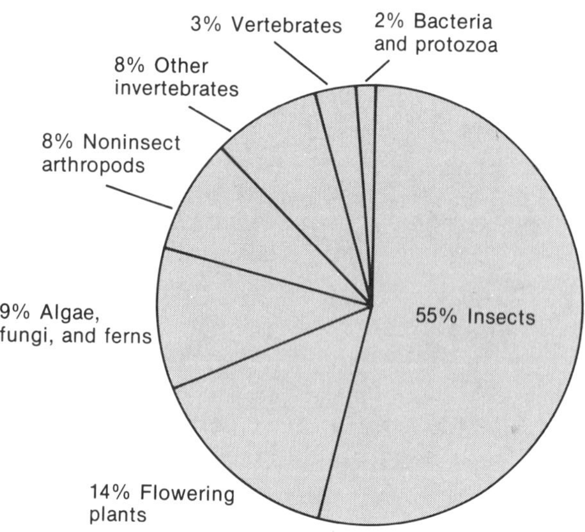

Of the currently described species, insects make up more than half the total. The number of flowering plants described are less than one-third of the percentage of insects.

NOTE: Percentages rounded to nearest whole number.

SOURCE: Office of Technology Assessment, 1986.

lands were deforested 1,000 years ago (22). Erosion from the deforested and overgrazed Armenian hills, which eventually led to the demise of productive agriculture in Mesopotamia, apparently began over 2,000 years ago (21). These kinds of changes undoubtedly caused local and regional losses of diversity.

What is new today is *the scale* on which resource degradation is occurring and thus *the rate* at which diversity is apparently being reduced. Fishing, hunting, and gathering beyond the capacity of ecosystems continues today, but the effects are being greatly exacerbated by degradation of ecosystems and significant reductions in natural areas. The decline of fisheries and sharp reduction of diversity in the Chesapeake Bay over the past two decades is an example well known to Congress, which has supported several initiatives to improve understanding of the complex syndrome of overfishing, pollution, and hydrologic changes related to the region's development.

Acceleration of resource degradation and diversity loss is partly a consequence of population growth, especially in rural areas of developing countries, where compound growth rates are often more than 1 percent per year. It is also a consequence of technologies developed over the past century that have enabled humans to devastate natural ecosystems even where population densities or growth rates are moderate. For example, modern drainage techniques and market conditions make accelerated wetland drainage possible in the United States, and veterinary drugs and modern well-drilling machinery enable African farmers to build livestock herds above the natural carrying capacity of their rangelands.

Accelerated loss of diversity is also caused by modern transportation, which reduces the effect of geographic barriers important in the evolution of diversity. Exotic species, diseases, and pests were for centuries carried across oceans, mountains, and deserts by hundreds of people traveling in ships and on foot, but now they are carried by hundreds of thousands of people traveling in trucks and airplanes.

Biologists and agriculturalists have thus become alarmed about the scale of plant and animal resource degradation during the last two to three decades. The alarm stems from observations of extensive reductions in habitat, coupled with a growing understanding of how such changes adversely affect diversity. (Key concepts that have aided this understanding are described in box 3-A.)

DIVERSITY LOSS

The problem of diversity loss is broader than extinction of species because diversity occurs at each level of biological organization.

- **Ecosystem diversity:** A landscape interspersed with agricultural fields, grasslands, and woodlands has more diversity than the

Box 3-A.—Biological Concepts

Trends in changing biological diversity cannot be measured directly; so many species exist that costs of inventories would be too high. Rather, trends must be inferred by applying biological concepts and by observing changes in habitats. Several biological concepts are relevant:

- **Species-area relationship:** Large sites tend to have more species than small sites. [So] When the areas of diverse natural ecosystems are reduced by land development or by degradation, diversity is reduced. From analysis of many sets of empirical data, scientists have derived a mathematical equation that can be used to predict the decrease in number of species that can be expected following a reduction in habitat area (5).
- **Provinciality effect:** Diversity of species and populations separated by geographic barriers, usually increases over time. But when species or varieties are carried across these barriers, as with the introduction of an exotic organism, provinciality is abruptly lost. Rapid loss of diversity can follow if native species have no defense against an exotic pathogen or pest, or if the exotic organism competes more aggressively for habitat (44). Examples include the introductions of Dutch elm disease to the United States, of cattle to California, of the paperbark tree to Florida, and of goats to many oceanic islands.
- **Narrow endemism:** Some species occur only within very restricted geographic ranges. This group includes many species that have evolved on islands, in mountaintop forests, in isolated lakes or other aquatic zones such as coral reefs, in areas with Mediterranean climates, including California, Western Australia, the Cape of South Africa, Chile, and the Mediterranean Basin countries. Areas with a high proportion of narrow endemic species contribute to global diversity more than other areas with similar numbers of species but less endemism. Thus, biological degradation in such areas reduces diversity more than it would elsewhere.
- **Species richness:** Some ecosystems have many more species than others. Generally, species richness is greatest in equatorial regions, and it decreases toward the poles. It is generally greater in warmer or wetter places than in colder or drier places. Thus, the hot, wet tropical forests, which cover only 7 percent of the Earth's land area, may have about half of the Earth's terrestrial plants and animals (30).
- **Species interdependence:** Interdependence can take a variety of forms. Symbiosis occurs when one or both of two species benefit from association. Mutualism occurs when neither species can survive without the other under natural conditions. Commensalism refers to associations in which one benefits and the other is unaffected.
- **Natural vulnerability:** Vulnerability to extinction varies with several factors. Narrow endemics are perhaps most vulnerable. Rare species may be less susceptible to catastrophe if widely dispersed, but dispersion may lessen their chances for successful mating. Other species relatively vulnerable to extinction include the following: top-level carnivores, species with poor colonizing ability, those with colonial nesting habits, migratory species, those that depend on unreliable resources, and species with little evolutionary experience with perturbations.

same landscape after most of the woodlands are converted to grassland and cropland.

- **Species diversity:** A rangeland with 100 species of annual and perennial grasses and shrubs has more diversity than the same land after grazing has eliminated or greatly reduced the frequency of the perennial grass species.
- **Genetic diversity:** Economically useful crops are developed from wild plants by selecting valuable inheritable characteristics. Thus, the wild ancestral plants contain many genes not found in today's crop

plants. A global agricultural environment that includes domestic varieties of a crop (such as corn) and the crop's wild ancestors has more diversity than the same environment after the wild ancestors are eliminated to make space for more domestic crops.

The quality of information used to assess the loss of biological diversity varies greatly for different ecosystems and different parts of the world. In general, both data and theories are better developed for temperate than for tropical biology; better for birds, mammals, and flowering and coniferous plants than for other classes of organisms; and better for the few major crop and livestock species used in modern agriculture than for the many species used in traditional agriculture.

Ecosystem Diversity

Natural ecosystem diversity has declined in the United States historically (26), and no evidence suggests that this long-term trend has been arrested. By comparing a nationwide evaluation of potential natural vegetation (PNV) with data on existing land uses from the 1967 Conservation Needs Inventory, scientists estimate what portion of land in the United States is still occupied by natural vegetation (26). This study estimates a percent change in area for each ecosystem type (each PNV) since presettlement times.

The greatest area reduction was 89 percent for the Tule Marsh PNV in California, Nevada, and Utah, mainly due to agricultural development. Twenty-three ecosystem types that once covered about half the conterminous United States now cover only about 7 percent. The agricultural States of Iowa, Illinois, and Indiana have lost the highest proportions of their natural terrestrial ecosystems (92, 89, and 82 percent, respectively).

States with the lowest losses were Nevada, Arizona, and New Hampshire (4, 7, and 12 percent, respectively). This assessment does not imply that 96 percent of Nevada is in the same condition that it was 400 years ago. The study did not assess degradation of areas still occupied by natural vegetation; rather, it indicated the areas whose uses remain unchanged. Also, the study was unable to consider some important ecosystem types, such as riparian and wetland areas, which are not included in the PNV categories (26).

Wetlands inventories are conducted by the U.S. Fish and Wildlife Service. The estimated total wetland area in the conterminous States in presettlement times was 87 million hectares. This amount was reduced to 44 million by the mid-1950s and to 40 million by the mid-1970s (figure 3-2). Thus, half the Nation's wetland area was lost in about 400 years, and another 5 percent was lost in the following two decades. Drainage for agricultural development has been the main cause of wetland ecosystem loss (48).

Riparian ecosystems are generally too small to be included as PNV types in major analyses. However, riparian areas contribute disproportionately to biological diversity, especially in the Western States, where they provide luxurious habitats compared with the adjacent uplands (9). Further, their maintenance is important to biological diversity in the streams and lakes they border. Natural riparian (mostly streamside) vegetation in the United States has

Figure 3-2.—Changes in Wetlands Since the 1950s (percentage of total)

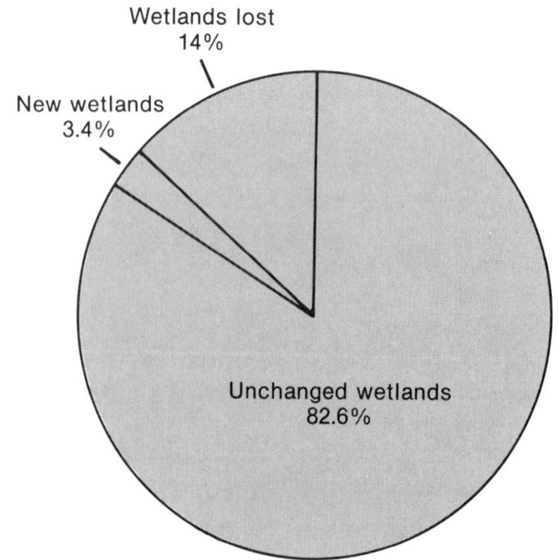

SOURCE: Data from Fish and Wildlife Service's National Wetland Trends Study, 1982.

been reduced some 70 to 90 percent during the last two centuries (46,54). In the Sacramento Valley of California, for instance, the estimated loss of riparian vegetation areas is 98.5 percent; for Arizona, the estimate is 95 percent (45).

The diversity of agricultural ecosystems, or agroecosystems, is also being reduced. System diversity is high in regions where agricultural land is divided into relatively small holdings and each farm uses a variety of crop and livestock species. As indicated in the preceding chapter, such landscapes support natural enemies of crop pests and are likely to contain species and varieties that can resist disease outbreaks and survive abnormal weather. However, on the fertile land of temperate-zone farming regions, where modern machinery and agricultural chemicals are used with crop varieties and where livestock are bred to maximize production, farmers can achieve substantial economies of scale on large holdings that specialize in relatively few crops or breeds. These less diverse agroecosystems are more productive and more profitable than the older systems (36). As yet, relatively little scientific effort is being made to determine how biologically diverse farming could be made more profitable. Thus, the continuing loss of agroecosystem diversity in the United States and throughout the world seems to be a function of both economic development and research priorities (10).

Time-series measurements of agroecosystem diversity are lacking, as is an understanding of the advantages and disadvantages of diversity. There is also a delay between the loss of diversity and consequent increased or decreased profits. Therefore, it seems likely that agricultural system uniformity may continue to increase beyond its economic optimum. Then, a period of restoring some diversity may occur. This process may be underway in some areas of the United States, where multiple cropping, crop rotations, and restoration of shelterbelts are becoming more popular practices (51).

Attempts to increase farm profits by methods that reduce diversity may fail where severe droughts or soil erosion are common and where hot temperatures and high rainfall have resulted in soils with little capacity to hold nutrients. Where such development failures occur, restoration of more diverse farming systems can be difficult, because topsoil, water resources, germplasm, or knowledge of traditional farming methods have been lost (11,25).

Most countries do not have detailed information on changes in ecosystem diversity. The greatest concern on a global scale is for reduction of natural areas in the tropical regions, where ecosystems are least able to recover from degradation. Data on deforestation from many tropical countries indicate that the closed-canopy tropical forests are being reduced by about 11 million hectares each year. (The deforestation rates are discussed in some detail in ref. 54.)

Few data are available for the developing countries on degradation of biological diversity and other resources within the areas that remain classified as forest. Nor are data available on changes in area or quality of grasslands, wetlands, open-canopy forests, riparian and coastal zones, or aquatic ecosystems. Nevertheless, compelling anecdotal evidence indicates widespread degradation of all types of ecosystems in developing countries. In Sri Lanka, for example, removal of coral reefs for production of lime has had several consequences:

- the disappearance of lagoons important as nursery areas for fish,
- the collapse of a fishery,
- reduction of mangrove areas,
- erosion of cultivated coconut land, and
- salination of wells and soil within half a mile of the shore (41).

Documents from development assistance agencies, such as the U.S. Agency for International Development (AID), the World Bank, the United Nations Development Programme, and the U.N. Food and Agriculture Organization abound with observations of resource degradation in developing countries. Evidence of ecosystem degradation is found in the environmental profile series that AID has been preparing since 1979. Usually the evidence is a description of problems caused by resource degradation rather than a report from careful monitoring of resource changes. At present, reports are available on about 67 developing countries, and nearly all describe ecosystem degradation. In some places the problems are the longstanding effects of unsustainable resource development; in others, the degradation has increased dramatically over the last decade and is constraining economic development (see box 3-B).

Species Diversity

Data to document changes in numbers and distribution of species are scarce. To document an extinction, the species must be named and described taxonomically and accurately observed at least once, then the loss must be recorded. Most documented extinctions have been of large terrestrial birds, mammals, and conspicuous flowering plants in the temperate zone and on tropical islands.

Modern taxonomic description goes back to 1753, but most recognized species were described much more recently, and the majority of species have yet to be described and named. For most of the estimated 385,000 living plant species, not much more is known than can be discerned from one or a few pressed, dried herbarium specimens. Nevertheless, personnel and financial support for the taxonomic work done in museums, herbariums, universities, and wildlife agencies around the world are being reduced (8).

Biologists estimate that at least two-thirds of all species live in the tropics. For example, a single tree in the Peruvian Amazon rain forest was found to harbor 43 species of ant belonging to 26 genera. That is a species richness about equal to the ant fauna of the entire British Isles (27). But two-thirds of the named species are in the temperate zone. This disparity reflects the historical distribution of taxonomists. In the United States, for example, about 500 plant taxonomists work with 18,000 species—a ratio of 36 species to 1 taxonomist. Tropical vascular plant species number about 190,000; about 1,500 taxonomists worldwide have expertise in tropical plants, yielding a ratio of 125 to 1 (8).

Even for conspicuous species that have been named, a considerable delay is involved in recording an extinction. For example, the U.S. Fish and Wildlife Service conducted a status review in 1985 for the ivory-billed woodpecker, whose last accepted sighting had been in the early 1950s. Had extinction been confirmed, then the lag between extinction and confirmation of loss could have been 30 years (16). The status of this species remains in doubt, however, because a sighting was reported in 1986 (2).

Indirect methods must be used to estimate changes in species diversity, because complete inventories of ecosystems would be too expensive, and because little is known of many species and the genetic attributes of populations. Methods include:

- preparing lists of species threatened with extinction and monitoring those species;
- monitoring populations of relatively well-known "indicator species" where habitats are being changed and inferring that other species in the same ecosystem are experiencing similar changes (indicator species are commonly trees, birds, large mammals, butterflies, or flowering plants); and
- using mathematical models of species-area relationships to project extinction numbers likely to result from various levels of habitat reduction.

Lists of Threatened Species

Lists of threatened animal species are prepared by the Species Conservation Monitoring Unit (SCMU) of the International Union for the Conservation of Nature and Natural Resources. For the United States, endangered animal lists are prepared by the Endangered Species Office of the U.S. Fish and Wildlife Service (FWS/ESO). For both the SCMU and the FWS/ESO, the information is better for temperate than for tropical species; better for terrestrial than for aquatic species; and better for birds and mammals than for reptiles, amphibians, fish, and invertebrates (16). Terms used in describing the status of threatened species are defined in box 3-C.

Box 3-B.—Typical Excerpts From Development Assistance Agency Reports Addressing Environmental Constraints to Development

India/Pakistan Border Lands in the Sind-Kutch Region (24)

The predominant natural terrestrial vegetation in the region appears to be a low, open-type of dry tropical thorn forest, interspersed in places with grassland. Due to the long and pervading influence of man and grazing stock, the present vegetation is frequently a highly degraded form of low and sparse xerophytic scrub.

The Indus delta is a critical area of high biological diversity and productivity. Mangrove zone creeks and mudflats hold crustaceans, are important to bird populations, and are a fisheries center of local and international significance. The delta appears to be subject to reduced freshwater input, increased salinity, overfishing, and environmental disturbances.

Honduras (49)

Deforestation by shifting cultivators in Yoro and Olancho is dramatic and well publicized. The human tragedy is even more serious for the many thousands of campasino families living on degraded lands in the Choluteca Valley and the areas bordering El Salvador. The forest cover has been peeled back leaving a sparse patchwork of grazed bush fallow and cultivated plots.

By far the greatest threat to natural area functions and wildlife is the indiscriminate destruction of all natural vegetation by shifting agriculturalists and cattlemen regardless of the inappropriateness of the sites for such uses.

Mauritania (35)

Rapid desertification of much of its territory and the consequent loss of land devoted to both grazing and subsistence agriculture is the major environmental problem facing Mauritania today. This process, although severely aggravated by the drought affecting the entire Sahelian area of Africa, has been made worse by certain human practices that have upset the delicate balance of the area's ecology.

Sudan (23,56)

Increased rural population has placed pressure on land resources used for agriculture, resulting in shortened fallow periods and inadequate restitution of fertility. Crop yields have been reduced over time, and exhausted land has been totally abandoned. With loss of the more productive areas, marginal land that is prone to erosion has been cultivated, often with disastrous consequences. Between El Fasher and Nyala, removal of the stabilizing effect of trees has set ancient and once stable sand dunes in motion.

Mechanized farming adversely affected the environment through encouraging soil erosion and desertification, especially in areas of inadequate and unreliable rainfall.

The present lack of a resource-use policy and the unbalanced use of land results in a loss of 5 million hectares from production annually. A total of 65 million hectares are affected by some form of environmental degradation, which is equivalent to 60 percent of Sudan's potentially useful land area.

The SCMU data are gathered from an international network of correspondents identified from research papers. For example, the person compiling data on mammals has about 5,000 informants and contacts worldwide (16). The lists, organized by classes of animals in geographic regions (e.g., New World mammals), are revised on roughly a 10-year cycle. Table 3-1 summarizes some of the data on threatened animals.

Since the mid-1970s, numerous lists of threatened plant species have been prepared. Because these are so new and most tropical regions do not have such lists, the data cannot yet indicate rates of extinction. For the temperate zone,

Box 3-C.—Definitions of Threatened Status

Two sets of definitions are used to classify the status of threatened species. Definitions based on the Endangered Species Act of 1973 are used in the United States. All other countries' lists of endangered species follow the definitions promulgated by the International Union for Conservation of Nature and Natural Resources (IUCN). The two sets of definitions are technically not compatible mainly because of differences in the concept of extinction and the IUCN inclusion of taxa rather than species.

Three technical definitions are used for classification of status in the United States:

1. An **endangered species** is in danger of extinction throughout all or a significant portion of its range.
2. A **threatened species** is likely to become an endangered species within the foreseeable future throughout all or a significant portion of its range.
3. **Critical habitats** are areas essential for the conservation of endangered or threatened species. The term may be used to designate portions of habitat areas, entire areas, or even areas outside the current range of the species.

The IUCN categories include five definitions:

1. **Extinct** taxa are species, or other taxa such as distinct subspecies, that are no longer known to exist in the wild after repeated search of their type localities and other locations where they were known or were likely to have occurred.
2. **Endangered** taxa are in danger of extinction and their survival is unlikely if the factors causing their vulnerability or decline continue operating.
3. **Vulnerable** taxa are declining and will become endangered if no action is taken to intervene.
4. **Rare** taxa are so rare that they could be eliminated easily but are under no immediate threat at present and have populations that are more or less stable.
5. **Intermediate** taxa belong in one of the above categories, but information is not sufficient to determine exactly which one (47).

SOURCES: M.J. Bean, *The Evolution of National Wildlife Law* (New York: Praeger Publishers, 1983); H. Synge, "Status and Trends of Wild Plant Species," OTA commissioned paper, 1985.

however, the numbers of threatened species do give some indication of the scope of the problem.

Nearly all industrial countries now have lists of threatened plant species. In Europe, all but five countries have such lists, and four of those five will have them soon (47). In the United States, many lists and reports cover both the Nation and individual States. Table 3-2 summarizes some information from the endangered plant species lists.

In North America, about 10 percent of the plant species are listed as rare or threatened. Many are plants endemic to small areas in California. A higher proportion of species are listed in Europe because of extensive threats to vegetation in the northern countries and the narrow endemism of many species in the Mediterranean countries. Data from the Soviet Union emphasize horticultural plants and are less complete than for Europe. For temperate-zone Southern Hemisphere countries, the lists are also dominated by narrow endemic plants from the Mediterranean climate regions (47).

Oceanic islands, because of geographic isolation during the millennia of evolution, typically have a very high proportion of endemic species. These areas are particularly vulnerable, because they are not adapted to animals and aggressive weeds that may be introduced by humans. Lists of threatened plants have been prepared for many such islands. Table 3-3 indicates how severe the threats are for islands with 50 to 1,000 endemic species.

Painting by George Sutton/Cornell Laboratory of Ornithology

The ivory-billed woodpecker is presumed extinct in the United States (last verified sighting occurred in 1971) and was thought to be extinct worldwide until the discovery of at least two specimens in the spring of 1986 in eastern Cuba.

Among tropical countries, Brazil has a listing project under way. Lists covering parts of India have been prepared and indicate about 900 threatened plant species. The Malayan Nature Society is preparing a database on threatened plants for the Malaysian peninsula. Listing projects are also complete or under way in some nontropical developing countries, such as Chile, Pakistan, and Nepal (47).

Monitoring Indicator Species and Habitats

Inferring trends in biological diversity from changes in the status of indicator species is a method that relies on time-series data assembled for management of economically significant species or species of special esthetic interest. For example, striped bass (known locally as rockfish) has been a highly valued species since precolonial times on the east coast of the United States, and commercial harvest data have been tabulated for areas like the Chesapeake Bay since 1924.

For 50 years, the trend in striped bass commercial harvest was upward, from around 2 million pounds landed in 1924 to 14.7 million pounds in 1973. Then, the trend reversed. By 1983, the catch had plummeted by 90 percent to 1.7 million pounds. The decline was believed to be due to a combination of overfishing and chemical contamination of the species' habitat (18). Thus, a reduction in populations of other species could be inferred from the striped bass trend.

Inference from this indicator species was confirmed in 1982 when a 7-year Environmental Protection Agency study indicated the extent of the decline in the bay. Subaquatic vegetation had declined 84 percent since 1971. Areas suffering from lack of dissolved oxygen had increased fifteenfold since 1950, and in Baltimore Harbor at least 450 organic compounds, mostly toxins, were identified. Corresponding dramatic declines were documented in native animal species of the bay, including oysters, shad, and yellow perch (53).

The spotted owl is an indicator species for diversity in old-growth forests of the Pacific Northwest. The owl feeds primarily on flying squirrels and other rodents of old-growth habitat. Its decline in a region is considered a sig-

Photo credit: Chesapeake Bay Foundation

Rockfish, once abundant in the Chesapeake, have been exploited and are now endangered in many areas of the estuary. The decline in population of this highly valued species has been attributed to overfishing and pollution.

Table 3-1.—Status of Animal Species in Selected Industrial Countries

	Mammals			Birds			Reptiles			Amphibians			Fishes		
	Species known	Threatened[a]		Species known	Threatened[a]		Species known	Threatened[a]		Species known	Threatened[a]		Species known	Threatened[a]	
		Number	Percent		Number	Percent		Number	Percent		Number	Percent		Number	Percent
Canada	94	6	6.4	434	10	2.3	32	2	6.3	54	2	3.7	800	15	1.9
United States[b]	466	35	7.5	1,090	69	6.3	368	25	6.8	222	8	3.6	2,640	44	1.7
Japan	186	4	2.2	632	35	5.5	85	3	3.5	58	1	1.7	3,144	4	0.1
Australia	320	43	13.4	700	23	3.3	550	9	1.6	150	6	4.0	3,200	X	X
New Zealand	68	14	20.6	282	16	5.7	37	7	18.9	6	X	X	777	3	0.4
Austria[c]	83	38	45.8	201	121	60.2	X	X	X	X	X	X	92	54	58.7
Denmark	49	14	28.6	190	41	21.6	5	0	0.0	14	3	21.4	166	17	10.2
Finland	62	21	33.9	232	15	6.5	5	1	20.0	6	0	0.0	58	4	6.9
France[d]	107	34	31.8	264	79	29.9	32	4	12.5	30	6	20.0	70	13	18.6
Italy	97	13	13.4	419	60	14.3	46	24	52.2	28	13	46.4	503	70	13.9
Netherlands[e]	60	29	48.3	257	85	33.1	7	6	85.7	15	10	66.7	49	11	22.4
Norway	71	10	14.1	280	28	10.0	5	1	20.0	5	1	20.0	X	X	X
Portugal	79	X	X	337	X	X	24	X	X	17	X	X	X	X	X
Spain[f]	100	53	53.0	389	142	36.5	49	20	40.8	23	18	78.3	137	12	8.8
Sweden	63	11	17.5	250	34	13.6	6	0	0.0	11	5	45.5	X	X	X
Switzerland	86	X	X	190	74	38.9	15	X	X	20	X	X	60	X	X
Turkey	31	11	35.5	217	36	16.6	X	X	X	X	X	X	X	X	X
United Kingdom[g]	51	12	23.5	200	24	12.0	6	2	33.3	6	2	33.3	37	8	21.6
West Germany[c e]	94	44	46.8	455	98	21.5	12	9	75.0	19	11	57.9	173	40	23.1

X = Not available.
[a]Threatened refers to the sum of the number of species in the endangered and vulnerable categories.
[b]Including Pacific and Caribbean Islands.
[c]Data for threatened animals include extinct and/or vanished species.
[d]Data for known freshwater fish species only.
[e]The number of bird species known includes occasional visitors.
[f]Peninsular Spain and the Baleares only.
[g]Data refer only to terrestrial mammals.

SOURCE: Organisation for Economic Cooperation and Development, *OECD Environmental Data Compendium 1985* (Paris: 1985).

Table 3-2.—Summary of Data From Endangered Plant Species Lists

Country/region	Species	Rare and threatened species	Extinct taxa	Endangered taxa
Australia	25,000	1,716	117	215
Europe[a]	11,300	1,927	20	117
New Zealand	2,000	186	4	42
South Africa	23,000	2,122	39	107
U.S.S.R.	21,100	653	≈20	≈160
United States[b]	20,000	2,050	90	?

[a]Excludes European U.S.S.R., Azores, Canary Islands, and Madeira.
[b]Excludes Hawaii, Alaska, and Puerto Rico.
SOURCE: S. Davis, et al., *Plants in Danger: What Do We Know?* (forthcoming), as cited in H. Synge, "Status and Trends of Wild Plants," OTA commissioned paper, 1985.

Table 3-3.—Data on Threatened Plant Species of Selected Oceanic Islands

Island	Number of endemic species[a]	Number listed as rare or threatened
Azores	55	30 (55%)
Canary Islands	569	383 (67%)
Galapagos	229	150 (66%)
Juan Fernandez	118	95 (81%)
Lord Howe Island	75	73 (97%)
Madeira	131	86 (66%)
Mauritius	280	172 (61%)
Seychelles	90	73 (81%)
Socotra	215	132 (61%)

[a]Endemic means the species occurs only on the island.
SOURCE: S. Davis, et al., *Plants in Danger: What Do We Know?* (forthcoming), as cited in H. Synge, "Status and Trends of Wild Plants," OTA commissioned paper, 1985.

nal that its prey and other species associated with the habitat are also in decline (3).

Diversity losses due to pollution may be indicated by animals' food chains, such as the bald eagle and other fish-eating birds. Plants susceptible to air pollution, such as lichens, may also be useful indicators. Extensive records of observations on smaller animals of long-standing interest to professional and amateur biologists can also gauge diversity change. The decline of Bay Checkerspot butterflies, for example, is taken as an indication of decline of many associated organisms in the San Francisco Peninsula area (13).

Models of Species-Area Relationships

The scale of diversity reduction can be estimated for most tropical countries only by inferences from the reduction of habitat. Nearly all attempts to estimate global extinction rates focus on the tropical moist forests. These ecosystems are exceedingly species-rich, contain areas of narrow endemism, and are undergoing extensive and rapid conversion to other uses. Because they typically have erosion-prone soils incapable of holding many plant nutrients and occur where rain and heat are especially intense, these forest ecosystems are highly susceptible to degradation. In fact, the undeveloped forests are so diverse, and the deforested, degraded landscapes to which they are often converted support so few species, that the models used to estimate extinction rates generally treat the diversity of deforested landscapes in the moist tropics as negligible (43).

A recent projection of bird and flowering-plant extinctions that could be caused by continued deforestation in tropical America is based on a mathematical model of the species-area relationship (see box 3-A). About 92,000 flowering plant species have been described for regions where the forested area for recent human-caused deforestation was about 6.9 million square kilometers (43). Over the next 15 to 20 years, the forested area will be reduced to about 3.6 million square kilometers if the rate of deforestation remains at the level of the 1970s. The mathematical relationship between area and species numbers, derived by analysis of some 100 empirical data sets (5), indicates that this reduced forest could be expected to support about 79,000 species. Thus, a 15-percent reduction in numbers of species is projected for the near future.

Deforestation is expected to continue for more than 20 years, however, and it seems likely that it will accelerate as the rapidly growing human populations of tropical American coun-

Photo credit: U.S. Fish and Wildlife Service

Northern spotted owl requires large tracts of Pacific Northwest old-growth forest as habitat. If harvesting of old-growth continues at current rate, the habitat for this species could disappear within the next two decades.

tries need more rapid resource development. A "worst-case" calculation indicates that if the forests were reduced until they covered only National Parks and their equivalents that had been established by 1979 (about 97,000 square kilometers), then the final effect could be as high as a 66-percent reduction (43).

About 704 species of land birds have been described in the Amazon region of tropical America. Using the same mathematical relationship as used for plants, a 60-percent reduction of the original forest area over the next 15 to 20 years could be expected to cause eventual extinctions of 86 bird species. The worst-case calculation, with reduction of the Amazon forest to the area of already established national parks, projected that 487 types of birds, or about 70 percent of the species, could become extinct (43).

Several assumptions tend to underestimate extinction rates. For example, extinctions resulting from reduced provinciality are ignored in the calculations, as are effects of the narrow endemism that occurs in several parts of tropical American forests. On the other hand, the assumption that none of the plant and bird species will find habitats they need after deforestation seems an exaggeration. Such projections may be helpful in stimulating institutional responses to prevent the worst cases from occurring. Many nations' governments have begun to take steps in the past decade to protect endangered habitats. The worst-case calculations are thus not predictions, but indications of the direction and scale of the projected trend.

Distracting Numbers Game

Projections of alarming losses in species diversity have attracted attention to this issue. But discrepancies among the estimated extinction rates have called into question the credibility of all such estimates. In one sense, the numbers themselves have become an issue, confusing policymakers and the general public and possibly detracting from efforts to deal with the causes and consequences of diversity loss (28). This numbers game also has defined the problem of loss mainly in terms of species extinction, which may be the most dramatic aspect of the diversity question, but it is only part of the problem. Further, global and national data and projections may mask the localized nature of resource degradation, diversity loss, and the consequences of both. Large inaccessible areas of forest, for example, may make the global deforestation rate seem moderate, but destruction of especially diverse forests in local areas of Australia, Bangladesh, India, the Philippines,

Brazil, Colombia, Madagascar, Tanzania, and West Africa proceeds at catastrophic rates (32).

Genetic Diversity

Ideally, concern about loss of biological diversity should be focused on genetically distinct populations, rather than on species (13,16,34). But with so little information available about the majority of wild species, this seems impractical.

For agricultural species, on the other hand, the concern is mainly about genetic diversity. The species do not seem to be in danger of extinction, but the variety of genes in many crops and livestock breeds is being reduced (39). Many distinct types are being eliminated as improved breeds and varieties that are genetically similar are gaining more widespread use. Ironically, success in exploiting genetic resources for agriculture threatens the genetic diversity on which future achievements depend.

With livestock, the principal diversity loss involves developing-countries' breeds being replaced by imported ones. The threat seems greatest for those species in which artificial insemination is widely used, and it is particularly a problem with cattle, for which over 270 distinct breeds exist. For farmers with only a few cows, artificial insemination is cheaper than keeping a bull. But developing countries lack facilities to collect and freeze semen from locally adapted breeds, so semen is usually imported from commercial studs in industrial countries. Threatened breeds include the criollo of Latin America and the Sahiwal and several others from Africa (see table 3-4) (15).

Llama and alpaca—as well as vicuna and guanaco, their wild relatives—are South American species used for meat, as beasts of burden, and for their hair and pelts. Numbers of all four species have declined sharply since the Spanish conquest of the Incan empire, and loss of genetic diversity has almost surely occurred, though it is unmeasured (15).

Poultry and swine breeds are also moving toward genetic homogeneity, because controlled breeding has been rapid and intensive to produce varieties suitable for modern commercial production. Poultry breeding has been

Table 3-4.—Endangered African Cattle Breeds

Breed	Location	Purpose	Reasons for decline in number	Advantages
Mutura	Nigeria	Meat, draft	Civil war, crossbreeding, lack of interest by farmers as tractors become available	Trypanotolerant,[a] hardy, good draft animal, low mortality
Lagune	Benin, Ivory Coast	Meat	Crossbreeding, lack of interest by farmers because of small mature size (125 kg) and low milk yields	Trypanotolerant, adapted to humid environment
Brunede l'Atlas	Morocco, Algeria, Tunisia	Meat	Crossbreeding to imported breeds	Adapted to arid zones
Mpwapwa	Tanzania	Milk	Lack of sustained effort to develop and maintain new breed	Adapted, dual-purpose
Baria	Madagascar	Milk, meat	Crossbreeding	Adapted, dual-purpose
Creole	Mauritius	Milk, meat, draft	Crossbreeding	Adapted, multiple-purpose
Kuri	Chad	Milk, meat	Political instability, numbers reduced by rinderpest and drought	High milk production, ability to float and swim in Lake Chad, tolerant of heat and humidity
Kenana	Sudan	Milk	Crossbreeding (artificial insemination) to imported dairy breeds, loss of major habitat to development scheme	High milk potential; adapted to hot, dry environment
Butana	Sudan	Milk	Crossbreeding	High milk potential; adapted to hot, semiarid environment

[a]Ability to survive Trypanosome infection (spread by tsetse fly), which normally causes African sleeping sickness in cattle.

SOURCE: Adapted from K.O. Adeniji, "Prospects and Plans for Data Banks on Animal Genetic Resources," *Animal Genetic Resources Conservation* (Rome: Food and Agriculture Organization, 1984), as cited in H. Fitzhugh, et al., "Status and Trends of Domesticated Animals," OTA commissioned paper, 1985.

Photo credit: H. Fitzhugh

Two Sahiwal cows on a government farm in Kenya. Sahiwal were originally developed in Pakistan and then imported to Kenya as an all-purpose breed (e.g., milk, meat). There is concern that inbreeding has sharply reduced the genetic variation in this breed, both in Pakistan and Kenya.

dominated by a few companies (probably fewer than 20 worldwide) (15). These firms typically retain a number of breeds from which to make selections and crosses, but they do not find it cost-effective to retain stocks that might prove useful more than 10 years in the future (7). Meanwhile, breeds adapted to traditional farm conditions are becoming rare in industrial countries, because fewer farmers want them and the number of small hatcheries producing them has declined sharply. Poultry breeds from industrial countries are being exported to urban markets in many developing countries, but no evidence exists that these have affected the genetic diversity of poultry in rural areas of developing countries.

Hundreds of plant species have been domesticated, and traditional farming systems continue to use many species. But modern agriculture produces most human sustenance, plant-derived fibers, and industrial materials from only a few species. Three-quarters of human nutrition is provided by just seven species: wheat, rice, maize, potato, barley, sweet potato, and cassava (31). Within the United States, the top 30 crops account for $57.7 billion in farm sales and imports annually, which is 60 percent of the combined value of all U.S. agricultural plant resources (see table 3-5) (39).

Within these 30 crops, modern varieties have replaced traditional ones, reducing diversity between and within agricultural sites and genetic populations. The narrow species and genetic base of modern agriculture generates two distinct concerns: 1) the extinction of genes, which reduces opportunities to produce new

Table 3-5.—Crops Grown or Imported by the United States With a Combined Annual Value of $100 Million or More (average 1976 to 1980)

Average annual value (U.S. $ millions)	Crop
11,278	Soybean
10,412	Corn
6,475	Wheat
4,233	Cotton
3,925	Coffee
2,851	Tobacco
1,723	Sugarcane
1,525	Grape
1,206	Potato
1,163	Rice
1,150	Sweet orange
1,147	Sorghum
1,054	Alfalfa
1,051	Tomato
1,016	Cacao
815	Apple
760	Beet crops
747	Peanut
706	Rubber
672	Barley
527	Lettuce
517	Common bean
393	Sunflower
368	Banana
365	Cole crops
355	Almond
349	Peach
314	Coconut
304	Oats
287	Onion
252	Strawberry
219	Grapefruit
198	Chrysanthemum
192	Cucumber
189	Melon
186	Pineapple
179	Roses
167	Celery
164	Walnut
158	Peppers/chilis
156	Jute
155	Plum/prune
148	Sweet cherry/sour cherry
146	Pear
144	Olive
143	Oil palm
142	Carrot
142	Pea
136	Lemon
130	Bermudagrass
128	Tea
116	Watermelon
116	Cashew
110	Sweet potato
102	Pecan
100	Azalea/rhododendron

SOURCE: C. Prescott-Allen and A. Prescott-Allen, *The First Resources: Wild Species in the North American Economy* (New Haven, CT, and London: Yale University Press, 1986).

varieties better suited for production at particular sites; and 2) the increased uniformity of crops, which makes them more vulnerable to pests and pathogens. Of these two, extinction of genes is the greater problem. For annual crops, uniform genetic vulnerability can be quickly corrected as long as a high degree of genetic diversity is maintained for the crop somewhere. Gene extinction, however, cannot be reversed.

Published information on status and trends of crop diversity usually consists of impressions by plant breeders and others on the loss of cultivated varieties or threats to wild relatives of crops, such as: "it may not be long before landraces are irretrievably lost" or "many locally adapted varieties have been replaced by modern varieties" (39). Such reports have been collected and evaluated by the International Board on Plant Genetic Resources (IBPGR). IBPGR's information has stimulated conservation action and has helped to determine general collecting priorities for protection of genetic resources.

Plant breeders and germplasm collectors generally concur that crop genes have been lost and that losses are still occurring rapidly and widely in many crops (39), in spite of progress with collection and offsite maintenance programs (see ch. 6). Three problem areas include:

1. crops that have low priority for IBPGR but are of major economic importance to the United States (e.g., grape, alfalfa, lettuce, sunflower, oats, and tobacco);
2. crops that despite being a high international priority still lack adequate provision for long-term conservation (including those maintained as living collections rather than as seeds, such as cacao, rubber, coconut, coffee, sugarcane, citrus, banana); and
3. wild relatives of major crops, which, except for sugarcane and tomato, are represented in collections by extremely small samples.

Detailed assessments of the status and trends of genetic diversity are lacking, even for crops whose collection is well advanced, such as rice, maize, potato, tomato, and bean. Such assessments are needed to understand the dynamics of crop genetic change and its relationship to social and economic change (39).

The status of diversity conservation for economically important timber trees lies between that of wild plants gathered for economic use and that of agricultural plants. Some commercial tree species are protected by parks and other protected natural areas, and the diversity of some is at least partially captured in offsite seed collections. In many extensively managed forests, commercial tree species regenerate naturally after logging, fire, or other disturbances, and local genetic diversity is maintained. However, replanting with stock propagated from selected parents and from tree-breeding programs is a common practice with some trees, such as Douglas fir, and gene pools for these species are being gradually altered (19).

The genetic resources of commercial trees and other economic plants and animals in developing countries are being eroded by conversion of forest areas to agriculture or grazing use. An international panel recently identified some 300 tree species or important tree populations (presumably with unique genetic characteristics) that are endangered (17). Thus, in the United States and developing countries where U.S. agencies provide assistance, protection of natural gene pools of commercial trees and other nondomesticated economic species could become an objective in development planning (see ch. 11). At present, economic species not used in agriculture or horticulture are poorly represented in genetic conservation programs.

CAUSES OF DIVERSITY LOSS

Forces that contribute to the worldwide loss of biological diversity are varied and complex, and stem from both direct and indirect pressures. Historically, concern has focused on the commercial exploitation of specific threatened or endangered species. But now attention is also being focused on indirect threats more sweeping in scope (30).

Development and Degradation

Economic development usually entails making sites more favorable for a manageable number of economic activities. Consequently, the changed landscape has fewer habitats and supports fewer species. Habitats may be affected by offsite development too—by pollution, for example, or changed hydrology. Some development, such as logging in a mosaic pattern through a large forested area, mimics natural processes and may result in a temporary increase in diversity.

But poorly planned or badly implemented development, such as agricultural expansion without investment in soil conservation, can severely disrupt biological productivity, and it can start a self-reinforcing cycle of degradation. For example, soil erosion reduces soil fertility, which in turn can reduce growth of plants for cover, leading to more soil erosion and to rapid depletion of diversity as the site becomes suitable for fewer and fewer species (51). Other causes of site degradation include grazing pressures, unnatural frequency or severity of fires, and excessive populations of herbivores (such as rabbits) where predators are eliminated. Arid and semiarid sites are especially susceptible to degradation from such pressures. Species may be reduced by one-half to two-thirds without outright conversion of the land use (33).

Modernization of farming systems is also a cause of diversity loss. Such systems often include many species of crops and livestock; genetic diversity is typically high, because cultivars and breeds adapted to the vagaries of site-specific conditions are used. To achieve productivity gains, however, traditional systems are being replaced with modern methods. Capital inputs, such as manufactured fertilizers and feeds, are used to compensate for site-specific differences. Thus, it is possible to replace traditional crops and livestock with fewer varieties

Photo credit: U.N. Photo 152,843/K. Muldoon

Overgrazing in Burkina Faso—one major cause of diversity loss.

bred to give high yields under more artificial conditions.

The loss of traditional agroecosystems is not restricted to developing countries. Native American farming systems that interplant corn with squash, numerous types of beans, sunflowers, and many semidomesticated species are reduced to isolated areas now and continue to be abandoned. These systems and crop varieties have been described in anthropological literature, but they are lost before being scrutinized by agricultural scientists.

Agricultural development may cause abrupt disappearance of traditional varieties, as with the replacement of traditional wheat in the Punjab region of India, or it may be gradual, as with fruit and vegetable varieties in the United States and livestock breeds in Europe. Locally adapted varieties may become extinct in a single year if germplasm for a traditional variety is lost because of a catastrophe or is destroyed to control a disease. Examples include traditional grain varieties that were replaced with modern ones in Africa when seed was eaten during the recent famine, and local swine populations that were exterminated in Haiti and the Dominican Republic to control a disease and then replaced with modern breeds (15).

Exploitation of Species

As noted earlier, concern with loss of biological diversity historically focused on extinctions or population losses that resulted from excessive hunting and gathering. Whales, cheetahs, passenger pigeons, bison, the North Atlantic herring, the dodo, and various orchids are all examples of organisms hunted or gathered in excess (30).

Today the direct threat to wildlife remains. Numerous monkeys and apes are endangered by overhunting, mainly to supply the demand

Photo credit: D. Ehrenfeld

The green turtle is an example of a species threatened by direct factors, such as exploitation of adults and eggs, and by indirect factors, including nesting beach destruction, ocean pollution, and incidental catch in shrimp trawls.

from medical research institutions. Some 108 primate species are hunted for an international trade worth about $4 million annually (30). For many, perhaps most of these, the capturing process is very destructive. Apes such as the gibbon are captured by shooting mother animals from the treetops and taking any infants that survive the fall. Many of the infants die while passing through the market system. Thus, the 30,000 primates sold in 1982 (30) actually compose a much higher number killed to support the trade.

The rhinoceros has declined more rapidly over the past 15 years than any other large mammal. From 1970 to 1985 there was an almost 80-percent decrease in the numbers of rhinos, from 71,000 to only about 13,500 today. The most spectacular decline has been that of the black rhino—from 65,000 to 7,000 in the past 15 years. Whole populations of black rhino have been almost totally eliminated over the past 10 years in Mozambique, Chad, Central African Republic, Sudan, Somalia, Ethiopia, and Angola. In the past 2 years, Mozambique has lost the white rhinoceros for the second time in this century—a dubious achievement indeed (29).

The main reason for this catastrophic decline since 1975 is due to the illegal killing of the animal, mostly for its horn. In the early 1980s about one-half of the horn put onto the world market went to North Yemen where it is used for the making of attractive dagger handles, while the remaining half went to eastern Asia where it is used mostly to lower fever, not—as often supposed—as an aphrodisiac (6).

Plant species are also subject to overharvest. A cycad plant species was reported eliminated in Mexico during just one year when 1,200 specimens were exported to the United States (55).

Vulnerability of Isolated Species

If the range of species is restricted to a relatively small area, such as an island or a mountaintop forest, a single development project or the introduction of competing or exotic species can lead to loss of diversity. Many recorded extinctions have been animals and plants from oceanic islands (see table 3-6) (52). Some of these areas, such as Haiti, are infamous for deforestation and rapid rates of soil erosion. It may be inferred that diversity loss has been and probably continues to be especially severe on such islands.

Complex Causes

Most losses of diversity are unintended consequences of human activity, and the species and population affected are usually not even recognized (30). Air and water pollution, for example, can cause diversity loss far from the pollution's source. Substantial gains in reducing these pressures have been achieved in industrial countries, particularly in the United

Table 3-6.—Oceanic Islands[a] With More Than 50 Endemic Plant Species

Island	Endemics	Percentage endemism
Madagascar	≈9,000	
Cuba	2,700	46
New Caledonia	2,474	76
Hispaniola[b]	1,800	36
New Zealand	1,618	81
Sri Lanka	≈900	
Taiwan	≈900[c]	
Hawaii	883	91
Jamaica	735	23
Figi	≈700	
Canary Islands	383	
Puerto Rico	332	12
Caroline Islands	293[c]	
Trinidad and Tobago	215	
Galapagos	175	25
Mauritius	172	
Ogasawara-Gunto	151[c]	
Reunion	≈150	
Vanuatu	≈150	
Tubuai	≤140	
Comoro Islands	136	
Socotra	132	
Bahamas	121	
Sao Tome	108	
Marquesas Islands	103	
Samoa	≥100	
Juan Fernandez	95	
Cape Verde	92	
Madeira	86	
Mariana Islands	81[c]	
Lord How Island	73	
Seychelles	73	

[a]Excludes Australia, New Zealand, Borneo, New Guinea, and Aldabra.
[b]Hispaniola comprises the nations of Haiti and the Dominican Republic.
[c]Omits monocoyledons.

SOURCE: Adapted from A.H. Gentry, "Endemism In Tropical Versus Temperate Plant Communities," *Conservation Biology,* M. Soulé (ed.) (Sunderland, MA: Sinauer Associates, Inc., 1986); and H. Synge, "Status and Trends of Wild Plants," OTA commissioned paper, 1985.

States, since passage of the National Environmental Policy Act.

Yet pollution remains a major threat to biological diversity, because abatement is often expensive and is sometimes a very complex organizational task, especially when it depends on international cooperation. Acid rain is an example. In Scandinavia, several fish species have declined in numbers because of acidification of lakes; in eastern Canada, a trout species has been placed in the severely threatened category (37). International pollution by acid rain has recently been reported to extend far from industrial regions into Zambia, Malaysia, and Venezuela, for example.

Climate change is apparently being caused by increased carbon dioxide and atmospheric dust, which result from fossil-fuel burning and from the release of carbon stored in vegetation when extensive areas are converted from forest to cropland or sparsely vegetated grassland. The expected consequences include significant changes in temperature and rainfall patterns. Temperature rises seem likely to occur rapidly, at least in evolutionary terms, so diversity will probably incur a net loss during the next century (38).

In both industrial and developing countries, diversity is lost as land is converted from forest, grasslands, and savanna to cropland or pasture. If the land being converted will support permanent agriculture with relatively high yields, the effect on diversity is contained. Moderate areas of such land can support substantial populations. But much of the newly cleared land is marginal or totally unsuitable for the cultivation or grazing practices being applied. As a result, extensive areas must be cleared, especially where the land is so poor that it degrades to wasteland and is abandoned after a few years, which is typical in the moist tropical forest regions and in semiarid areas of both the temperate and tropical zones (30).

The underlying causes of inappropriate land clearing are many and exceedingly complex. Population growth, poverty, inappropriate agricultural technologies, and lack of alternative employment opportunities are all problems far too complex for biologists and conservationists alone to resolve.

Population growth in itself may not seem intrinsically threatening to biological diversity. In some industrial countries, such as the United States and Japan, disruption of ecosystems has been mitigated by urbanization, establishment of parks, and land use regulation (30). But the connections between affluent populations and their impacts on biological diversity are obscured by the complexity of commerce. The Japanese, for example, carefully protect the diversity of their own remaining forests, but they use large quantities of timber from forests in other countries where controls are lax. Much has been written about the "hamburger con-

Figure 3-3.—Past and Projected World Population

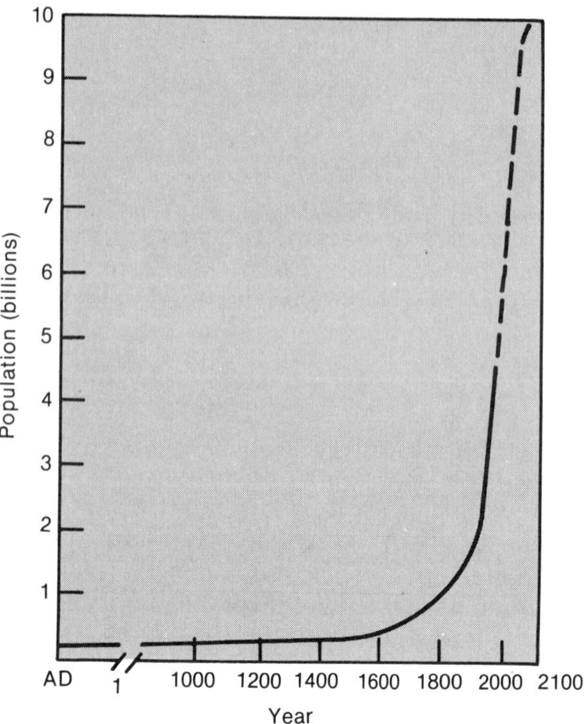

If the current growth in population continues, by the year 2000 more than 6 billion people will inhabit the world. With this growth, irreversible environmental degradation and loss of biological diversity can be expected.

SOURCE: World Resources Institute and International Institute for Environment and Development, *World Resources 1986* (New York: Basic Books, 1986).

nection" by which U.S. and European beef consumers are said to be causing loss of diversity in tropical countries where forests are converted to pasture. The very difficult task of identifying, measuring, and mitigating such negative economic-ecologic links between nations is increasingly important as the world economy becomes more and more international (30).

The causal link between human population size and diversity loss is clearer in developing countries where population growth in rural areas continues to be rapid, and land-use regulations do not exist or are poorly enforced. Between 1980 and 2000, rural populations are expected to increase by 500 million in the developing world (57) (figure 3-3). Where these people continue to rely on extensive agriculture, resource degradation and diversity loss can be expected to accelerate. The harmful impacts of population growth are also likely to be exacerbated by development programs that encourage large resettlements of landless people into deserts or tropical lowlands without providing the means to sustain agricultural productivity in such difficult sites (42).

CONCLUSION

Circumstantial Evidence

Biological diversity is abundant for the world as a whole. More than 10 million species may exist, but after more than 200 years of study, scientists have only named and described some 1.7 million. Many of these species contain numerous genetically distinct populations, each with a different potential for survival and utility.

The abundance and complexity of ecosystems, species, and genetic types have defied complete inventory or direct assessment of changes. But from events and circumstances that can be measured, it can be inferred that the rate of diversity loss is now far greater than the rate at which diversity is created.

The circumstantial case is based on the knowledge that each wild species and population depends on the habitat to which it is adapted. Diverse natural habitats are being converted to less diverse and degraded landscapes. On those sites, diversity has been reduced. The sites that remain in a natural or nearly natural condition are often fragmented patches that will not support the diversity of larger areas.

For domesticated agricultural plants and animals, the concern is genetic diversity, which must be maintained by active husbandry. Farming systems with high genetic diversity are being replaced by new systems with much lower diversity, so husbandry of many genetic types is abandoned. Thus, gene combinations that reproduce particular characteristics and took decades to develop may be lost in a single year. The rapid rate and large scale of agricultural modernization imply that genetic diversity losses are great, though quantitative estimates have not been made.

Data for Decisionmaking

In recent decades, inventories and monitoring of ecosystems, species, and genetic types have improved, and the knowledge of what exists has greatly enhanced abilities to maintain diversity. Biologists, resource managers, and conservationists concur that information available now is adequate in virtually every country to guide programs to maintain diversity.

The circumstantial case for diversity loss in the United States and other industrial countries is bolstered by abundant site-specific data as well as by regional survey data on ecosystems and species. This information has moved public and private organizations to allocate substantial resources to the establishment and management of nature reserves, abatement of pollution, and other programs that sustain biological diversity. Opportunities to improve the use of these data are discussed in chapter 5.

The situation is quite different in developing countries. Circumstantial evidence of diversity loss is compelling, and many countries have designated parks and natural areas in recent years. But the available data are not adequate to support policy decisions to allocate enough funds and other resources to maintain diversity. Both money and trained personnel are needed to develop the necessary information.

Public and private funds that might be used for conservation are extremely scarce. Therefore, a great need exists for good data and comprehensive planning, so that whatever funds can be allocated will be used effectively. Organizations such as The Nature Conservancy and the IUCN are working to develop the data and local planning expertise needed to adequately assess the status of biological diversity and prospects for its conservation. More concerted support from public institutions is needed, however, both in the United States and abroad.

CHAPTER 3 REFERENCES

1. Adeniji, K.O., "Prospects and Plans for Data Banks on Animal Genetic Resources," *Animal Genetic Resources Conservation by 44/1* (Rome: Food and Agriculture Organisation of the United Nations, 1984). *In*: Fitzhugh, et al., 1985.
2. Anonymous, "Ivory-billed Woodpecker Found," *Bioscience* 36(10):703, 1986.
3. Barton, K., "Wildlife on Bureau of Land Management Lands," *Audubon Wildlife Report, 1985* (New York: National Audubon Society, 1985).
4. Bean, M.J., *The Evolution of National Wildlife Law* (New York: Praeger Publishers, 1983).
5. Connor, E.F., and McCoy, E.D., "The Statistics and Biology of the Species-Area Relationship," *American Naturalist* 113:791-883, 1979.
6. Crawford, M., "Rhinos Pushed to the Brink for Trinkets and Medicines," *Science* 234:147, Oct. 10, 1986.
7. Crawford, R.D., "Assessment of Poultry Genetic Resources," *Canadian Journal of Animal Science* 64:235-251, 1984. *In:* Fitzhugh, et al., 1986.
8. Crosby, M.R., and Raven, P., "Diversity and Distribution of Wild Plants," OTA commissioned paper, 1985.
9. Crumpacker, D.W., "Status and Trends of Natural Ecosystems in the United States," OTA commissioned paper, 1985.
10. Dahlberg, K.A. (ed.), *New Directions for Agriculture and Agricultural Research: Neglected Dimensions and Emerging Alternatives* (Totowa, NJ: Rowman & Allanheld, 1986).
11. Dasmann, R.F., Milton, J.P., and Freeman, P.H., *Ecological Principles for Economic Development* (New York: John Wiley & Sons, Inc., 1973).
12. Davis, S., et al., *Plants in Danger: What Do We Know?* (Gland, Switzerland: International Union for the Conservation of Nature and Natural Resources, forthcoming). *In*: Synge, 1985.
13. Ehrlich, P., and Ehrlich, A., *Extinction* (New York: Ballantine, 1981).
14. Erwin, T.L., "Tropical Forest Canopies: The Last Biotic Frontier," *Bulletin of the Entomological Society of America* 29(1):14-19, 1983. *In*: Myers, 1985.
15. Fitzhugh, H.A., Getz, W., and Baker, F.H., "Bio-

logical Diversity: Status and Trends for Agricultural Domesticated Animals," OTA commissioned paper, 1985.
16. Flesness, N., "Status and Trends of Wild Animal Diversity," OTA commissioned paper, 1985.
17. Food and Agriculture Organization of the United Nations, *Report of the Fifth Session of the FAO Panel of Experts on Forest Gene Resources* (Rome: 1984).
18. Fosburgh, W., "The Striped Bass," *Audubon Wildlife Report, 1985* (New York: National Audubon Society, 1985).
19. Franklin, J.F., Chief Plant Ecologist, U.S. Forest Service, personal communication, May 26, 1986.
20. Gentry, A.H., "Endemism in Tropical Versus Temperate Plant Communities," *Conservation Biology*, M.E. Soulé and B.A. Wilcox (eds.) (Sunderland, MA: Sinauer Associates, Inc., 1986).
21. Gill, V.G., and Dale, T., *Topsoil and Civilization* (Norman, OK: University of Oklahoma Press, 1974).
22. Grant, C.J., *The Soils and Agriculture of Hong Kong* (Hong Kong: The Government Printer at the Government Press, 1960).
23. Institute of Environmental Studies, *Preassessment of Natural Resources Issues in Sudan*, University of Khartoum and the Program for International Development, Clark University, Worcester, MA, 1983.
24. International Union for Conservation of Nature and Natural Resources, *A Preliminary Environmental Profile of the India/Pakistan Border Lands in the Sind-Kutch Region*, a report for the World Bank prepared at the IUCN Conservation Monitoring Center, October 1983.
25. Klee, G.A. (ed.), *World Systems of Traditional Resource Management* (New York: John Wiley & Sons, Inc., 1980).
26. Klopatek, J.M., et al., *Land-Use Conflict With Natural Vegetation in the United States*, Publication No. 1333 (Oak Ridge, TN: U.S. Department of Energy, Oak Ridge National Laboratory, Environmental Services Division, 1979). *In*: Crumpacker, 1985.
27. Lewin, R., "Damage to Tropical Forests, or Why Were There So Many Kinds of Animals?" *Science* 234:149-150, Oct. 10, 1986.
28. Mares, M.A., "Conservation in South America: Problems, Consequences, and Solutions," *Science* 233:734-739, Aug. 15, 1986.
29. Maruska, E.J., Executive Director, Zoological Society of Cincinnati, presentation to the House Committee on Science and Technology, Subcommittee on Natural Resources, Agriculture Research, and Environment, Sept. 25, 1980.
30. Myers, N., "Causes of Loss of Biological Diversity," OTA commissioned paper, 1985.
31. Myers, N. (ed.), *GAIA: An Atlas for Planet Management* (Garden City, NJ: Doubleday & Co., 1984).
32. Myers, N., *Conversion of Tropical Moist Forests* (Washington, DC: National Academy of Sciences, 1980).
33. Nabhan, G., Assistant Director, Desert Botanical Garden, Phoenix, AZ, personal communication, May 16, 1986.
34. Namkoong, G., "Conservation of Biological Diversity by *In-Situ* and *Ex-Situ* Methods," OTA commissioned paper, 1985.
35. National Research Council, *Staff Report: Environmental Degradation in Mauritania*, Board on Science and Technology for International Development, Commission on International Relations, National Research Council, National Academy Press, Washington, DC, 1981.
36. Oldfield, M.L., *The Value of Conserving Genetic Resources* (Washington, DC: U.S. Department of the Interior, National Park Service, 1984).
37. Organisation for Economic Cooperation and Development, *Report on the State of the Environment* (Paris: 1985).
38. Peters, R.L., and Darling, J.D.S., "Greenhouse Effect and Nature Reserves," *Bioscience* 35(11): 707-717, 1985.
39. Prescott-Allen, C., "Status and Trends of Agricultural Crop Species," OTA commissioned paper, 1985.
40. Prescott-Allen, C., and Prescott-Allen, A., *The First Resources: Wild Species in the North American Economy* (New Haven, CT, and London: Yale University Press, 1986).
41. Salam, R.V., *Coastal Resources in Sri Lanka, India, and Pakistan: Description, Use, and Management*, U.S. Fish and Wildlife Service, International Affairs Office, Washington, DC, 1981.
42. Saunier, R.E., and Meganck, R.A., "Compatibility of Development and the *In-situ* Maintenance of Biotic Diversity in Developing Countries," OTA commissioned paper, 1985.
43. Simberloff, D., "Are We On the Verge of a Mass Extinction in Tropical Rain Forests?" *Dynamics of Extinction*, D.K. Elliot (ed.) (New York: John Wiley & Sons, Inc., 1986).
44. Simberloff, D., "Community Effects of Introduced Species," *Biotic Crises in Ecological and Evolutionary Time*, M.H. Nitecki (ed.) (New York: Academic Press, 1981).
45. Smith, F.E., "A Short Review of the Status of

Riparian Forests in California," *Proceedings: Riparian Forests in California—Their Ecology and Conservation Symposium*, A. Sands (ed.), University of California, Davis, Division of Agricultural Sciences, May 14, 1977 (Berkeley, CA: 1980). *In*: Crumpacker, 1985.
46. Swift, B.L., and Barclay, J.S., "Status of Riparian Ecosystems in the United States," presented at 1980 American Water Resources Association National Conference, U.S. Fish and Wildlife Service, Eastern Energy and Land Use Team, 1980.
47. Synge, H., "Status and Trends of Wild Plants," OTA commissioned paper, 1985.
48. Tiner, R.W., Jr., *Wetlands of the United States: Current Status and Recent Trends* (Newton Corner, MA: U.S. Fish and Wildlife Service, Habitat Resources, 1984). *In*: Crumpacker 1985.
49. U.S. Agency for International Development, *Honduras Country Environmental Profile: A Field Study*, AID Contract No. AID/SOD/PDC-C-0247 (Arlington, VA: JRB Associates, August 1982).
50. U.S. Congress, Office of Technology Assessment, *Technologies To Sustain Tropical Forest Resources,* OTA-F-214 (Washington, DC: U.S. Government Printing Office, 1984).
51. U.S. Congress, Office of Technology Assessment, *Impacts of Technology on U.S. Cropland and Rangeland Productivity*, PB 83-125 013 (Springfield, VA: National Technical Information Service, 1982).
52. U.S. Department of the Interior, Fish and Wildlife Service, *Recovery Plan for the Endangered and Threatened Species of the California Channel Islands* (Portland, OR: 1984).
53. U.S. Environmental Protection Agency, *The Chesapeake Bay Program: Findings and Recommendations*, September 1983. *In*: Fosburgh, 1985.
54. Warner, R.E. (recorder), *Proceedings: Fish and Wildlife Resource Needs in Riparian Ecosystems Workshop*, Harpers Ferry, WV, U.S. Fish and Wildlife Service, Eastern Energy and Land Use Team, Resources Analysis Group, May 30-31, 1979. *In*: Crumpacker 1985.
55. Wildlife Trade Monitoring Unit, *A Perception of the Issue of High-Trade Volume* (Cambridge, UK: Conservation Monitoring Centre, 1984).
56. World Bank, *Sudan Forestry Sector Review*, Report No. 5911-SU, 1986.
57. World Resources Institute and International Institute for Environment and Development, *World Resources 1986* (New York: Basic Books, 1986).

Chapter 4
Interventions To Maintain Biological Diversity

CONTENTS

Tables

Figure

Chapter 4
Interventions To Maintain Biological Diversity

HIGHLIGHTS

- Four management systems are used to conserve diversity: 1) managing ecosystems, 2) managing populations and species in natural or seminatural habitats, 3) maintaining and propagating living organisms offsite as in zoos or botanic gardens, and 4) storing seeds or other germplasm, usually with refrigeration or freezing.
- The four systems for maintaining diversity are complementary, but linkages between the strategies (e.g., between zoos and nature reserves) are less well developed than they could be to maximize conservation efforts.
- Biological, political, and socioeconomic factors must be evaluated to choose the best mix of management interventions. Because the importance of maintaining diversity has only recently begun to attract widespread recognition, scientific methods for evaluating trade-offs are at an early stage of development. Methods for evaluating socioeconomic factors seem to lag behind development of biological methods.

The majority of plants, animals, and microbes survive without any specific human interventions to maintain them. However, as natural areas continue to be modified—through fragmentation of habitats, for example—their survival and, in turn, maintaining biological diversity will increasingly depend on active management. A spectrum of technologies—broadly defined to include management systems and other means by which knowledge is applied—can be used to maintain diversity.

MANAGEMENT SYSTEMS TO CONSERVE DIVERSITY

Two general approaches are followed in maintaining diversity: 1) onsite maintenance, which conserves the organism in its natural setting; and 2) offsite maintenance, which conserves it outside its natural setting. Onsite maintenance can focus either on an entire ecosystem or on a particular species or population. And offsite maintenance can focus on living collections or on stored germplasm. These four broad management systems are necessary components of an overall strategy to conserve diversity. Conservation objectives can be enhanced by any combination of the four systems and by improving the linkages between them to take advantage of their potential complementariness.

Table 4-1 lists some technology programs associated with each management system. In general, the technologies on the right side of the table entail more human intervention.

Onsite Ecosystem Maintenance

Where the conservation objective is to maintain as much biological diversity as possible, the only practical and cost-effective approach

Table 4-1.—Examples of Management Systems To Maintain Biological Diversity

Onsite		Offsite	
Ecosystem maintenance	Species management	Living collections	Germplasm storage
National parks	Agroecosystems	Zoological parks	Seed and pollen banks
Research natural areas	Wildlife refuges	Botanic gardens	Semen, ova, and embryo banks
Marine sanctuaries	*In-situ* genebanks	Field collections	Microbial culture collections
Resource development planning	Game parks and reserves	Captive breeding programs	Tissue culture collections

Increasing human intervention →

← Increasing natural processes

SOURCE: Office of Technology Assessment, 1986.

is to maintain ecosystem diversity. Offsite maintenance cannot accomplish this objective because many species cannot live outside their natural habitats. An ecosystem approach allows processes, such as natural selection, to continue. Survival, for some species, depends on complex interactions with other species in their habitats. Maintaining diverse ecosystems also continues ecological processes, such as nutrient cycling, that typically depend on the interaction of numerous species (5).

Programs to maintain a diversity of ecosystems usually identify different ecosystem types and then attempt to preserve a sample of each type (see ch. 5). Some types, such as cloud forests, are rare and confined to small areas. These are especially vulnerable and receive special emphasis in some conservation programs.

The ecosystem approach is used not only for natural areas but also for traditional agricultural ecosystems. Pressures to modernize these "agroecosystems" threaten the characteristically high levels of crop and livestock genetic diversity these systems represent—and upon which modern agricultural systems continue to depend.

Onsite Species Maintenance

When the objective is to maximize direct benefits from diversity, such as production of an optimal mixture of game, fish, timber, and scenic values, then the preferred approach is often to manage particular species and their habitats. Managing at the population level is preferred when the objective is to avert extinction of a rare or threatened species or subspecies.

Because managing all species would be impossible, biological, political, and economic factors determine which species will receive direct attention. Preference is given to species with recognized economic value, for example.

Noncommercial species that are rare and endangered also are given management attention to ensure survival of wild populations. Similarly, species that provide important indirect benefits, such as pollination or pest control, may receive attention. Ideally, management should also focus on keystone species, i.e., those with important ecosystem support or regulatory functions.

Offsite Maintenance in Living Collections

Zoological parks, botanic gardens, arboretums, and field collections are common homes for living collections. Living collections serve several conservation objectives. Zoos and botanic gardens can propagate species threatened with extinction in the wild, sometimes enabling the repopulation of a newly protected or restored habitat. Arboretums and living collections kept at places like agricultural research stations maintain the genetic diversity of plants not amenable for germplasm storage as well as numerous livestock varieties that are not commercially popular but are culturally significant or are needed for research and breeding programs.

The number of species maintained in living collections is limited by the size of the facilities and the relatively high maintenance cost per species. Managers of offsite collections face the dilemma of maintaining populations large enough to ensure viability and at the same time providing refuge to as many species as possible.

Historically, the primary objectives of living collections have been research and display. However, growing concern over the loss of biological diversity is leading to greater efforts to develop collections for their conservation potential. Offsite facilities are also used to breed and propagate organisms, so they no longer rely solely on collecting from the wild to replenish their stocks. Instead, they can make a positive, direct contribution to species' survival.

Offsite Maintenance in Germplasm Storage

Storage of dormant seeds, embryos, and clonal materials, or germplasm storage, is the most cost-effective method to preserve the genetic diversity of the thousands of agricultural varieties and their wild relatives when biological factors allow (5). As farmers increasingly abandon the traditional varieties in favor of genetically uniform, modern ones, the preservation of diverse, locally adapted crops will depend heavily on offsite storage (1).

The need to maintain a convenient source of germplasm for breeding purposes and the ability to draw on germplasm from different geographic areas are important objectives met by the offsite storage systems (see ch. 7). Germplasm storage is also the principal method for maintaining identified strains of microbes (see ch. 8). And it is increasingly used to store wild plant species and a few wild animal species.

DECIDING WHICH MANAGEMENT SYSTEM TO APPLY

The efficacy of onsite and offsite technologies depends on biological, political, and economic factors. The following four chapters in this report examine how these various considerations determine which technologies are applied. General observations on how these factors help match management systems to conservation objectives are considered here. (See table 4-2 for a summary of objectives.)

Biological considerations are central to the objectives and choice of systems. Because not all diversity is threatened, the task of maintaining it can focus on the elements that need special attention. Biological uniqueness is important in setting priorities for conservation programs. A unique species—one that is the only representative of an entire genus or family, for example, or a species with high esthetic appeal—may be the focus of intensive conservation management either onsite, offsite, or both.

Biological uniqueness can present problems in applying conservation technologies, because species-specific research is often required to develop management or recovery plans (ch. 5). Species with unique reproductive physiology, for example, often cannot be maintained offsite until a considerable investment has been made in developing propagation techniques (ch. 6).

Political factors also influence conservation objectives and management systems. Commitments of government resources, policies, and programs determine the focus of attention and, to a large extent, such commitments reflect public interests and support. For example, in the United States a disproportionate share of resources is devoted to conservation programs for a select few of the many endangered species. Substantial sums have been spent in 11th-hour efforts to save the California condor and the black-footed ferret, while other endangered organisms such as invertebrate species receive little notice.

National instability may also threaten biological resources either directly in the cases of civil strife or warfare or indirectly through encouraging neglect. Such cases warrant special efforts

Table 4-2.—Management Systems and Conservation Objectives

Onsite		Offsite	
Ecosystem maintenance	Species maintenance	Living collections	Germplasm storage
Maintain:	Maintain:	Maintain:	Maintain:
• a reservoir or "library" of genetic resources	• genetic interaction between semidomesticated species and wild relatives	• breeding material that cannot be stored in genebanks	• convenient source of germplasm for breeding programs
• evolutionary potential	• wild populations for sustainable exploitation	• field research and development on new varieties and breeds	• collections of germplasm from uncertain or threatened sources
• functioning of various ecological processes	• viable populations of threatened species	• offsite cultivation and propagation	• reference or type collections as standard for research and patenting purposes
• vast majority of known and unknown species	• species that provide important indirect benefits (for pollination or pest control)	• captive breeding stock of populations threatened in the wild	• access to germplasm from wide geographic areas
• representatives of unique natural ecosystems	• "keystone" species with important ecosystem support or regulating function	• ready access to wild species for research, education, and display	• genetic materials from critically endangered species

SOURCE: Office of Technology Assessment, 1986.

to collect an endangered species or germplasm and maintain it outside the country to ensure survival and to facilitate access.

The applicable management systems and technologies also depend largely on economic factors. Costs of alternative management systems and the value of resources to be conserved may be relatively clear in the case of genetic diversity. For example, the benefits of breeding programs compared with the cost of seed maintenance easily justify germplasm storage technologies (see ch. 7). However, cost-benefit analysis is more difficult when benefits are diffuse and accrue over a long period (7). This problem is particularly acute for onsite maintenance programs where competition for land exists. Current threats to biologically rich tropical forests by land seeking peasant agriculturalists illustrate these conflicting interests.

COMPLEMENTARINESS OF MANAGEMENT SYSTEMS

Each of the four management systems serves different objectives. Historically, the two offsite approaches have developed independently from onsite approaches. However, some links have developed between the different management programs (see figure 4-1). Improvement of such links will contribute substantially to the cost-effectiveness of each management system and will help to achieve the overall goal of maintaining biological diversity.

Biological Linkages

Transfers of biotic material among the four management systems can enhance diversity. Exchanges between onsite systems occur, for example, when genetic material from wild plants becomes incorporated into locally cultivated varieties. Exchanges between offsite systems occur, for example, when seeds and clones of agricultural varieties are taken from storage and grown out in living collections for use in breeding programs. Similarly, a zoo may collect animal semen from its living collection and place it in cryogenic storage to expand the number of individuals it can maintain—in a sense creating a "frozen zoo."

Exchanges of species or germplasm between wild areas and living collections are most evident when wild specimens are taken for zoos,

Figure 4-1.—Transfers of Biotic Material Between Management Systems

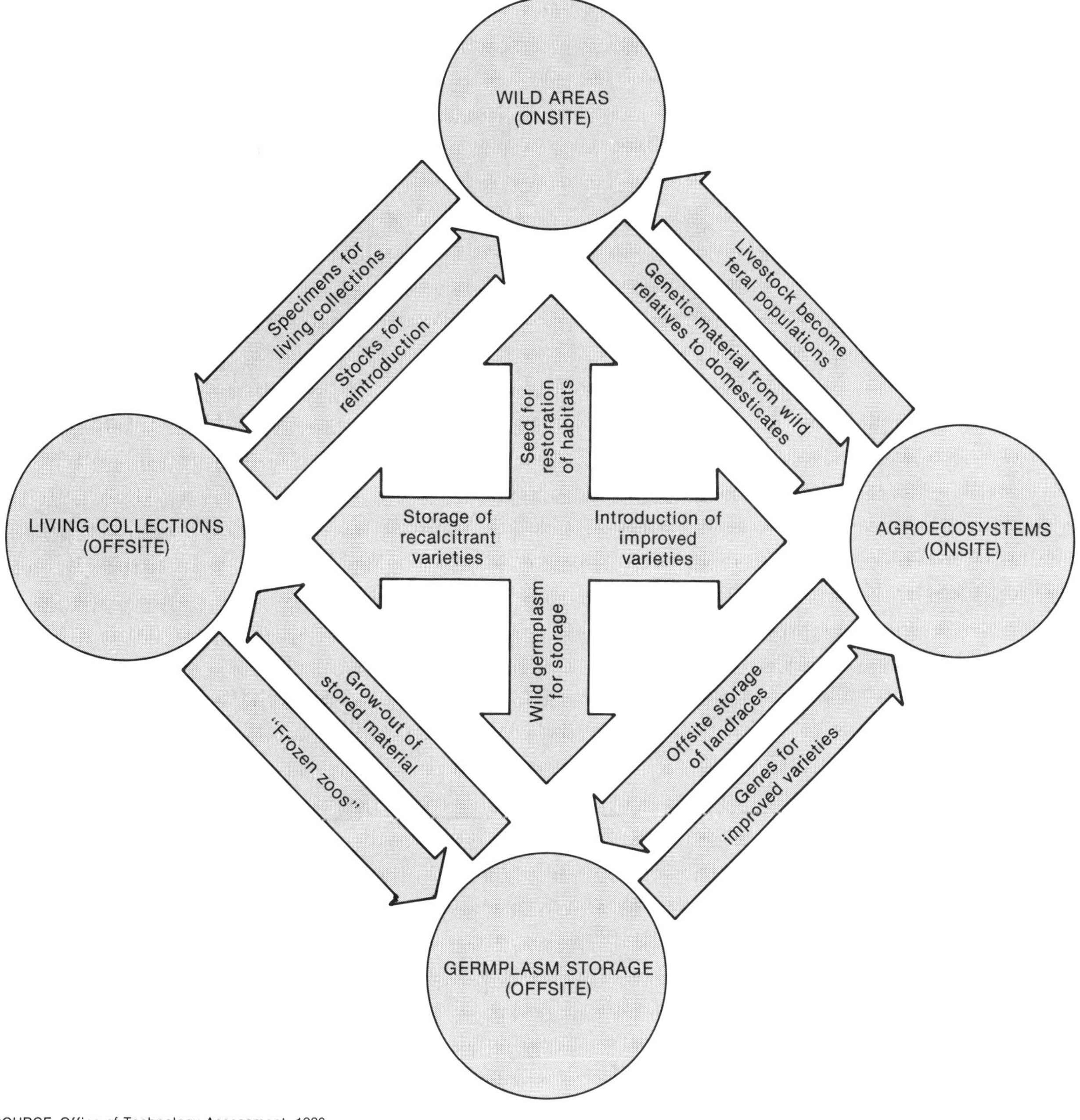

SOURCE: Office of Technology Assessment, 1986.

botanic gardens, or private collections. Taking wild specimens for living collections may provide material for research and public education, may prevent captive populations from inbreeding, and may even enhance wild populations through later reintroduction. However, these activities can—and have—threatened wild populations of a number of species.

It is often possible to take only germplasm—seeds, cuttings, and semen—rather than entire organisms. This approach has the advantage of being less destructive to rare or endangered populations (6). In the interest of preserving endangered populations, mammal germplasm collection is increasingly being attempted. Semen has been collected from wild populations of cheetahs, rhinoceroses, and elephants (10). But collecting mammal germplasm, without keeping the animal in captivity, is more difficult and costly. And many of the wildlife specimens maintained in zoos are the survivors of destructive capturing procedures.

Efforts increasingly are being made through captive breeding to produce stocks for reintroduction into the wild. The golden lion tamarin program in Brazil (11) has been successful for example. Less attention has been focused on reintroducing threatened plant species, but the recent reintroduction of a wild olive tree species on the island of St. Helena suggests that this approach is possible (6).

Perhaps the most important transfer of genetic material occurs between agroecosystems and germplasm storage: Landraces produced in traditional farmers' fields are the result of thousands of years of natural and human selection from thousands of different crop varieties. Many varieties can no longer be maintained by the farmers, who abandon them to plant higher yielding crop varieties. But the genetic diversity of traditional varieties is needed to create improved varieties. Thus, collecting expeditions to transfer these varieties into offsite storage are critically important to maintenance of the world's agriculture. Germplasm flows from storage back to agroecosystems, via research and breeding programs, as new varieties are introduced into agricultural systems.

Transfers are not always beneficial, however. Livestock that escape captivity can become feral animals with populations so high that they threaten native wild plants and animals. Feral goats on Pacific islands and feral horses on some rangelands of the United States are well-known examples. Similarly, exotic plants introduced as ornamentals or agricultural crops sometimes escape to become weeds that crowd out native species. Efforts to capture specimens for living collections can also be destructive. The challenge is to manage the transfers among sites and programs to enhance the positive contributions to diversity maintenance and minimize the negatives ones.

Technological Linkages

Research and technology transfers between diversity management programs can increase the efficiency, effectiveness, and capacity for maintaining biological diversity. Some technologies developed for domesticated species can be adapted for use with wild species. For example, technologies for offsite maintenance of wild species—particularly germplasm storage and captive breeding—have benefited substantially from the research and experience in agriculture. Perhaps the most dramatic linkage is embryo transfer technologies developed for livestock that are now being adapted to endangered species (ch. 6). Similarly, storage technologies developed for agricultural varieties, such as cryogenics and tissue culture, may become valuable tools for maintaining collections of rare or threatened wild plant species.

Like biological linkages, technological linkages work both ways. For example, research on living collections has provided information that can be applied to maintaining populations in the wild (2). Likewise, research on wild populations supplied information on a number of species' special reproduction requirements, which led to successful results with breeding in captivity.

Technological linkages among institutions engaged in researching, developing, and applying technologies have been limited. Researchers and resource managers in this area have historically worked in relative isolation, dealing almost exclusively with others in their fields of activity. The few interactions that have occurred have had a positive impact. Thus, the potential for benefits from increased cooperative work seems apparent, but institutions are slow to make such changes.

Institutional Linkages

Exchanges of organisms and technologies have occurred because they have been considered necessary for success of the different programs. However, most programs focus on relatively narrow subsets of diversity. Some groups devote their attention exclusively to maintaining certain agricultural crops, while others focus on specific wild species—e.g., whales or migratory waterfowl. The result is that much of the work is done in isolation, and the scope and effectiveness of overall diversity maintenance effort becomes difficult to monitor. And while particular concerns may be well-addressed, other concerns receive little or no attention.

Institutional problems that impede overall maintenance of biological diversity include:

- overlap and interinstitutional or intrainstitutional competition,
- gaps between goals and the human and financial resources available to achieve them, and
- lack of complementariness or cooperation between initiatives (4).

Institutional links can identify common interests, strengths, and weaknesses of various organizations as well as gaps and opportunities to address overall concerns. Not all activities should be operationally linked, however. A diversity of approaches in conservation activities is beneficial, and interaction should occur principally with those programs closest in purpose and approach (4).

Useful technologies emerge through a series of steps. Basic research provides an understanding of the nature of biological systems. Drawing on this knowledge, researchers define requirements and develop techniques to manage ecosystems, species, or genetic resources. Once the techniques are developed, however, researchers must synthesize techniques into technologies, then transfer and apply the technologies to site-specific circumstances.

In practice, the process of technology development is impeded by institutional constraints. Research is undertaken by scientists in many institutions, including universities, botanic gardens, zoological institutions, and government agencies responsible for natural resources. These scientists commonly emphasize the theoretical. At the other end of the spectrum are resource managers who apply particular techniques. Although the transfer of basic research to applied research is a problem in developing useful technologies, the principal weakness seems to be the failure of institutions to support synthesis of scientific information into useful management tools.

The problem of technology development is more pronounced for onsite than for offsite maintenance. This difference perhaps reflects the more pragmatic nature of offsite maintenance, where institutions (most with agricultural interests) emphasize research *and* development. Focus on technology development is commonly lacking in onsite maintenance. Institutions may deter scientists from translating research into practical techniques (8). To apply ecological studies to onsite maintenance, greater emphasis needs to be placed on comparative and predictive science, which implies less emphasis on descriptive studies (4).

Forces working against diversity are largely social and economic. Therefore, human dimensions need to be included in the scientific investigations, and natural and social sciences must be involved in conservation initiatives. There is, however, a paucity of social science research for the development of technologies to conserve biological diversity. This lack is partly because of the complexity and difficulty of such work, and partly because the potential for social science to make important contributions has been overlooked.

Greater support is needed for inventory and monitoring of diversity in natural systems and in agricultural systems. Some of this information is already available, but most of it has not been assimilated or made available to decisionmakers.

Finally, science needs to be applied to provide policymakers and the general public with better information on the scope and ramifications of diversity loss. Such information needs to be accurate, compelling, and digestible by a lay audience. To produce such information, scientists and scientific institutions need to become more directly involved and accommodating within the public policymaking process (9). At the same time, the information they provide should meet the four criteria for effective public policy:

1. **Adequacy**—meets the accepted standards of objectivity, completeness, reproducibility, and accuracy, and is appropriate to the subject and the application.
2. **Value**—addresses a worthwhile problem; neither too narrow nor too broad.
3. **Effectiveness**—able to influence policy in a constructive way and linked with the network of decisionmakers responsible for the issue.
4. **Legitimacy**—carries a widely-accepted presumption of correctness and authority (3).

CHAPTER 4 REFERENCES

1. Arnold, M.H., "Plant Gene Conservation," *Nature* 39:615, Feb. 20, 1986.
2. Benvischke, K., "The Impact of Research on the Propagation of Endangered Species in Zoos." *Genetics and Conservation*, C. Schonewald-Cox, et al. (eds.) (Menlo Park, CA: Benjamin/Cummings Publishing Co., Inc., 1983).
3. Clark, W.C., and Majone, G., "The Critical Appraisal of Scientific Inquiries With Policy Implications," *Science, Technology, and Human Values*, forthcoming.
4. diCastri, F., "Twenty Years of International Programmes on Ecosystems and the Biosphere: An Overview of Achievements, Shortcomings, and Possible New Perspectives," *Global Change*, T.F. Malone and J.G. Roederer (eds.) (New York: Cambridge University Press, 1985).
5. Frankel, O.H., and Soulé, M.E., *Conservation and Evolution* (Cambridge, MA: Cambridge University Press, 1981).
6. Lucas, G., and Oldfield, S., "The Role of Zoos, Botanical Gardens and Similar Institutions in the Maintenance of Biological Diversity," OTA commissioned paper, 1985.
7. Randall, A., "Human Preferences, Economics, and the Preservation of Species," *The Value of Biological Diversity,* B.G. Norton (ed.) (Princeton, NJ: Princeton University Press, 1986).
8. Salwasser, H., U.S. Forest Service, personal communication, September 1986.
9. Shapiro, H.T., et al., "A National Research Strategy," *Issues in Science and Technology*, spring 1986, pp. 116-125.
10. Wildt, D.E., National Zoological Park, Smithsonian Institution, Washington, DC, personal communication, July 1986.
11. World Wildlife Fund—U.S., *Annual Report 1984* (Washington, DC: 1984).

Part II
Technologies

Chapter 5
Maintaining Biological Diversity Onsite

CONTENTS

Tables

Figures

Boxes

Chapter 5
Maintaining Biological Diversity Onsite

HIGHLIGHTS

- Maintaining plant, animal, and microbial diversity in their natural environment (onsite) is the most effective way to conserve maximum biological diversity over the long term.
- Strategies to maintain diversity onsite have evolved from strict preservation to multiple use. More recently, attention is being given to integrating conservation with development in areas outside protected zones.
- Guidelines for optimum biological design of protected areas are improving. But decisions on design are determined more often by socioeconomic and political factors than by scientific principles.
- Techniques for restoring diversity on degraded sites are being improved as knowledge of natural plant and animal succession increases. However, complete restoration is often not feasible, and partial restoration is usually slow and expensive.
- Opportunities for improving national and global conservation of diversity onsite include 1) promoting an ecosystem approach to protected area establishment and management, 2) encouraging innovative resource development methods that treat conservation as a form of development, 3) supporting multidisciplinary research on the many factors to consider when designing nature reserves, and 4) developing training and job opportunities for experts in all these areas.

INTRODUCTION

Plants and animals can be maintained where they are found, that is, onsite, either by protecting certain sites from change or by managing change to support some portion of the natural biota. Most biological diversity can only be maintained in a natural condition for three reasons:

1. For most species, technologies have not been developed to keep substantial numbers of individuals alive outside their natural environments.
2. For species that can be kept alive in artificial conditions, preserving genetic diversity usually entails maintaining numerous individuals from genetically distinct populations. Such preservation is financially and logistically feasible for only a few of the hundreds or thousands of species of many ecosystems.
3. Species survive gradual changes in their natural environments by continuous evolution and adaptation—processes that are arrested in offsite collections.

Strategies for maintaining biological diversity onsite range from single-species management to protection of complete ecosystems in designated natural areas. The various approaches are complementary. For example, a European nature reserve system established with broad conservation objectives contains some 10,000 sites of plant species that also are useful for breeding and for research into the chemistry of natural substances.

CLASSIFYING AND DESIGNING PROTECTED AREAS

Maintenance of biological diversity per se has often not been the primary objective of protected natural areas. Instead, many such areas have been set aside and managed for other conservation values, such as preservation of scenic landscapes or protection of watersheds (11). More recently, however, the U.S. Congress and other policymakers have begun to authorize actions to address the maintenance of biological diversity directly. With this new mandate, biologists, agricultural scientists, and conservation program managers have started to develop new ways to apply science to the problem of maintaining biological diversity onsite.

The development of techniques for onsite maintenance of biological diversity has so far focused mainly on protected areas. This section is concerned, therefore, largely with where these protected areas should be established and how biological principles can be used in the design and management of protected areas. The technologies appear to be scientifically sound, yet too little implementation has occurred thus far for a conclusive assessment of effect.

Even if the biological techniques are demonstrated to be correct, the actual location, design, and management objectives for protected areas will be determined mainly by social (including economic and political) factors. For example, costs will usually be a stronger consideration than biological criteria in choosing whether to have one large reserve or several small ones. Boundaries usually reflect what area has been made available rather than what would provide the best habitat for flora and fauna.

Development activities other than conservation may also take precedence in decisions to change the boundaries of protected areas. In tropical countries, where diversity is most threatened, many natural areas are occupied by farmers, hunters, gatherers, and fishermen (see ch. 11). Strategies to safeguard biological diversity must recognize that development of natural resources is imperative and must incorporate socioeconomic and political considerations. However, conservationists and resource developers should also view conservation as a necessary component of economic development.

In spite of the powerful influence of social factors, social sciences are applied less often than natural sciences in efforts to maintain biological diversity. Development planning techniques that do use social science data and principles have been proposed, however, and used occasionally to integrate natural resource conservation with other forms of economic, cultural, and social development. Resource development planning is discussed in some detail in the OTA report, *Technologies To Sustain Tropical Forest Resources* (83). A variation of resource development planning, integrated development planning, is described briefly later in this chapter.

Classification Systems for Protected Areas

Strategies to develop a system of protected areas typically begin by classifying and mapping types of ecosystems using data on plant and animal distributions and on climate and soil parameters. This information is compared with the locations of already-protected areas to approximate priorities for allocating the resources available for site protection.

Descriptions of threatened ecosystems are now adequate in every country to undertake effective programs for conserving biological diversity. In nearly all regions, however, continued improvements in ecosystem classification and assessment would facilitate better decisions on where protection is most needed. Preservation priorities need to be based on knowledge of which ecosystems:

- have high diversity,
- have high endemism (a high proportion of the species having a limited natural range),
- are threatened by resource development or degradation patterns,

- are located where social and economic conditions are conducive to conservation, and
- are not adequately represented in existing protected areas.

The major patterns of nature can be described for most terrestrial areas with existing data. Several biogeographic systems have been developed that relate data on distribution of plant and animal species to factors such as climate and natural barriers like oceans, deserts, and mountain ranges. These systems classify the Earth into zones, with each zone containing distinctive ecosystems and life forms.

Much less information is available on aquatic sites, such as lakes and streams. Aquatic ecosystems are difficult to map on a large scale, and the way to integrate them into land classifications is poorly understood. The same is true of riparian vegetation, mountain meadows, and other azonal ecosystems.

Classification systems take two broad approaches. "Taxonomic" methods establish land units by grouping resources or sites with similar properties. "Regionalization" methods subdivide land into natural units on the basis of spatial patterns that affect natural processes and the use of resources (1). The two approaches can be integrated to identify ecosystem diversity in considerable detail.

The taxonomic approach is typified by the Society of American Foresters (SAF) Cover Type Classification system, which aggregates similar stands of forest trees on the basis of the kind, number, and distribution of plant species and the dominance by tree species (19). The basic taxonomic units—forest cover types—are named after the predominant tree species. The Renewable Resources Evaluation of the U.S. Forest Service further aggregates many of the SAF categories into 20 "major forest types," which are the basis for the only map of forest cover types available for the United States as a whole.

The regionalization approach, on the other hand, begins with a nation or continent and subdivides it into progressively smaller, more closely related units. An example of this is the ecoregions classification system, used extensively by U.S. Federal land-managing agencies (1). Ecosystem regions for North America are defined as domains on the basis of climate. The domains are subdivided into divisions, which are subdivided into provinces on the basis of what plant communities can be expected to develop if the natural succession of species is not interrupted by human activity. Provinces are subdivided into sections on the basis of the composition of the vegetation types that eventually would prevail. Extending this ecoregion classification system to cover the world on a scale of 1:25 million is being considered.

A recently developed system for classification of the world's marine and coastal environments combines physical processes with biotic characteristics (34). This classification system will be used as a basis for selecting U.S. coastal biosphere reserves (13).

Each classification system has advantages and disadvantages for programs to maintain biological diversity. The taxonomic approach identifies and classifies each component. For example, separate taxonomies are used to identify flora, fauna, and soils. This separation facilitates location of natural areas that will conserve concentrations of high-priority components, such as a vegetation type or animal species. The regionalization approach allows scientists to determine whether the same type of ecosystem in distinct biogeographic regions actually represents two different ecosystems (2).

Biogeographic classification maps indicate what ecosystems would be found under natural conditions, but the discrepancy between expected and actual features is often great because of human intervention. Sparse grasslands may occur where climate, physical features, and species distribution records suggest tropical moist forest should grow. Furthermore, the major classification systems cover only broad zonal features of the environment. Azonal features—e.g., wetlands, riparian areas, and coral reefs—cannot be included. So conservation strategies must take a different approach to identify priorities for these ecosystems. Typically, plans

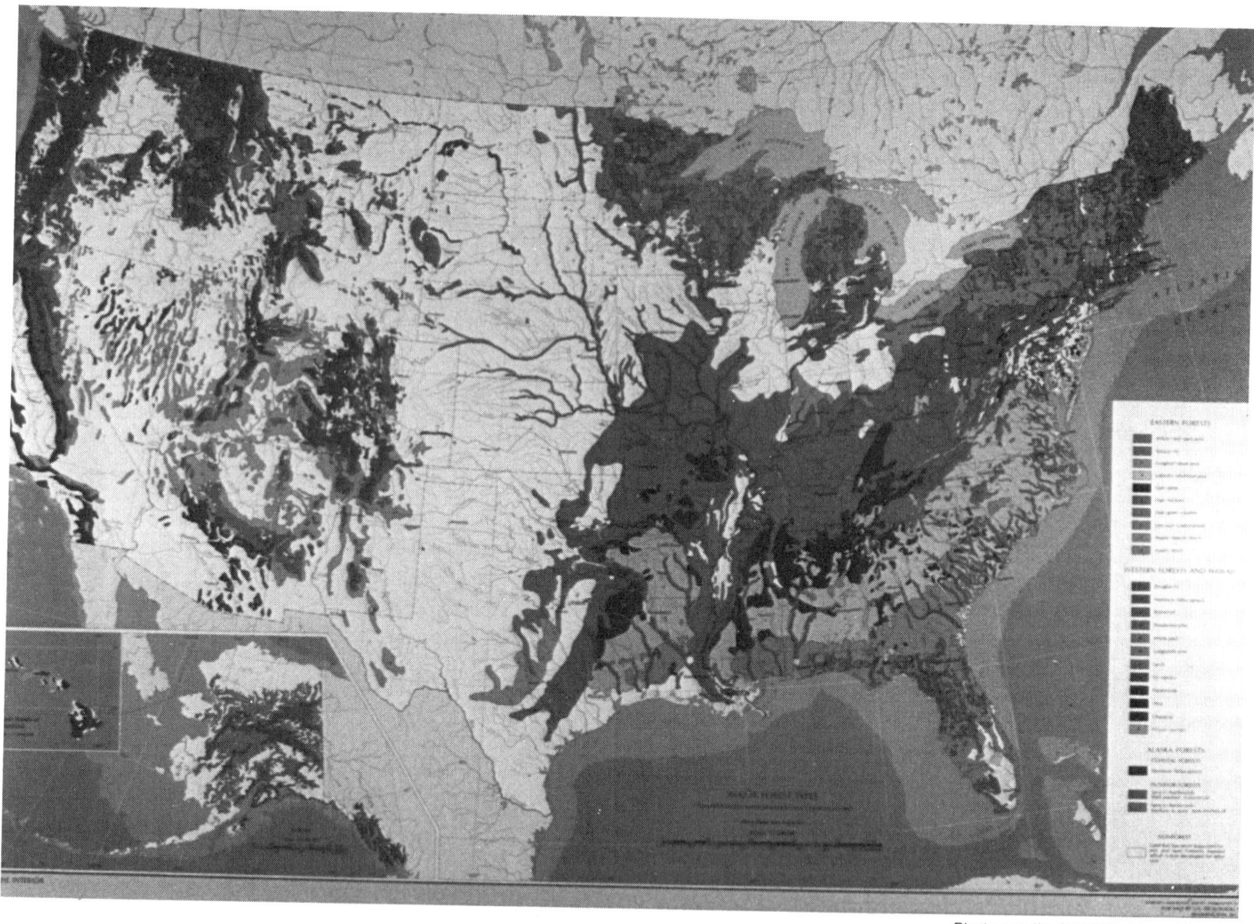

Photo credit: SAF/U.S. Forest Service

The Society of American Foresters Cover Type Map, an example of an ecosystem classification system, aggregates similar stands of forest trees on the basis of the kind, number, and distribution of plant species and the dominance by tree species.

for conservation of azonal ecosystems are based on surveys that cut across the biogeographic zones.

Biogeographic classification systems also need to be supplemented with information on endemism. Patterns of endemism vary among taxa and among regions. Some species with restricted distribution are quite common locally, whereas others are extremely rare (26). On the broadest scale, taxa may be endemic to a continent or subcontinent; on a narrow scale, many plant species seem to be restricted naturally to areas as small as a few square kilometers.

Identifying centers of endemism has been an ongoing effort of conservationists, especially tropical ecologists. An area such as an island or a mountain forest may not have an unusually high number of species present, but it may have a high proportion of species not found elsewhere (i.e., high endemism). Such areas are considered valuable for maintaining biological diversity, because they contribute substantially to diversity on a global scale.

In sum, a variety of ecosystem classification systems are currently being used by many organizations with different objectives. Although these maps do not indicate the extent of existing ecosystems (e.g., how much forest actually remains), they do correspond roughly to the boundaries of species distributions. Thus, they can be compared with maps of natural areas already protected, and planners can then choose which sites to focus on for more detailed assessment of an ecosystem's contribution to diversity, its vulnerability, and its social and economic significance.

Design of Protected Areas

The sizes and locations of protected areas are determined first by political and financial constraints. Within those limits, the designs of nature reserves have usually been based on natural history characteristics of the particular species of greatest interest. Recently, however, scientists have begun to develop theories for designing nature reserves to optimize protection of biological diversity rather than protection of particular species. These theories are still based mainly on inferences from general scientific principles and are largely untested. Thus, they are the subject of much academic debate among scientists (53).

Criteria for optimum size and shape for protected areas have been based on information from insular ecology (e.g., refs. 15,16,74,90). These criteria, however, are widely viewed as too simplistic, and the theories are being further developed with information from ecological-evolutionary genetics (24,70,73,79) and from theoretical population dynamics (28,74, 80). These theories focus mainly on terrestrial protected areas and probably have limited use for the design of coastal marine reserves. The great dispersive abilities of marine organisms and the interconnections of adjacent communities thus complicate decisions concerning the proper size and spacing of reserves. (See box 5-A for discussion contrasting terrestrial and coastal-marine systems.)

Islands and Boundaries

Information on the occurrence and natural distribution of species on islands has been used to formulate theoretical size and location criteria for protected areas intended to maintain diversity. The equilibrium theory of island biogeography (52) maintains that greater numbers of species are found on larger islands because

Box 5-A.—Differences Between Terrestrial and Coastal-Marine Systems

It is difficult to gauge the relative differences in biological diversity in terrestrial and coastal-marine environments. Dry land contains approximately four times the number of species found in the sea; on that consideration alone, terrestrial ecosystems seem inherently more diverse. Differences in faunal diversity between marine and terrestrial environments are primarily due to insects. Without them, marine fauna would be more diverse than terrestrial fauna. However, terrestrial flora clearly exhibit greater diversity than marine flora (51).

Viewed from a different perspective, in which the number of higher taxa (particularly animal) indicate degree of diversity, the sea would appear more diverse because many higher taxa (i.e., phyla, classes, orders) are exclusively marine. Implicit in this view is the notion that higher levels reflect greater genetic differences—i.e., a single species may be the sole representative of an order, class, or phylum, and the loss of one of these species might cause a far greater genetic loss than would the loss of a species in a taxon made up of several hundred or thousand members (51).

Another difference is that many fish and invertebrates that make up the bulk of marine species pass through several life stages from egg to adult. In many of the life stages, the organisms seem unrelated to that of the adult form. These different forms can live in different ecosystems or in distinctly different niches within the same ecosystem. Maintaining one species may therefore require maintenance of several different ecosystems (51).

Movement of organisms and materials between different community types—seagrass, coral reef, and mangrove—means that terrestrial and marine communities sometimes cannot be defined simply by their physical boundaries. Effectiveness of efforts to protect one community type may be diminished by failure to protect neighboring communities as well as adjacent watersheds (40).

the populations on smaller islands are more vulnerable to extinction. That vulnerability is due to probabilistic nature of individual births, deaths, occurrences of disease, and changes in habitats. Also, islands farther from continents have fewer species, because colonists from large land masses are less likely to reach them. This theory was extended from true islands to their terrestrial analogs (e.g., forest patches in agricultural or suburban areas), and the field of study become known as "insular ecology" to reflect this broader perspective. Scientists do not concur that the theory accurately explains natural patterns of species diversity, and research has been initiated to test the theory. (See Gilbert (27) and Simberloff (76) for reviews of studies that confirm or refute the equilibrium theory.) In any case, the island analogy—that much of the natural diversity is being reduced and confined to small, often isolated areas—is not in dispute.

Nature reserves serve as islands for species incapable of surviving in human-dominated habitats. Isolated natural areas are likely to experience declining numbers of species when their size is reduced by deforestation or similar habitat changes. The analogy between islands and nature reserves was reinforced by findings from some of the early tests of equilibrium theory. These findings led to proposed design criteria for nature reserves intended to minimize the loss of species over time (53). The designs called for large nature reserves near each other, to reduce the effects of small areas and distances on species survival. Other design elements not explicitly derived from equilibrium theory but thought to maintain a greater number of species at equilibrium also exist. However, these are rather academic "all other things being equal" principles, and on the ground, complex habitat differences among areas should weigh more heavily in pragmatic choices of which sites to conserve.

The applicability of insular ecology to conservation is being tested by the World Wildlife Fund's Minimum Critical Size of Ecosystems Project (49). Biologists took inventories of plant and animal life in an Amazon forest area before it was fragmented by development. Various-sized patches of forest were kept intact through coordination with the deforestation and development program, and biologists now monitor plant and animal populations in each patch. Although the project is only 20 years old, changes in the biota are already evident (47,48).

The guidelines for optimum biological design still have many limitations (72). Most of the relevant research has focused on animals, particularly forest-dwelling birds; too little research has been conducted on plants or on other types

Photo credit: R.O. Bierregaard, Jr.

The Minimum Critical Size of Ecosystems Project (above) of the World Wildlife Fund and Brazil's National Institute for Amazon Research is a long-term study of the effects of fragmentation on the Amazon forest as it is cleared to create pasture. This study is providing data on which to base design and management recommendations for national parks and reserves of particular relevance to the Amazon forest.

of habitats. Also, the occurrence and persistence of a species on any particular site may be governed not only by the populations on that site but also by whether groups of loosely connected populations can survive in the region (46,71). This sort of scientific question requires long-term study, which is only beginning to be conducted.

As the Earth increasingly becomes a patchwork of natural and developed areas, the effects of activities on or near the boundaries of protected areas are becoming more important. Small areas and those with angular boundaries have a higher proportion of boundary-to-interior than larger or more circular areas. Nature reserves seldom have sharply defined natural boundaries like oceanic islands. Instead they have political boundaries that can do only so much to prevent movements in and out. Many species can migrate across nature reserve boundaries, and the results of human activities (e.g., pollution) may enter by air, water, or land. Consequently, another theory on the optimum design of protected areas, "the boundary model," has been proposed. It accounts for the boundary effects, including the effects of human activities (69).

Designated protected areas include both political and biological boundaries. Some of the biological boundaries are the natural edges between ecosystems; others result from human activities, most of which originate outside political boundaries. Those biological boundaries that fit the ecological definition of an "edge" (box 5-B) may increase local diversity as edge-adapted species prosper. Over time, however, survival of species in the interior may be reduced if edges are enlarged, because the habitat for species adapted to less-disturbed conditions is reduced. Poor protection at the political boundaries generally shifts the biological boundaries toward an area's interior.

Zones where resource-conserving development activities are encouraged have been tested to buffer the boundary effect (e.g., the United Nations Educational, Scientific and Cultural Organization's (UNESCO) Man and the Biosphere Program). Such buffer zones can help reserves by increasing the habitat area and minimizing the potential exposure to harm. The idea of buffer zones is not new, but implementation has been slow; few evaluations have been done yet to develop guidelines about the necessary character and width of the zones or the shifting nature of boundaries.

Box 5-B.—The Edge Effect

Natural boundaries between ecosystems, or edges, are considered to be ecologically diverse areas. Edges can be created by human manipulation of vegetation in an attempt to encourage maximum local diversity (14). Along an edge, animals from each of the abutting vegetation types may be found, together with animals that make frequent use of more than one vegetation type and those that specialize on the edge itself (41). Game animals commonly are edge-adapted, as are animals of many urban, suburban, and agricultural areas (e.g., birds) (8).

Corridors of habitat to connect nature reserves have been proposed for sites where reserve sizes are below-optimum. These corridors should facilitate gene flow and the dispersal of individuals between protected areas, which should, in turn, increase the effective size of populations and thus raise the chance of survival for semi-isolated groups (6). Also, corridors could increase the recolonization rate if species are eliminated locally (78). Corridors, however, are another theoretical concept, and they may not be appropriate for all sites. As noted earlier, geographic isolation is a cause of genetic diversity. Thus, corridors where none previously existed might cause locally adapted genotypes to be lost due to gene flow. The applicability of corridors is another aspect of design theory now being actively researched.

The use of corridors and boundary zones has been proposed for protected areas in the Western Cascades region of the United States. This area contains the largest tract of uncut forest in the conterminous United States as well as natural riparian habitats (32). The proposal suggests surrounding islands of old-growth forest with zones of low-intensity, long-rotation timber harvesting and then linking the islands by corridors of old-growth vegetation. This design would presumably provide mobility for species like the cougar and bobcat—far-ranging carnivores that would have populations too small for survival and continued evolution if confined to a single habitat island. Proposals like this must be considered planning hypotheses, suggested by general theory; and as such must be subjected to close, case-by-case scrutiny before implementation.

Genetics

Genetic considerations are another dominant concern in the literature on population viability and conservation. Attempts are being made to determine the smallest number of interbreeding individuals that will enable a species to survive indefinitely—adapting to changing environmental conditions without suffering the negative effects of a small population size (population instability, erosion of genetic variability, inbreeding). Because each individual carries only part of the genetic variation characteristic of its species, the size of a population—and thus, the amount of genetic variation—may determine how much and how fast a population can evolve.

Application of genetics to the issue of population size and viability has led to theoretical estimates of minimum populations for successful conservation of birds and mammals. One such estimate, known as the 50/500 rule, is that effective population size (in genetics sense) of 50 breeding adults is the minimum needed to sustain captive breeding programs over decades or a century (e.g., as in zoos), but a population 10 times as large is needed to sustain a species in its natural habitat as it evolves over millennia to survive changing environmental stresses (25,45).

The 50/500 rule is an approximation based on studies of only a few species. But the effect of population size depends on several factors that differ for various species, such as sex ratio, age structure, mating behavior, and behaviors such as feeding. Thus the rule, when applied to a particular species, could project a need for populations larger than 50/500—perhaps orders of magnitude larger. Empirical or experimental evidence is lacking to determine how resilient a "genetically viable" population would be when confronted with other pressures (e.g., demographic, environmental, or catastrophic uncertainty) (72).

Population Dynamics

Scientists have long recognized that, in general, smaller populations are more susceptible to extinction than larger ones, because death for individual organisms is an event determined largely by change, and populations are collections of individuals. Models of the impact of change on individual births and deaths were developed decades ago (e.g., ref. 21), and these have been applied to estimate the extinction time for particular species under various circumstances. Models also have been developed to evaluate the effect of chance environmental variations and chance population-wide catastrophes.

Recently, more sophisticated models of stochastic population dynamics have been formulated specifically to investigate questions of population viability. These models do not give specific prescriptions for minimum population size to assure survival, but they are leading to a better understanding of the role of chance in populations. They indicate that to avoid extinction resulting from the impact of chance on individual births and deaths may require only a few hundred breeding individuals. But larger, perhaps much larger, population sizes are necessary if the condition of the species' environment varies, and still larger populations are needed for species that are susceptible to catastrophes (72).

The modeling approach is useful but has significant limitations. First, data for population models encompassing both environmental and genetic factors exist for only a few species. Also, species experience the effects of chance at individual, environmental, population, and genetic levels. But models that could simultaneously simulate all these factors would be too complex for existing analytical capabilities.

Even if such models were developed, they could prove very costly to use (72).

The theoretical population models are yielding other plausible hypotheses, some of which have important implications for conservation. Extinction probabilities depend critically on population growth rates, on environmentally induced variability in this rate, and on particular catastrophic scenarios to which the species are subject. One recent analysis employs a stochastic population model and the general relationships between body-size and population growth rates and between body-size and population density to estimate the sizes of populations and habitats necessary for mammals to have a 95 percent probability of persistence for 1,000 years.

The preliminary results from this analysis are startling. For larger animals, the viable population size is smaller, but the necessary habitat must be larger to support the requisite populations. Thus, smaller mammals can have a viable population size of a million individuals but a habitat requiring only tens of square kilometers. The largest mammals, on the other hand, may have a viable population with only hundreds of individuals but may need a million square kilometers of habitat (3).

These are preliminary analyses. But even if subsequent work reduces the estimates by two orders of magnitude, larger mammals may need contiguous habitats of tens of thousands of square kilometers to survive indefinitely. Few protected natural areas are that large, implying that conservation strategies for certain species should not depend as much on protected reserves as on monitoring and managing larger areas (24).

ESTABLISHMENT OF PROTECTED AREAS

Since the world's first two national parks were established in the 1870s, some 3,500 protected areas have been set aside for conservation, covering some 4.25 million square kilometers (1,050 million acres) (35). (See figure 5-1 for rate of growth.)

Growth in the number and size of protected areas was slow at first. It accelerated during the 1920s and 1930s, halted during World War II, and regained momentum by the early 1950s. The number doubled during the 1970s, but growth has slowed over the past few years (33). Before 1970, most protected areas were located in industrial countries. But for the past 15 years, the Third World has led in both numbers added and rates of establishment.

Designation as a protected area does not necessarily mean that protection is effective, of course. The extent of actual protection in the 3,500 areas has not been determined, but anecdotal evidence indicates that illegal or unmanaged hunting, fishing, gathering, logging, farming, and livestock grazing are common problems (83). Thus, data on designated areas exaggerate the scope of conservation actually being achieved.

Acquisition and Designation

Most protected areas are established by official acts designating that uses of particular sites will be restricted to those compatible with natural ecological conditions. At the Federal level in the United States, designating a land area or water body for conservation involves making a formal declaration of intent to assign a certain category of protection and then providing an opportunity for extensive public comment on the proposed action. Other governments use similar processes, although the extent of public participation varies.

The degree of protection depends partly on the objectives of the acquisition or designation. There are many different types of designations. Kenya, for example, has national parks, national reserves, nature reserves, and forest reserves. The wildlife sanctuaries in Kiribati in the South Pacific are very different in conser-

Figure 5-1.—Growth of the Global Network of Protected Areas, 1890-1985

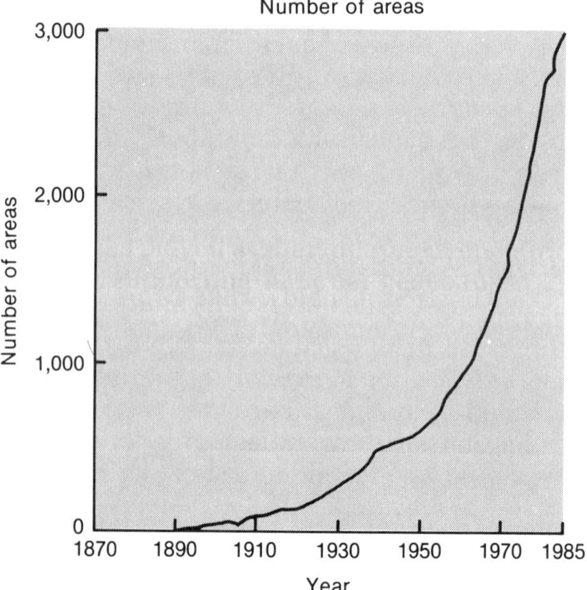

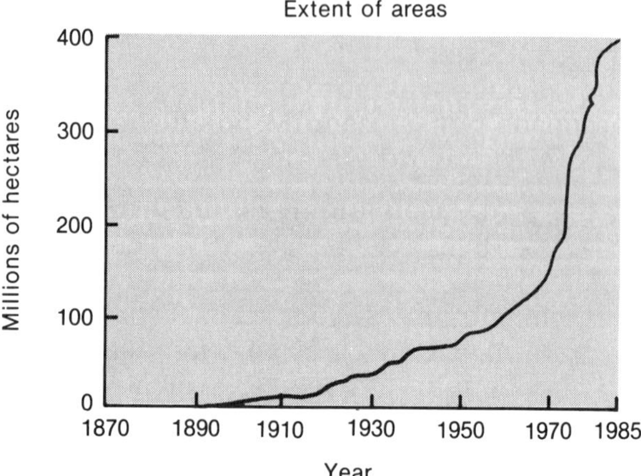

SOURCE: International Union for the Conservation of Nature and Natural Resources, *The United Nations List of National Parks and Protected Areas* (Gland, Switzerland: 1985).

vation terms from wildlife sanctuaries in India. Designated national parks of the United Kingdom are quite different from national parks in the United States. And in Spain, national parks, nature parks, and national hunting reserves indicate different types of protection.

To clarify this situation and to promote the full range of protected area options, the International Union for the Conservation of Nature and Natural Resources (IUCN) provides a series of 10 management categories (37,38). Protected areas are categorized according to their management objectives, rather than by the name used in their official designations (see table 5-1). Thus, the national parks of the United Kingdom are placed under category V (protected landscape or seascape), rather than under category II (national parks). Standardization of the categories also facilitates international compar-

Table 5-1.—Categories and Management Objectives of Protected Areas

I. *Scientific reserve/strict nature reserve:* To protect and maintain natural processes in an undisturbed state for scientific study, environmental monitoring, education, and maintenance of genetic resources.

II. *National park:* To protect areas of national or international significance for scientific, educational, and recreational use.

III. *Natural monument/natural landmark:* To protect and preserve nationally significant features because of their special interest or unique characteristics.

IV. *Managed nature reserve/wildlife sanctuary:* To assure the conditions necessary to protect species, groups of species, biotic communities, or physical features of the environment that require specific human manipulation for their perpetuation.

V. *Protected landscape or seascape:* To maintain nationally significant landscapes characteristic of the harmonious interaction of humans and land, while allowing recreation and tourism within the normal lifestyles and economic activities of these areas.

VI. *Resource reserve:* To protect the natural resources of the area for future use and prevent or contain development activities that could affect the resource, pending the establishment of objectives based on knowledge and planning.

VII. *Natural biotic area/anthropological reserve:* To allow the way of life of societies living in harmony with the environment to continue.

VIII. *Multiple-use management area/managed resource area:* To provide for the sustained production of water, timber, wildlife, pasture, and outdoor recreation, with conservation oriented to the support of the economic activities (although specific zones may also be designed within these areas to achieve specific conservation objectives).

IX. *Biosphere reserve:* To conserve an ecologically representative landscape in areas that range from complete protection to intensive production; to promote ecological monitoring, research and education; and to facilitate local, regional, and international cooperation.

X. *World heritage site:* To protect the natural features for which the area was considered to be of world heritage quality, and to provide information for worldwide public enlightenment.

SOURCE: J.W. Thorsell, "The Role of Protected Areas in Maintaining Biological Diversity in Tropical Developing Countries," OTA commissioned paper, 1985.

isons and provides a framework for all protected areas.

Criteria for Selection of Areas To Protect

Protected areas can be located and managed to protect biological diversity at three levels:

1. **at the ecosystem level:** by protecting unique ecosystems, representative areas for each main type of ecosystem in a nation or region, and species-rich ecosystems and centers of endemism;
2. **at the species level:** by giving priority to the genetically most distinct species (e.g., families with few species or genera with only one species), and to culturally important species and endemic genera and species; and
3. **at the gene level:** by giving priority to plant and animal types that have been or are being domesticated, to populations of wild relatives of domesticated species, and to wild resource species (those used for food, fuel, fiber, medicine, construction material, ornament, etc.).

Ecosystem Approach

Conserving ecosystem diversity maintains not only a variety of landscapes but also broad species and genetic diversity. Indeed, it may be the only approach to conserving the many types of organisms still unknown to science.

A strategy to maintain ecosystem diversity generally begins with the biogeographic classification system described earlier. The system can be used to identify which ecosystem types need to be acquired or designated to achieve more complete protection of biological diversity.

The extent to which diverse U.S. ecosystems are represented within protected areas is being assessed on a State-by-State basis by the natural heritage inventory programs of the different States (see ch. 9). Recent estimates of the proportion of major terrestrial ecosystem types that are not protected in the Federal domain vary from 21 to 51 percent, depending on the size and number of each type thought to be needed for adequate protection (13).

The extent to which the world's terrestrial ecosystems are included in protected areas has been crudely estimated using the Udvardy biogeographic classification system, which divides the world's land into 193 biogeographical provinces. Since each province typically contains many distinct types of ecosystems, the degree to which province locations correlate to protected area locations gives only an approximation of where greater protection is needed. The 3,514 protected areas listed by IUCN are located in 178 provinces. The coverage is patchy: several provinces have few protected areas, which implies that numerous unique ecosystems have yet to be included in the worldwide network of protected areas (see table 5-2) (33). An estimate of the cost of completing this network is $1 billion (17).

Ten provinces have fewer than 1,000 square kilometers protected but more than five protected areas, while another 29 have more than 1,000 square kilometers but only five or fewer separate protected areas. Determining the extent of the patchiness requires better figures for analysis, such as accurate estimates of province sizes. In addition, aquatic and azonal ecosystems (e.g., wetlands and coral reefs) do not fall easily within this system.

A U.S. effort that helps maintain representative aquatic ecosystems is the Marine Sanctuary Program conducted by the National Oceanic and Atmospheric Administration of the Department of Commerce. Potential marine sanctuary sites were listed after consultation with scientific teams familiar with the different ecological values of sections of the coastal zone (86). All current and future designations into the marine sanctuaries will be made from the site-evaluation list. Maintenance of community or ecosystem diversity is not a specific objective of the Marine Sanctuaries Program, but if all sites on the list were designated sanctuaries, coastal ecosystem diversity would be significantly protected.

An international effort that contributes to conserving representative ecosystems is UNESCO's

Table 5-2.—Coverage of Protected Areas by Biogeographic Provinces

Provinces lacking any protected areas:
Arctic Archipelago, Arctic Ocean
Argentinean Pampas, Argentina
Ascension/St. Helena, South Atlantic Ocean
Baikha, U.S.S.R.
Burman Rainforest, Burma
Greenland Tundra, Greenland
Laccadive Islands, Laccadive Sea
Lake Ladoga, U.S.S.R.
Lake Tanganuika, Africa
Lake Titicaca, Peru
Lake Turkana, Kenya
Maldives/Chagos Archipelago, Indian Ocean
Pacific Desert
Revillagigedo Island, Alaska
South Trinidad
Provinces with five or fewer protected areas and a total protected area of less than 1,000 km² (247,000 acres):
Aldabra, Seychelles
Amirante Isles, Seychelles
Aral Sea, U.S.S.R.
Araucania Forest, Chile
Atlas Saharien Steppe, Algeria-Morocco
Cayo Coco, Cuba
Central Polynesia, Pacific Ocean
Cocos (Keeling) Islands, Christmas Island, Australia
Comoros, Indian Ocean
East Melanesia, South Pacific Ocean
Fernando de Noronha Archipelago, Brazil
Guerrero, Mexico
Hindu Kush Highlands, Afghanistan-Pakistan
Insulantarctica
Kampuchea
Lake Malawi, Africa
Lake Ukerewe (Victoria), Africa
Malagasy Thorn Forest, Indian Ocean
Mascarene Islands, Indian Ocean
Micronesia, North Pacific Ocean
Patagonia, Argentina
Planaltina, Brazil
Ryukyu Islands, Japan
Sichuan Highlands, China
Sri Lankan Rainforest, Sri Lanka
Taiwan, ROC
Tamaulipas, Mexico
West Anatolia, Turkey

SOURCE: J. Harrison, "Status and Trends of Natural Ecosystems Worldwide," OTA commissioned paper, 1985.

Man in the Biosphere (MAB) Program. MAB has established a network of biosphere reserves in a global system of protected areas (see ch. 10). The objective is to have a comprehensive system covering all 193 biogeographic provinces. The MAB program exists in 66 countries, and approximately 256 biosphere reserves have been designated thus far (61).

Species Approach

Natural areas are also selected to conserve the habitats of rare or endangered species or to protect areas with high species endemism. Using species presence as the criteria for protected area location and management is useful for several reasons (62,82):

- Certain species can be used to indicate the effectiveness of management. If the more conspicuous rare species cannot survive, then the design and management of the reserve should be changed.
- Species provide a focal point or objective that people can readily understand.
- Some species have an appeal that wins sympathy, an important factor in raising funds and public awareness.

Protection of an area to conserve a rare or endangered species should be based on the best existing evidence on its location and habitat needs. The United States has accumulated a great deal of such information as a result of the Endangered Species Act and the work of The Nature Conservancy. For other regions of the world, information on endangered species ranges from precise (in northwestern Europe) to nonexistent (in the Amazon Basin). At the international level, the IUCN's Conservation Monitoring Center tracks the status of species and publishes its findings in the Red Data Books (10).

Genetic Resources

Genetic variation within species needs to be conserved because it enables species to adapt to changing conditions and provides the raw material for domestication of plants and animals and the continued improvement of already domesticated crops and livestock.

Protected areas designated specifically to protect genetic variability of particular species are often called *in-situ* genebanks. They may be established as separate areas for particular crop relatives, timber trees, animal species, and so on. Or, the maintenance of genetic diversity of important species may be one of several objectives of a protected area (63).

India and the Soviet Union have expressed commitment to *in-situ* conservation of the wild relatives of crop species (63). India has designated the first gene sanctuary, for citrus, and some Indian biosphere reserve areas are expected to have genetic conservation as a major objective. For example, a reserve area has been proposed for the Nilgiri Hills area, which is rich in wild forms of ginger, tumeric, cardamom, black pepper, mango, jackfruit, plantain, rice, and millets. The Soviet Union has reportly designated 127 reserves for protection of wild relatives of crops and has proposed an additional 20 areas for protection. Expeditions to a region known as the Central Asian gene center have found 249 species of wild crop relatives (63).

In East Germany, an inventory is being made of important genetic resources within the country's nature reserves, including 24 forage crop species, 51 medicinal plants, and 27 fruit species. As noted earlier, the inventory is expected to identify about 10,000 places within the country's reserve system where protection is afforded for plants relevant to breeding, breeding research, and study of the chemistry of natural substances (68).

Trade-Offs

In selecting areas for onsite maintenance of biological diversity, trade-offs occur when any of the above criteria are given priority. If the strategy is to protect areas where rare and endangered species are found, then the diversity of ecosystems that exists may not be maintained adequately because only certain types include identified rare species. Concentrating on biogeographic categories for broad coverage of ecosystem types may not protect habitats for rare or endangered species sufficiently or for centers of endemism. The third criterion, protecting genetic variability, includes consideration of economic and social factors that may contribute less to the objective of maintaining maximum diversity but aid the larger goal of conserving resource opportunities for human welfare.

In practice, other objectives and various social and economic constraints prevail in the decisions on where to locate protected areas. Other objectives include preservation of scenic resources, provision of recreation opportunities, and protection of watersheds. Constraints include budgetary feasibility, competing demands for use of the area, and opportunity for local support of protected status.

The literature on conservation strategies contains few objective methods to evaluate these trade-offs, except to note that the three biological approaches—ecosystem, species, and gene pool—are both complementary and necessary. Decisions are often initially made by the intuitive judgment of conservationists but ultimately by the political processes that lead to the official protected area designation.

PLANNING AND MANAGEMENT

Planning and management strategies for onsite maintenance seek to conserve either the species and genetic diversity within a given area or the diversity of ecosystems across a geographic region. Planning tools range from mathematical models that simulate how an area's biological resources are likely to respond given different management options to written plans for natural area management. Management is concerned both with managing external pressures affecting a protected area and with managing the natural succession of plant and animal communities within the area. Management activities range from no intervention to active manipulation of an ecosystem.

Planning Techniques

Planning for protected areas can begin before designation is finalized. Biologists generally agree that plans to maintain diversity need to begin with site surveys to determine the following information (65):

- the number, abundance, and distribution of species, and the interactions between species and community types;

- the types, extent, locations, and effects of human uses, the degree of dependence of local inhabitants on these uses, and the possible alternatives for activities that are harmful to the site;
- the present and potential threats from activities outside the immediate area of concern;
- the opportunities for making the site more useful to local inhabitants; and
- the best approach for law enforcement on the site.

Agency budgets and policies for management planning often omit some of these surveys, considering them fundamental research rather than pragmatic planning activities. For example, the U.S. Bureau of Land Management's Resource Management Plan process does not include collection of detailed site data if no deleterious human impact or other problem is known. Biologists argue, however, that the problems cannot be fully identified without the surveys.

Historically, the specific plans to conserve biological diversity were left to the area manager to devise and implement. This approach still prevails in many regions of the world. In the United States, conflicts in land and water management and the increasing need to justify all management activities to a governing institution have resulted in specialized tools for planning the conservation of biological resources. Much of this development of planning techniques has occurred in the Federal land management agencies.

Modeling

A recent innovation in planning techniques is the use of mathematical models. The models are highly simplified versions of natural environments. Biological data are used to develop equations that represent assumptions about cause-and-effect interactions between plant and animal populations and their habitats. Numerous equations interact, and the outcome projects responses of the biological resources to different management options. The accuracy of the projections depends on how well the equations and the data reflect the situation in the natural environment.

Various kinds of wildlife-habitat models have been used, and recently, the population simulation models described earlier have begun to be used widely. These population models predict how management activities would affect population size, structure, and recovery rate. They can describe, for example, the probable size of a fish population before and at various times after a specified fishing season.

Wildlife-habitat models are built from natural history data on species distribution and abundance in various habitats, from which cause-and-effect relationships are deduced to predict how wildlife populations will change as a result of changed habitat conditions. Indicator Species Models, for instance, focus on one or a few species known to reflect broader ecosystem qualities. Another example is the Habitat Evaluation Procedure used by the U.S. Fish and Wildlife Service to describe the responses of vegetation and, hence, wildlife habitats to certain management options such as timber harvesting (75).

Geographic Information System models also account for the changes in vegetation or wildlife habitats that result from different management options, but the output is presented on maps, which facilitates evaluation of cumulative impacts by area. A complementary technique being developed by U.S. National Park Service personnel, the Boundary Model, is intended to assess not only management activities but also the effects of human actions outside the protected areas (69).

Biologists warn that the accuracy of models is constrained by the need to reduce complex, often poorly understood interactions to assumptions simple enough to be represented with mathematical equations. Often, data are too limited to test all the assumptions. None of the natural area models can predict all the possible ways that biological resources might respond to habitat changes. Thus, models are best used to make the assumptions and logic of scientists, managers, and natural-area users explicit, so that final plans and management decisions can be based on clear, thorough, and objective understanding of all perspectives.

Management Plans

Management plans can help avoid typical protected area problems, such as inappropriate development; sporadic, inconsistent, and ad hoc management; and lack of clearly defined management objectives. Management planning also serves to review existing databases, to encourage resource inventories, and to identify other needed research and monitoring activities. Unfortunately, such plans do not exist for many of the world's protected areas, which constitutes a major constraint on maintaining diversity (82).

Species-specific management plans identify actions for maintaining healthy, reproductive populations of a particular species. Often, the species are either economically valuable or are rare, endangered, or sensitive to certain land- or water-management practices. The Office of Endangered Species of the U.S. Fish and Wildlife Service is the lead agency for recovery plans to restore populations listed on the Federal Threatened and Endangered Species List. For example, two Federal agencies, two State agencies, one university, and two agencies from British Columbia cooperated in development of the Selkirk Mountain Caribou Management Plan/Recovery Plan. This plan provides details on caribou population dynamics, behavior, and habitat in Idaho, Washington, and British Columbia. It describes past and present caribou management activities, specifies management goals and objectives to recover the species, indicates priorities for action, and assigns these to specific agencies (87).

Site-specific management plans outline the options for maintaining biological resources within given locations, commonly parts of natural areas. For example, the Bureau of Land Management developed a plan to maintain resources within the Burro Creek Section of the Kingman Resource Area in Arizona (88). The plan has clearly stated management objectives. It describes the resources of the site, presents the management issues pertaining to the area, details the management practices that will be used on the site, and indicates what other resource activities will be allowed (e.g., mining) (88).

Large-area planning documents can include maintaining diversity as one objective to be balanced with others, but they generally do not recommend site-specific actions. Examples include the plans prepared by the U.S. Forest Service for national forest management, plans by the Bureau of Land Management for resource area management, and plans by the National Marine Sanctuary Program for the management of marine sanctuaries. These planning processes generally involve numerous experts from various disciplines who identify and weigh management options. The planning document then describes resources, the options available for managing those resources for various uses, the trade-offs that would be made in various resource-use scenarios, and finally, the proposed management strategy.

National and subnational conservation strategies (NCSs) tend to be generic documents that may include but are not limited to conserving biological diversity. Some 30 countries had begun to develop national conservation strategies by the end of 1985 (62) (see figure 5-2). To date, only a few NCSs have been completed. The United States, for example, does not have a plan for conservation of biological diversity.

One example of a completed countrywide plan is the Zambia National Conservation Strategy, which identifies the major environmental issues and ecological zones that need immediate attention in that country (29). Objectives of the strategy are to maintain the essential life support systems, maintain genetic diversity of both domestic and wild species, promote wise use of natural resources, and maintain suitable environmental quality and standard of living. To accomplish these objectives, plans and policies for sustainable management of natural resources are to be integrated with all aspects of the country's social and economic development. The strategy outlines schedules of action for the major agencies and identifies necessary interagency linkages to assure cooperation. The official status of this plan and the extent to which it is being implemented is not clear.

Management plans vary in geographic scale and levels of specificity. Plans at the more general levels require less detailed information on

Figure 5-2.—National Conservation Strategy Development Around the World, July 1985

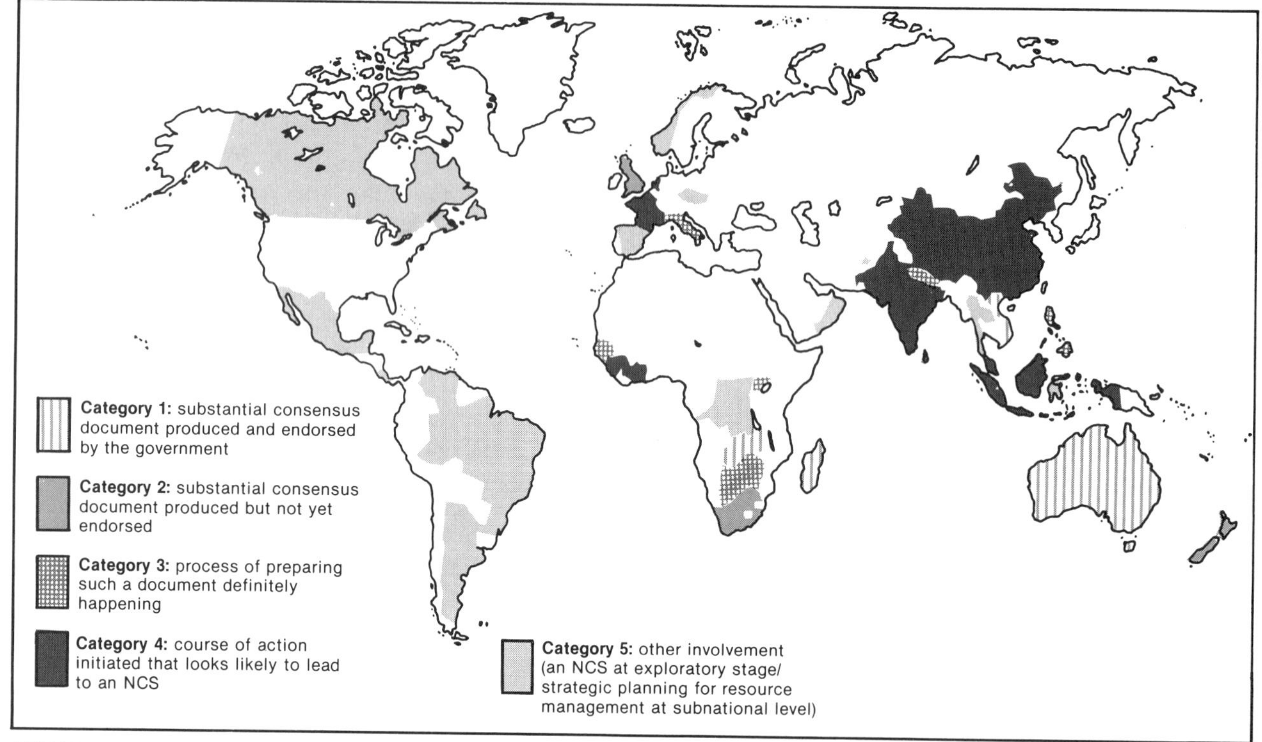
Category 1: substantial consensus document produced and endorsed by the government
Category 2: substantial consensus document produced but not yet endorsed
Category 3: process of preparing such a document definitely happening
Category 4: course of action initiated that looks likely to lead to an NCS
Category 5: other involvement (an NCS at exploratory stage/ strategic planning for resource management at subnational level)

SOURCE: *IUCN Bulletin Supplement* (Gland, Switzerland: IUCN, 1985).

the characteristics of species but greater understanding of larger cause-and-effect relationships and of social, economic, and political factors.

Management Strategies

Increasingly active management of factors affecting biological diversity will be needed to overcome the effects of human activity and the gradual fragmentation of natural areas (89). Natural areas change over time, as various plant and animal communities succeed one another, and gradual change in the components and quantity of biological diversity occurs. To sustain particular components, such as game animals or songbirds, protected-area managers therefore need to intervene in the natural processes. The interventions vary with objectives, and conflicts may occur. For example, developing optimum habitat for a particular species may not be compatible with maximizing the diversity of community types.

Manipulating habitats to manage particular species sometimes involves controlling populations of certain animals—removing an exotic fish from a lake, for example. More often, the intervention involves modifying vegetation. If the target species are grazing or browsing animals such as deer, intervention might mean cutting trees to prevent woodlands from evolving to the climax stage; for prairie birds such as cranes, it could mean burning grasslands to prevent encroachment by woody plants. Certain plants may be propagated for food or cover for the target species. The U.S. Fish and Wildlife Service uses such management techniques in national wildlife refuges, which are the only extensive federally owned lands managed chiefly for conserving wildlife.

Management to maximize the diversity of community types involves similar interventions. Again, a basic consideration is the variety of plant succession stages to be maintained within an ecosystem. Manipulation management is likely to be needed to preserve communities representing early stages of succession. For example, savanna ecosystems are maintained by fire, wildlife, and human influence. Management techniques to conserve savanna systems include regulating animal numbers and species and using controlled burning. Rain forests in a mature successional stage require little intervention, but they are likely to need active protection because they are not generally resilient if cleared in large areas (82).

Where the U.S. Forest Service manages land with wildlife diversity as a goal, it attempts to provide an appropriate mix of successional stages within each plant community (84). The approach of the U.S. National Park Service is to maintain natural processes to the extent possible, including catastrophic changes such as localized fire, to allow a relatively natural mix of succession stages to occur.

Management strategies have evolved from strict preservation and protection to multiple-use approaches and, more recently, to integrated approaches. Strict preservation strategy entails setting aside large blocks of natural areas where designation and protection alone would be expected to achieve conservation objectives. Protection would mean severely restricting the uses of, and the changes within, an area to ensure the continued natural condition of its biological resources and regular policing of boundaries to prevent trespassing or poaching. Where possible, fences would be erected to restrict access by humans and livestock.

Moderate versions of this strategy may be effective in some locations, particularly where the land is owned by an individual or nongovernmental organization. In many areas, strict controls are impractical. It has not been very successful in developing countries. Moreover, neither fences nor patrols can prevent all external influences from damaging a protected area. Regular patrols of a marine sanctuary could not stop the effects of water pollution, for example.

Strict preservation of biological diversity is not an explicit objective of any federally protected area in the United States. The objective closest to it is protection of "biological resources" or "ecological processes" on lands in the National Wilderness Preservation System, which is an evolving system of public lands relatively undisturbed by humans and large enough to have potential for wilderness recreation. (Most wilderness areas contain at least 5,000 acres, although some in the Eastern United States are smaller.)

Other countries also have extensive areas set aside for preservation while allowing some human access. Examples include large segments of Antarctica and isolated parts of the Amazonian forest. Some natural areas, such as Wood Buffalo National Park in Canada and Salonga National Park in Zaire, have wardens to guard the boundaries and prevent trespassing (61). But increasingly, countries cannot afford to designate large areas for strict preservation. Particularly in developing countries, adequate fences, patrols, or other means to deny access to designated areas are seldom logistically, economically, or socially possible. In addition, preservation strategies have exacerbated perceived conflicts between conservation and development.

Another strategy for protected areas is to incorporate multiple uses or objectives. This strategy is usually based on one or two approaches: developing an optimum mix of several uses on a local parcel of land or water; or creating a mosaic of land or water parcels, each with a designated use, within a larger geographic area.

Developing an optimum mix of uses in an area requires careful incorporation of each objective so that all can be met. This approach is used by the U.S. Forest Service on national forest lands and by most States on wildlife areas and State forests. In national forests, the potential of each subsection is evaluated for recreation, grazing, timber production, wildlife or fisheries habitat, mineral development, and other uses. Management objectives for each site usually include more than one use. Thus, an area that is managed for timber production may also provide sites for grazing livestock or foraging wildlife. Sometimes, mining or another use will be exclusive, at least temporarily.

The California Desert Conservation Area, managed by the Bureau of Land Management, is an example of the second approach to multiple use, in which protected areas are managed primarily as distinct parcels with different primary uses. The conservation area is broken into different land units and classified according to the level of human activity allowed in each. Research areas, wilderness areas, areas of critical environmental concern, areas of geologic or archeologic significance, and critical habitats of endangered or sensitive plant or animal species are mapped and sometimes identified by markers posted at the sites. The rest of the land is classified for various levels of use ranging from restricted to extensive human use and alteration. Management of the area evolves as human needs for resources of the California Desert change.

The biosphere reserve concept is another example of multiple-use based on buffer zones that would moderate the extent that activities affect the core. The UNESCO Man and the Biosphere Program (see also ch. 10) champions this idea. An idealized scheme includes three areas:

1. The core areas strictly protect ecological samples of natural ecosystems that can serve as benchmarks for measuring long-term changes in ecosystems.
2. The buffer zones have land-use controls, which allow only activities compatible with protection of the core area, such as research, environmental education, recreation, and tourism.
3. The transition areas surround the core and buffer zone and are usually not strictly delineated. In these areas, researchers, managers, and the local population are to cooperate in rehabilitation, traditional use, development, and experimental research on natural resources (30).

The areas should facilitate management by reducing conflicts, because the more incompatible uses would be physically distant from one another. And effectiveness of protection should be enhanced, because conflicting uses could be detected before they spread into the core (see box 5-C).

This approach has not yet been implemented sufficiently to assess its worldwide effect, but plans for such development now exist and await political commitment and implementation in several nations. One example is the development plan for the San Lorenzo Canyon area in Mexico (60). Multiple-use development is indicated for a 225,000-acre chaparral and desert area where watershed protection is a primary objective. The plan delineates four zones:

1. a core *scientific* area to be used for research and watershed protection,

Box 5-C.—Cluster Concept for Biosphere Reserves

In the United States, a promising development of biosphere reserves is the cluster concept. The approach is intended to link complementary areas administered by different agencies so they can cooperate in monitoring research, educational, and management activities.

A particularly promising multiple-unit biosphere reserve is emerging in the Southern Appalachians. Efforts are underway to link the existing Great Smoky Mountains National Park, the Forest Service's Coweeta Hydrological Station, the Department of Energy's Oak Ridge National Environmental Research Park, and other nearby State and Federal agencies managing natural resources to form a Southern Appalachian Biosphere Reserve. The existence of a permanent association of Federal agencies and regional universities has served as a useful mechanism to help coordinate regional research and management activities involving the biosphere reserve.

Another promising example is on St. John, the Virgin Islands, where the National Park Service manages the V.I. National Park. A cooperative effort involving agencies and research institutions from Puerto Rico, the U.S. Virgin Islands, and the British Virgin Islands has coordinated a major research program focused on developing a biosphere reserve on St. John. As the only U.S. national park in a developing region, the area provides opportunities in the transfer of research and resource management technologies suitable for small islands of the region.

Figure 5-3.—Design of a Coastal or Marine Protected Area

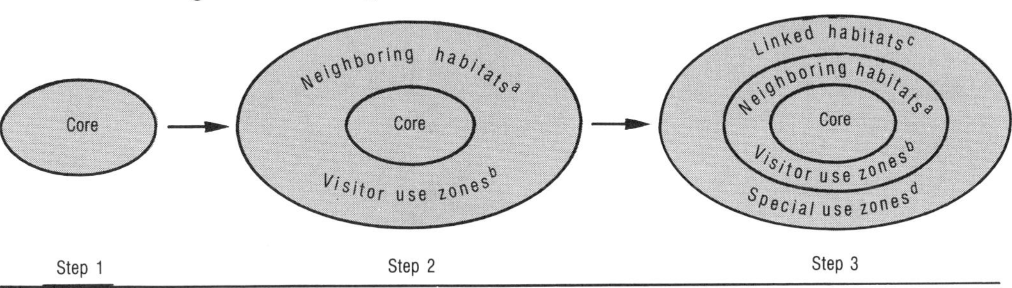

[a]Habitats adjacent and linked to habitats of interest by species movements or flows of nutrients.
[b]Headquarters; ranger stations; and traditional fishing, recreational, research, and education zones.
[c]Watersheds, agricultural lands, urban and industrial developments, rivers, and estuaries.
[d]Shipping lanes, commercial fishing grounds, and intensive use zones.

SOURCE: R.V. Salm, *Marine and Coastal Protected Areas: A Guide for Planners and Managers* (Gland, Switzerland: International Union for the Conservation of Nature and Natural Resources, 1984).

2. a *primitive* area to be used for watershed protection and recreation,
3. an *extensive use* area for recreation and education, and
4. a *natural recovery* area eventually for agricultural and commercial use.

Zoned development seems an especially important concept for marine and coastal areas, which are particularly vulnerable to events outside their boundaries even when they are protected. Figure 5-3 is an idealized design for a coastal- or marine-protected area.

A more recently developed integrated approach holds potential for resolving many of the problems that arise in onsite maintenance of biological diversity. An example is the integrated regional development planning, which is discussed later in this chapter.

Ecosystem Restoration

As degraded ecosystems become more common, restoration will play an increasing role in conserving biological diversity. Underlying most of the discussion in this chapter has been the assumption that protected areas are designated where ecosystems are in a relatively natural condition. Another important approach, however, is to protect and sometimes manipulate degraded ecosystems in order to restore some degree of biological diversity. Restoration techniques are being used by conservation organizations, such as The Nature Conservancy and the Audubon Society, and by government agencies, such as the National Park Service to enlarge or adjust the shape of reserves (43).

Reclamation is action intended to restore damaged ecosystems to productive use (43). Restoration is the re-creation of entire communities of organisms, closely modeled on communities that occur naturally. Reclamation gradually becomes restoration as more and more naturally occurring species are used and as natural plant and animal succession occurs. Restoration technologies, which depend heavily on the knowledge gained from reclamation experience, attempt to accelerate natural succession processes while assuring that indigenous rather than exotic species dominate.

Restoration is an onsite method that provides links with offsite activities to preserve species. Zoos and botanic gardens conserve rare species offsite for reintroduction onsite (see ch. 6). Nurseries and seed facilities provide plants and seeds for a variety of revegetation efforts, although materials for most native plants still must be gathered from the wild (42,44). Reintroductions of animals from captive populations are few but include the Arabian oryx, the golden lion tamarin (a recent effort), and plans to reintroduce Przewalski's horse in Mongolia (see ch. 6, box 6-E). A few plant reintroductions from offsite collections also exist. The Knowlton's cactus (*Pediocactus knowltonii*) has been returned to the natural environment from cuttings by the Fish and Wildlife Service (55).

Some States, such as Florida, require the use of native species in reclamation, but reclamation work generally falls short of restoration. Reclamation is generally task-oriented, and the objective is usually to establish productive plant cover, such as pasture or stands of trees. Relatively little attention is given to species not directly related to the objective, and relatively few species are used. Consequently, efforts to reclaim land have largely focused on the use of common plant and animal species that are easily propagated and multiply rapidly. Often these are nonnative species; rare or difficult-to-establish species are seldom used. It is easiest and most cost-effective to use those few species that have been shown to be adequate for particular uses, such as for stabilizing beaches.

Tree planting is one of the most frequently used techniques for reclaiming degraded lands, and a wealth of literature on various forms of reforestation exists (23,84). The potential for reforesting degraded forest land is especially great in the tropics (83). But restoring forests with diverse native species is seldom attempted. Instead, most programs use one or a few exotic species, partly because of a lack of seeds and techniques to propagate native trees and partly because of the cost-effectiveness of planting fast-growing tree species known to have commercial value.

The Santa Rosa National Park in Costa Rica is one of the few forest restorations that has been undertaken. The area was a cattle ranch for 400 years, but since designation as a protected area, a dry forest ecosystem of native species has been reestablished from seed sources on nearby mountain slopes. The principal management technique has been to stop the human-caused fires, allowing woody species to reinvade the pure grass pastures. The restorative effect has now been proved and is to be used in the proposed Guanacaste National Park (39).

Prairie restoration offers a kind of prototype for the development and use of ecosystem restoration. Restoration of prairies began early, motivated by concerns such as diversity and community authenticity (43). Techniques developed to restore prairies to moderately high levels of native plant diversity borrow extensively from agriculture techniques used in prairies. One approach to restoring 2 to 40 hectares recommends plowing, followed by disking at intervals of a year to reduce weeds, followed by seeding with a mixture of prairie species (56). In Crex Meadows, WI, restoration of prairie plants and animals was possible with little intervention other than protection and controlled burning, because many native prairie species had apparently continued to grow, unobserved, for decades while the site was forested. Little information on the cost of prairie restoration is available. Up to now, much of the effort that has gone into restoring the highest quality prairies has depended heavily on dedicated volunteers (4).

Photo credit: D. Franzen

Planting prairie plants in a restoration project at the University of Wisconsin-Madison Arboretum. The purpose of the experiment is to study competition between species by planting various combinations. The results will be useful in developing techniques for introducing "difficult" species into prairies as they are being restored and managed.

Although the technology of reclamation has developed rapidly in recent years, partly as a result of legislation such as the Surface Mine Control and Reclamation Act of 1977, restoration has not yet become an established discipline. Restoration research and technology development vary tremendously from one natural community to the next.

The availability of seed and plant stock for varieties adapted to local conditions is a problem. Use of local seeds is not required by law, and the high cost of small, special seed collections often precludes use of local seeds in favor of cheaper, nonlocal ones of relatively few species (31,54). Western nurseries and the Soil Conservation Service's Plant Material Centers have responded to the demand for more native plants, but many species still are not available commercially. For many that are available, germplasm is limited to specialized ecotypes or registered cultivars with limited value for restoration.

The cost of reclamation varies greatly, depending on the extent of disturbance, the extent of restoration, and the type of ecosystem. The average cost of seeding for reclamation of surface mines in seven Western States has been estimated at $620 per hectare (in 1977 dollars) (59). This estimate included fertilization, mulching, and irrigation (the most expensive component). The cost of earth-moving brought the total bill to $10,000 per hectare. Mechanical planting of shrubs costs from $500 to $2,000 per hectare in Utah, depending on whether bareroot or containerized stock was used (20). Hand-planting to simulate natural vegetation patterns would further increase costs.

Establishing the same community that occurred on a site prior to disturbance is often not feasible because of the high cost and a lack of information regarding, for example, necessary conditions for seed germination and other aspects of survival and reproduction of native species. Although restoration technologies cannot quickly restore the diversity that existed before degradation, they can be used to break the cycle of resource degradation and to reestablish a community of indigenous organisms. Normal plant and animal succession may eventually lead to a self-sustaining and relatively natural ecosystem that provides most of the values of biological diversity.

OUTSIDE PROTECTED AREAS

Most of the discussion thus far has dealt with protected areas where maintaining biological diversity is a management objective. But the majority of biological resources are found outside these areas. Few strategies have been designed yet for conserving diversity in nondesignated areas. Various resource conservation techniques with other objectives serve to enhance biological diversity, however.

Genetic Resources for Agriculture

Conservation of genetic variability outside protected areas is especially important because so many crop varieties and livestock species and many of their wild relatives are not found in areas designated for protection. In addition, evolutionary processes, such as crop-pest and crop-pathogen interactions, can continue. This type of conservation occurs when farmers have chosen to maintain traditional crop varieties and livestock breeds.

Crop varieties with a broad genetic base and wild relatives of crop plants are mainly located where traditional farming practices prevail. Large proportions of these resources (50 percent or more for many species) have not yet been evaluated or collected for preservation offsite (50). Germplasm collection programs focus on the world's major staple crops, so many species that are not yet widely grown are unlikely to be preserved offsite. Both these and local varieties of major crops are threatened with extinction as they are replaced by modern varieties, which the economics of agriculture favor.

A diverse mix of local varieties theoretically protects traditional farmers from catastrophic crop losses. And, with locally adapted varieties, farmers depend less on subsidized inputs, such as pesticides, fertilizers, irrigation equipment, and processed animal feed. In many countries, traditional farmers appear to be motivated more by avoiding risk than by maximizing profit. Nevertheless, as agricultural development occurs, farmers are shifting to fewer varieties, modern methods, and higher profits.

One response to this trend is the recently established program to monitor the remaining collections of teosinte, a wild relative of maize found only in Mexico, Guatemala, and Honduras. The habitats for teosinte include some of Mexico's best agricultural land, where it survives in narrow strips of untilled soil along stone fences bordering maize fields. As land use intensifies, these strips are brought into cultivation. And isolated stands of teosinte that interbreed with maize can be genetically "swamped" by the maize and lose their ability to disperse seed. Thus, teosinte populations with unique genetic characteristics of potential value for maize breeding are threatened with extinction.

Fortunately, the international agricultural research institute that focuses on maize, Centro Internacional de Mejoramiento de Maíz y Trigo (CIMMYT), is located near many of the sites where teosinte still grows. The CIMMYT maize staff and colleagues in the Mexican and Guatemalan national maize research programs have begun a monitoring program. The status of each teosinte population is checked annually. The intention is to take preservation action whenever a recognized population is placed in immediate danger of extinction (9).

Another way to safeguard genetic resources outside protected areas would be to preserve traditional agriculture systems in selected regions. To do this, farming systems must become more productive and produce more cash income. Presumably, higher productivity means applying scientific methods for crop production and genetic development but keeping the local varieties and livestock breeds. Some farming systems research does attempt this. For example, the Centro Agronomico Tropical de Investigaciones y Ensenanza has consulted with the Kuna people of northeastern Panama about agricultural development of their 60,000-hectare, indigenous-reserve area in the context of a park project (91). Similarly, the International Council for Research in Agroforestry in Africa trains researchers to identify opportunities for marginal improvements in traditional farming systems. But such work is outside the mainstream of agricultural development and is at best a modest and poorly funded effort.

A complementary approach is for modern farmers to maintain diverse varieties while relying on other income sources. They generally must turn to off-farm income or other, more modern areas of their farms. In developing countries, where traditional farming is still extensive and crop and livestock diversity are greatest, continued planting of nonprofitable traditional varieties would probably have to be subsidized.

Such an approach is not without precedent. Native American farmers are paid to produce seed of traditional cultivars in Arizona by a nonprofit organization, Native Seeds/SEARCH (58). In developing countries, similar programs might be administered by some of the same agricultural research organizations that maintain offsite germplasm collections. But the agricultural research community has not identified this as a priority for their limited funds. At best, only a very small sample of diversity could be maintained on subsidized traditional farms. It seems that such subsidies would be as cost-effective as marginal improvements in offsite collections.

Conservation As A Type of Development

Maintaining biological diversity by establishing parks is becoming increasingly difficult because of demographic, economic, and political pressures. The preservation approach to conservation may become less common in the future, especially in tropical developing countries where diversity seems to be most threatened. As a consequence, conservation organizations and conservationists within development organizations, such as the Agency for International Development (AID), the World Bank, and the Organization of American States (OAS), have begun to promote the concept that biological diversity can be conserved where natural resources are being developed if conservation is considered a development activity.

A recently published paper of the IUCN Commission on Ecology (64) supports this concept:

> The idea of basing conservation on the fate of particular species or even on the maintenance of a natural diversity of species will become even less tenable as the number of

threatened species increases and their refuges disappear. Natural areas will have to be designed in conjunction with the goals of regional development and justified on the basis of ecological processes operating within the entire developed region and not just within natural areas.

Conservation has long been a criterion in carefully planned development of agriculture, forestry, fisheries, grazing land, and other primary-industry development. But maintenance of biological diversity is relatively new as an explicit development objective. Some innovative approaches are beginning to be implemented, including the use of conflict resolution and systems analysis techniques in resource development planning.

Integrated Regional Development Planning (IRDP), being used by the OAS (60,66), subdivides a region into small spatial units and analyzes the sectoral interactions in each, in contrast to approaches that subdivide issues into sectoral components. IRDP addresses interactions, like competition for the same goods or services by two or more interest groups, and analyzes changes that occur in the mix of available goods and services as a result of activities in one sector that are detrimental to another sector.

IRDP uses systems analysis and conflict resolution methods to distribute the costs and benefits of development activities throughout affected populations or sectors. Integration of all the sectors—including maintenance of biological diversity—is necessary because individual sectoral activities may help, but often hinder, activities of other sectors aimed at appropriating goods and services from the same or allied ecosystems. Decisions about which activities are appropriate or how each can be adjusted to reduce conflict are made through negotiation by parties representing all the sectors that are involved (67).

A major constraint to considering diversity maintenance as a development activity is that the benefits of diversity are hard to calculate. No economic valuation techniques exist that can capture its full value. Thus, biological diversity has not fared well under the standard cost-benefit analyses applied to development activities. Although some efforts have been made to better account for biological diversity values, the results have been unsatisfactory and not widely applied. (See ch. 11 for further discussion of this topic.)

DATA FOR ONSITE MAINTENANCE OF BIOLOGICAL DIVERSITY

To set priorities and to allocate funds and other resources, decisionmakers need to know how various ecosystems contribute to biological diversity, how vulnerable they are to degradation, how well protected they are by existing programs, and what the social and economic prospects are for local cooperation. Management programs need details on the nutrition, space, and reproductive requirements of organisms. Most such information comes from taxonomy, biogeography, natural history, ecology, anthropology, and sociology. For agricultural species, information is also needed on genetics, microbiology, seed technology, and physiology.

Generally, enough is known to improve substantially the programs for maintaining diversity. But more and better data on many aspects of this subject are badly needed, and funding for conservation falls far short of the needs implied by the apparent rates and consequences of diversity loss (see ch. 3). So investments must be concentrated on the most cost-effective approaches possible, which implies the need to thoroughly understand the ecological, social, and economic aspects of biological diversity (77).

Uneven Quality of Information

The quality of data on biological diversity is uneven for different ecosystems and different parts of the world. For some places, such as tropical South America, data are rudimentary and theories are very tentative. For others, such as temperate North America, information is well developed and theories have been extensively tested. The unevenness is in part due to data being collected for different purposes, stored in different forms, and scattered among different institutions.

In general, both data and theories regarding biological diversity are better for temperate than for tropical biology; better for terrestrial than for aquatic sites; better for birds, mammals, and vascular plants than for the lower classes of organisms; and better for the few major crop and livestock species used in modern agriculture than for the many species used in traditional agriculture. Taxonomic coverage is increasing, but the pace is slow relative to the quantity of unknown organisms. Each year about 3 species of birds, 11 mammals, up to 100 fish, and dozens of amphibians and reptiles are identified for the first time (22,57). Insects are the largest order of organisms, and hundreds of species are newly identified annually (57); nevertheless, estimates of the number of insect species not yet identified range from 1 to 30 million (18).

Information needed to maintain diversity is even more limited on the ecosystem and community levels, partly because ecology is a younger discipline than taxonomy. Moreover, species interactions within ecosystems are so subtle that laborious, time-consuming field research is necessary to understand them. For example, the endangered red-cockaded woodpecker (*Dendrocopus borealis*) requires old-growth longleaf and loblolly pine trees for nesting. These pines persist in forest communities where occasional fires destroy the seedlings of other, more competitive species (5). Such fires require accumulation of appropriate fuel to carry the kinds of fires that favor the two pine seedlings. Conservation of red-cockaded woodpeckers, therefore, entails conserving appropriate species to generate the right kind of vegetation and litter on the forest floor.

Databases

Efforts to collect biological information have increased during the last two decades as a result of growing awareness of the importance of services provided by natural ecosystems and of the need for better use and management of natural resources. Biological data are now collected and analyzed at the international, national, and local levels. Databases—the collections of data that are organized for further analyses—can be used to make onsite diversity maintenance efforts substantially more effective.

International databases provide overviews that can identify potential gaps, status, and trends of biological diversity worldwide. The main international organizations involved in collecting biological data are the United Nations Environment Programme (UNEP); the Food and Agriculture Organization (FAO); United Nations Educational, Scientific, and Cultural Organization (UNESCO); the Conservation Monitoring Center (CMC) of the International Union for the Conservation of Nature and Natural Resources (IUCN); the World Wildlife Fund/Conservation Foundation; The Nature Conservancy International (TNCI); and the International Council for Bird Preservation (10). (See ch. 10 for further discussion of international databases.)

The utility of international databases has been limited because they are not readily available to resource planners and other analysts who might use them to advise development decisionmakers. To resolve this problem, the UNEP's Global Environment Monitoring System (GEMS) program is establishing a computerized Global Resource Information Database (GRID). This program will centralize access to numerous environmental databases and will include training in data analysis for developing-country participants.

Data for Management of Diversity

Biological data needed to plan management of diversity and other natural resources at the national level are collected by government agencies, academic institutions, and research centers. Completeness of the information varies from country to country. Several countries, such as Australia and Sweden, have compiled comprehensive biological surveys of their flora and fauna. Other countries, such as the Soviet Union and China, and some international regions, such as North, East, and West Africa, have completed or have made significant progress toward completing surveys of their flora. North America is the only part of the north temperate zone that has neither synthesized the data on its plant and animal resources nor created a national biological database (81).

In fact, the United States has abundant information on its biota at a regional or broad ecosystem level. But data acquisition is designed to serve the specific objectives of various organizations. As a result, many of the databases relevant to biological diversity are widely scattered, are often incompatible, and are in effect inaccessible to numerous potential users. The objective of maintaining biological diversity has been only a tangential consideration in most data-collection efforts. However, files on endangered species assembled by the Smithsonian Institution and the U.S. Fish and Wildlife Service do address an important aspect of species diversity directly.

The only comprehensive nationwide information system dealing directly with both species and ecosystem diversity is the national aggregation of State Natural Heritage Program data. This system is extensively used for decisionmaking on acquisition, designation, and management of protected areas.

The heritage program inventories are continually updated through a system of information gathering and ranking. They begin with a broad information search of secondary sources for rare species and ecosystems. These are then ranked, and further search, including field work, takes place for the rarest ones. The inventory is made up of a series of manual and computer files containing the species and ecosystem's classification, location, site where it occurs, land ownership of the site, and sites located on already protected land. Inventory data are plotted on U.S. Geological Survey maps to analyze which lands are most important to protect and what impacts specific development projects will have on diversity.

In recent years a great deal of attention has been given to the use of computers for managing biological data. Data management is facilitated by the flexibility of the hardware and by the many types of software on the market today. For example, Geographic Information Systems (GIS) are being used by the Forest Service and the National Park Service to integrate databases with spatial information. This technique produces overlay maps that have great potential to aid efforts to maintain biological diversity (figure 5-4). At present such overlay maps are used to assess the extent to which ecosystem diversity is being protected by combining details on ecosystem and species distribution with information on boundaries of various types of protected areas. International and nongovernment agencies are also finding the technique useful: GIS are a basic tool for GRID, The Nature Conservancy (TNC) has recently begun using GIS in its international program, and IUCN's Conservation Monitoring Center plans to acquire a system, once funding is secured (33).

The data on biological diversity generated at the State level are being aggregated at the national level by TNC. The quality and quantity of information varies from State to State, a few States do not yet have programs, and inventory of species and communities that are not threatened is just beginning. In spite of these limitations, this is the most comprehensive national database on biological diversity. In many geographic areas, TNC is the only institution collecting data on rare, sensitive, or endemic resources that may require special management to maintain their integrity as populations. In these areas, the heritage programs help to fill an important gap in biological data needed for the onsite maintenance of biological diversity.

Selection among such data management technologies as the GIS depends on the financial resources and the objectives of the organiza-

Figure 5-4.—Representation of a Geographic Information System Function Overlaying Several Types of Environmental Data

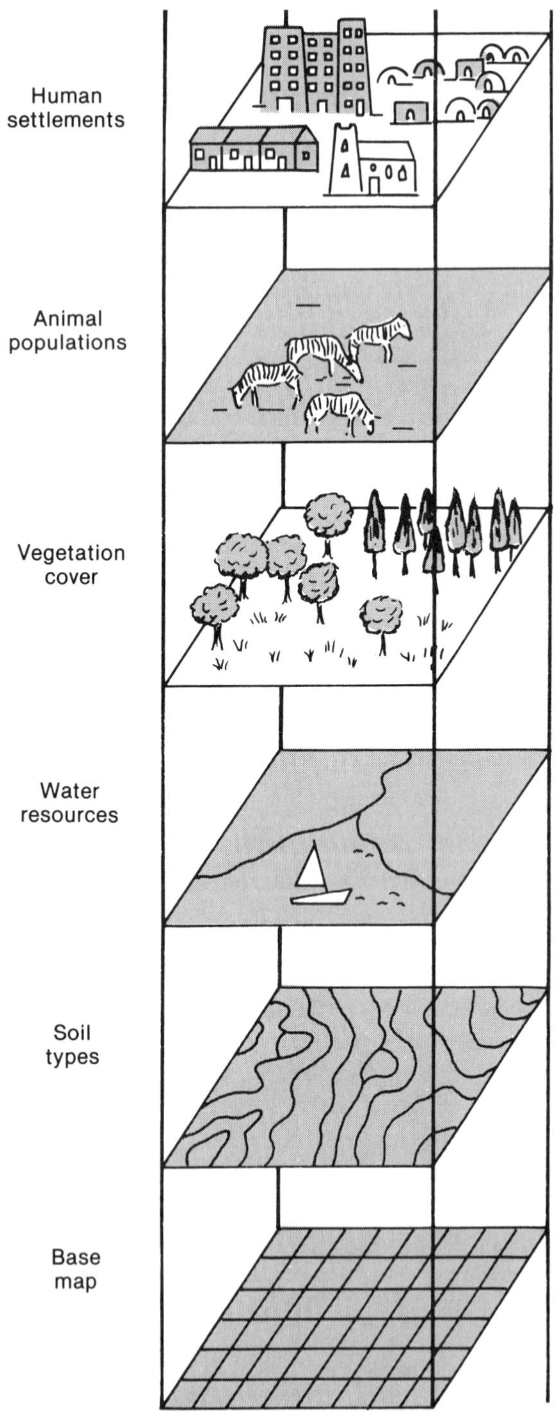

SOURCE: United Nations Environmental Programme/Global Environment Monitoring Systems, *Global Resource Information Databases* (Nairobi: GEMS Publication, 1985).

tion sponsoring the data collection. If the objective is to provide an overview of the status and trends of biological diversity in large areas, then remote sensing with sample surveys on the ground for verification and analysis with GIS may be the most cost-effective approach. If the objective is to design a management plan for a particular area, detailed field surveys are necessary, but tools such as GIS may still prove valuable.

For implementation of resource development, information on biological diversity at a local, site-specific level is most important. Yet this is the level at which the quality of biological information is most variable. For some heavily studied areas, detailed field inventories and analyses of ecosystem interactions have been completed, whereas for others, especially the remote areas, often little detail of biological diversity is known. Development of needed site-specific diversity data is constrained by the common attitude of land managers that diversity assessment is fundamental research that should be limited mainly to land areas where research is the designated major use. This is a problem even for agencies that are sensitive to the issue of biological diversity, such as the U.S. Fish and Wildlife Service.

Coordination

The quantity of biological data may increase as information becomes easier to handle and less costly to acquire and maintain. Linking databases developed for different purposes can greatly increase their utility and thus their cost-effectiveness. Data incompatibility hinders such linking, however, making it necessary to reenter data manually at great cost, or more often to forgo the improved analysis that linked databases would allow. Data sharing in the United States among and within Federal agencies frequently is constrained by a lack of standards. For example, different agencies generally use different terminology to define ecosystem types. This problem also exists at the international level, especially where classification schemes used to aggregate data are not standardized.

Coordination of data-collection efforts can reduce incompatibilities, lessen duplications, and identify gaps in collection. For example, CMC and UNESCO plan to feed information into the GRID system. TNC's regional databank has incorporated the classification system used by CMC to improve compatibility between the two data systems (33).

Coordination efforts at the U.S. Federal level have involved formal interagency cooperative agreements. (See OTA Background Paper #2, *Assessing Biological Diversity in the United States: Data Considerations*, for a description of these Federal interagency efforts to coordinate data collection and maintenance.) These efforts have resulted in recommendations and guidelines for standardization of databases. Most agencies would have to invest some personnel and funding to make their databases compatible with those of other agencies, however, which may not occur without specific congressional mandates.

Social and Economic Data

Human activities are the main cause of the accelerated loss of biological diversity, and successful implementation of onsite maintenance methods described in this chapter depends on cooperation of people living on and near the land that is affected. Collection and analysis of social and economic data, therefore, are essential to understanding the changing patterns of biological diversity and to planning and implementing conservation strategies (7).

The complexity of natural ecosystems rivals the complexity of social and economic processes that affect them. Thus, socioeconomic research should be no less rigorous than the biological research. Unfortunately, social and economic data are often the weak link in conservation planning.

Demography is a well-established social science with reliable data sources, theories, and methods to describe population patterns. Theories on how population growth under various circumstances affects biological diversity are lacking, however.

The status of biological diversity is greatly affected by the supply and demand of raw materials, agricultural commodities, and natural products. Natural-resource economic data and analytical methods have been developed for other fields of resource management, such as forestry, fisheries, and agriculture, but application of economics to issues of biological diversity has hardly begun. Some biologists and geographers have started to do economic analyses, but few professional economists are interested in biological diversity.

Data on technological change, especially in agriculture and pollution-causation and abatement, are sometimes assessed as part of the environmental-impact assessment process required when Federal funding is involved in resource development in the United States. Improved methods for such assessment have developed in the years since the National Environmental Policy Act became law. But methods for technology-impact assessment are sorely lacking for other parts of the world, especially for the tropical regions where diversity is most threatened.

Social and political processes influencing how biological diversity is perceived and valued are probably the least well-understood and, in the long run, the most important factors affecting success of onsite diversity maintenance. Geographers, sociologists, anthropologists, historians, and biologists who have ventured outside their field of technical expertise have developed important descriptions of social factors affecting diversity maintenance at specific sites. But the analysis needed to develop a broader understanding and theories from which to generalize has yet to be undertaken.

NEEDS AND OPPORTUNITIES

Onsite management of natural areas has usually been focused on limiting the impacts of outside pressures. In multiple-use protected areas, the technologies used to maintain biological diversity are mainly based on manipulation of habitats or populations to favor particular species. These methods, many of which derive from the fields of natural history and wildlife management, are effective for the target species. Biologists generally agree, however, that broad biological diversity values are not adequately served by species management alone. This approach necessarily concentrates on species with immediate commercial or recreational value and lets too many others, with less obvious values, perish if they do not happen to live in the type of environment maintained for the target species. Thus, technologies are needed to maintain diversity at the ecosystem level.

Onsite maintenance technologies commonly have been developed in relatively well-known temperate zone ecosystems. Plant and animal communities in these ecosystems generally can recover from moderate human disturbances if they are protected for years or decades. But biologists are not sanguine about adapting these technologies to tropical and other ecosystems, such as coral reefs, that are poorly known and that have much less natural ability to recover from disturbances.

Although most existing onsite technologies are focused on natural areas where development is restricted, attention is beginning to be directed beyond simple protected area programs. Resource development planning methods that treat conservation as an integral part of economic and social development have been devised and tested. These strategies hold promise, but they need to be taken from the conceptual stage to practical implementation.

The remainder of this chapter addresses these and other opportunities to improve the research, development, and application of onsite technologies to maintain biological diversity.

An Ecosystem Approach

An ecosystem approach is necessary to maintain biological diversity onsite for many reasons: 1) because the numbers of threatened species and genetically distinct populations is so high, 2) because so little is known about life histories or even the identity of many species, and 3) because many species are interdependent. Yet attempts to develop and implement ecosystem approaches are few.

In the United States, development of onsite maintenance technologies is largely the task of Federal land-managing agencies, such as the National Park Service, the Forest Service, the Fish and Wildlife Service, and the Bureau of Land Management. The mandates of these agencies emphasize species- and habitat-oriented technologies. A shift toward more ecosystem-oriented management would require policy changes within the agencies. Most, for example, consider an inventory of an area's biological diversity and the investigation of species interactions to be appropriate activities for basic research programs but not appropriate as pragmatic resource management activities. Changes in policies to encourage an ecosystem approach to protected areas may not occur without a congressional mandate directing agencies to manage lands and bodies of water in a way that maintains ecosystem diversity.

An important strategy for maintaining diversity is to safeguard representative samples of ecosystems from changes that would reduce their diversity. The United States lacks a comprehensive program for ecosystem diversity maintenance, although some efforts are being made through existing programs. The U.S.-MAB program is attempting to establish samples of ecosystems in the United States. Because the areas are identified on the basis of ecological criteria rather than political boundaries, various Federal, State, and private organizations must cooperate to implement the program successfully, which may explain the sluggish pace. Federal agencies could be directed to give more support to interagency and Federal, State, and private initiatives to support the MAB agenda.

The number and size of additional protected areas required for ecosystem diversity maintenance are unknown. Extensive inventory programs (e.g., the State heritage inventories of The Nature Conservancy) have been initiated to determine how to enhance coverage. But support has been sporadic and progress is slow. TNC's State-level approach and mobilization of private sector support has been effective, so the Federal Government could continue to support this and similar programs.

International organizations, led by IUCN and the World Wildlife Fund/Conservation Foundation, are promoting conservation of samples of the world's ecosystems. The coverage of ecosystems, indicated by comparing protected areas to Udvardy's biogeographical classification system, is encouraging but still incomplete. The next major step will be to survey the degree of actual protection in the designated natural areas. Such surveys could also identify gaps in ecosystem protection at a finer biogeographic level than Udvardy's classifications. Better information is needed, especially on aquatic ecosystem types such as coral reefs, to develop and implement strategies for international ecosystem conservation efforts. A U.S. Government interagency task force could identify personnel for this task and ways in which their work might serve the objectives of international conservation.

Innovative Technologies for Developing Countries

Many onsite technologies have been developed in industrialized, temperate zone countries, and thus, they may not be appropriate for developing countries' ecosystems, which are mostly tropical and where the biological, sociopolitical, and economic situations are fundamentally different. Hence, innovative technologies are especially needed in these areas.

The biosphere reserves concept is one such approach that appears to merit scrutiny and support. Continued U.S. Government support of UNESCO's MAB program and ways to increase support for MAB in developing countries could be explored in congressional committee hearings.

Integrated land management that includes conservation in development activities is another approach that should be encouraged. The OAS Integrated Regional Development Planning could provide a model for other development assistance agencies, such as AID or the World Bank.

Long-Term Multidisciplinary Research

The most important problems affecting implementation of biological diversity maintenance efforts are not amenable to resolution by any one field of biology, or indeed by the natural sciences alone. Biological diversity is so broad that its maintenance requires methods from numerous disciplines, such as natural history, population biology, genetics, and ecology. In addition, many other factors—economic, political, and social—contribute to decisions about the sizes, shapes, and locations of protected areas. Application of social sciences to diversity maintenance, for instance, to help communicate the issue's importance to decisionmakers at all levels, is probably the most needed research area.

The formation of a discipline called conservation biology is an encouraging sign of the scientific community's effort to start breaking down traditional barriers among disciplines and in ways of approaching problems. The goal is to provide principles and tools for maintaining biological diversity. Signs that the new discipline is gaining momentum include establishment of a Center for Conservation Biology at Stanford University, the creation of a department of conservation biology at Chicago's Brookfield Zoo, and the development of programs of study at the University of Florida and at Montana State University. More recently, a professional Society for Conservation Biology with its own journal, *Conservation Biology*, has been established.

As yet, the National Science Foundation and other research funding organizations have not recognized the status of conservation biology as a discipline by according it a separate funding category. Its impact on resource management should increase as it gradually becomes recognized, encouraged, supported, and broadened to include professional social scientists.

Personnel Development

A major constraint to maintaining diversity onsite is the shortage of personnel—taxonomists, social scientists, resource managers, and technicians with adequate training, motivation, and work experience. These individuals are needed to plan, manage, and explain the need to maintain biological diversity to decisionmakers. Training and institutional development to provide employment opportunities for these kinds of experts are sorely needed, particularly in developing countries.

The number of plant taxonomists working in the world today is estimated at 3,000, for example; but twice as many would probably be needed for an adequate study of the world's flora (12). Moreover, most taxonomists reside in the temperate zone and only study species there.

Even if money were available to train new taxonomists, job opportunities would have to be provided to attract people to the field. The number of taxonomic positions in museums, herbaria, universities, and resource-managing agencies currently is low and may be falling, as research funds are directed at more popular disciplines (e.g., molecular biology). It may be time for the museums and botanic gardens to explore innovative ways to promote the field of systematic biology. These institutions could help by defining systematic biology's role in the maintenance of biological diversity as a way of making the discipline more appealing to potential specialists.

Data To Facilitate Onsite Protection

Decisions on where and how to apply various methods for onsite maintenance of diversity need to be based on accurate data and correct theories on the interactions among numerous biological and human factors. Abundant data exist, especially for the temperate zone regions of the world. The data are being used both to develop and improve theories regarding biological diversity and to make specific decisions regarding resource management. However, use of the existing information is inefficient when data are not collected into readily accessible databases at the scale on which decisionmakers operate.

Thus, a significant opportunity to improve onsite maintenance of diversity, both within and outside protected areas, is to support accelerated development of comprehensive databases, which would include, for example, TNC's State Natural Heritage Programs and its international conservation data center program. It could also include development of a nationwide description and evaluation of all flora and fauna species in the United States, possibly under the auspices of TNC.

Large gaps in knowledge of tropical species and ecosystems constrain the effectiveness of diversity maintenance efforts in developing countries. Opportunities include increased development assistance support to build institutions and train scientists capable of accelerating progress in the fundamental sciences of taxonomy, natural history, and ecology.

Possibly the most severe information deficiencies relate to the poor understanding of how social, political, and economic factors interact with biological diversity. A great need exists for social scientists trained and employed to develop information on how social and economic conditions can be made conducive to onsite maintenance of biological diversity. Unfortunately, this is a need difficult for biologists and natural resource managers to address. It requires new levels of interest and commitment from social science institutions.

CHAPTER 5 REFERENCES

1. Bailey, R.G. (compiler), *Description of the Ecoregions of the United States* (Washington, DC: U.S. Department of Agriculture, Forest Service, Intermountain Region, 1978). *In*: Crumpacker, 1985.
2. Bailey, R.G., Pfister, R.D., and Henderson, J.A., "Nature of Land and Resource Classification—A Review," *Journal of Forestry* 76(10):650-655, 1978.
3. Belovsky, G., "Extinction Models and Mammalian Persistence," *Viable Populations for Conservation*, M. Soulé (ed.) (Cambridge, MA: Cambridge University Press, forthcoming).
4. Betz, R., Northeastern Illinois University, personal communications, 1985. *In*: Jordan, et al., 1986.
5. Bonner, F., U.S. Forest Service, Starkville, MS, letter to OTA, May 1986.
6. Brown, J.H., and Kodric-Brown, A., "Turnover Rates in Insular Biogeography: Effect of Immigration on Extinction," *Ecology* 58:445-449, 1977. *In*: Soulé and Simberloff, 1986.
7. Brush, S., International Agricultural Development Program, University of California, Davis, letter to OTA, June 1986.
8. Butcher, G.S., Niering, W.A., Barry, W.J., and Goodwin, R.H., "Equilibrium Biogeography and the Size of Nature Reserves: An Avian Case Study," *Oecologia (Berl.)* 49:29-37, 1981. *In*: Noss, 1983.
9. Centro Internacional de Mejoramiento de Maíz y Trigo (CIMMYT), "Conservation of the Wild Relatives of Maize," *CIMMYT Research Highlights, 1985* (El Batan, Mexico: 1986).
10. Collins, M., "International Inventory and Monitoring for the Maintenance of Biological Diversity," OTA commissioned paper, 1985.
11. Conservation Foundation, *National Parks for a New Generation: Visions, Realities, Prospects* (Washington, DC: Conservation Foundation, 1985).
12. Crosby, M., and Raven, P., "Diversity and Distribution of Wild Plants," OTA commissioned paper, 1985.
13. Crumpacker, D.W., "Status and Trends of the U.S. Natural Ecosystems," OTA commissioned paper, 1985.
14. Dasmann, R., *Wildlife Biology* (New York: John Wiley & Sons, Inc., 1964).
15. Diamond, J.M., "Normal Extinctions of Isolated Populations," *Extinctions*, M.H. Nitecki (ed.) (Chicago, IL: University of Chicago Press, 1984). *In*: Shaffer, 1985.
16. Diamond, J.M., "The Island Dilemma: Lessons of Modern Biogeographic Studies for the Design of Natural Reserves," *Biological Conservation* 7:129-146, 1975. *In*: Shaffer, 1985.
17. Eidsvik, H., Director of Parks, Canada, personal communications, June 1986.
18. Erwin, T.L., "Tropical Forests: Their Richness in *Coleoptera* and Other Arthropod Species," *Coleopterists' Bulletin* 36(1):74-75, 1982.
19. Eyre, F.H. (ed.), *Forest Cover Types of the United States and Canada* (Washington, DC: Society of American Foresters, 1980).
20. Fairchild, J., Utah Division of Wildlife Resources, Salt Lake City, UT, personal communication, 1985. *In*: Jordan, et al., 1986.
21. Feller, W., "Die Grundlagen der Volterraschen Theorie des Kampfes ums Dasein in Warscheinlichkeistheoretischer Behandlung," *Acta Biotheoretica* 5:11-40, 1939. *In*: Shaffer, 1985.
22. Flesness, N., "Status and Trends of Wild Animal Diversity," OTA commissioned paper, 1985.
23. Food and Agriculture Organisation of the United Nations, *Tree Growing by Rural People*, FAO Forestry Paper No. 64 (Rome: 1986).
24. Frankel, O.H., and Soulé, M.E., *Conservation and Evolution* (Cambridge, MA: Cambridge University Press, 1981).
25. Franklin, J.A., "Evolutional Change in Small Populations," *Conservation Biology: An Evolutionary-Ecological Perspective* M.E. Soulé and B.A. Wilcox (eds.) (Sunderland, MA: Sinauer Associates, 1980).
26. Gentry, A.H., "Chapter 8: Endemism in Tropical versus Temperate Plant Communities," *Conservation Biology*, M. Soulé (ed.) (Sunderland, MA: Sinauer Associates, 1986).
27. Gilbert, F.S., "The Equilibrium Theory of Island Biogeography: Fact or Fiction?" *Journal of Biogeography* 7:209-235, 1980. *In*: Shaffer, 1985.
28. Ginsburg, L., Slobodkin, L.B., Johnson, K., and Bindman, A.G., "Quasi-extinction Probabilities as a Measure of Impact on Population Growth," *Risk Analysis* 2:171-181, 1982. *In*: Shaffer, 1985.
29. Government of Zambia and IUCN, *National Conservation Strategy for Zambia* (Gland, Switzerland: 1985).
30. Gregg, B., U.S. National Park Service, Washington, DC, letter to OTA, Sept. 19, 1986.

31. Harrington, J., University of Wisconsin, personal communications, 1985. *In:* Jordan, et al., 1986.
32. Harris, L., *The Fragmented Forest* (Chicago, IL: University of Chicago Press, 1984).
33. Harrison, J., "Status and Trends of Natural Ecosystems Worldwide," OTA commissioned paper, 1985.
34. Hayden, B.P., Ray, G.C., and Dolan, R., "Classification of Coastal and Marine Environments," *Environmental Conservation* 11(3):199-207, 1984.
35. International Union for the Conservation of Nature and Natural Resources, *The United Nations List of National Parks and Protected Areas* (Gland, Switzerland: 1985).
36. International Union for the Conservation of Nature and Natural Resources, *National Conservation Strategies: A Framework for Sustainable Development* (Gland, Switzerland: 1985).
37. International Union for the Conservation of Nature and Natural Resources, "Categories, Objectives, and Criteria for Protected Areas," *National Parks, Conservation, and Development: The Role of Protected Area in Sustaining Society*, Proceedings of the World Congress on National Parks, Bali, Indonesia, Oct. 11-22, 1982, J.A. McNeely and K.R. Miller (eds.) (Washington, DC: Smithsonian Institution Press, 1984).
38. International Union for the Conservation of Nature and Natural Resources, *Categories, Objectives, and Criteria for Protected Areas* (Morges, Switzerland: 1978).
39. Janzen, D.H., "Guanacaste National Park: Tropical Ecological and Cultural Restoration," a report to Servico de Parques Nacionales de Costa Rica from Department of Biology, University of Pennsylvania, Philadelphia, 1986.
40. Johannes, R.E., and Hatcher, B.G., "Chapter 17: Shallow Tropical Marine Environments," *Conservation Biology*, M. Soulé and B. Wilcox (eds.) (Sunderland, MA: Sinauer Associates, 1986).
41. Johnston, V.R., "Breeding Birds of the Forest Edge in East-Central Illinois," *Condor* 49:45-53, 1947. *In*: Noss, R., "A Regional Landscape Approach to Maintain Diversity," *BioScience* 33(11):700-706, 1983.
42. Jones, J.O., Espey, Houston & Associates, Inc., Austin, TX, personal communications, 1985. *In*: Jordan, et al., 1986.
43. Jordan, W., Peters, R., and Allen, E., "Ecological Restoration as a Strategy for Conserving Biodiversity," OTA commissioned paper, 1986.
44. Kollar, S.A., Jr., Hartford Community College, Bel Air, MD, personal communication, 1985. *In*: Jordan, et al., 1986.
45. Lande, R., and Barrowclough, G.F., "Effective Population Size, Genetic Variation and Their Use in Population Management," *Viable Populations for Conservation*, M.E. Soulé (ed.) (Cambridge, MA: Cambridge University Press, forthcoming).
46. Levins, R., "Extinction," *Some Mathematical Questions in Biology, Lectures of Mathematics in the Life Sciences*, vol. 2, M. Gerstenhaber (ed.) (Providence, RI: American Mathematical Society, 1970). *In*: Shaffer, 1985.
47. Lovejoy, T.E., Bierregaard, R.O., Jr., Rankin, J.M., and Schubert, H.O.R., "Ecological Dynamics of Forest Fragments," *Tropical Rain Forests: Ecology and Management*, S.L. Sutton, T.C. Whitman, and A.C. Chadwick (eds.) (Oxford, England: Blackwell Scientific Publications, 1983). *In*: Shaffer, 1985.
48. Lovejoy, T.E., Rankin, J.M., Bierregaard, R.O., Jr., Brown, K.S., Jr., Emmons, L.H., and Van der Voort, M.E., "Ecosystem Decay of Amazon Forest Remnants," *Extinctions*, M.H. Nitecki (ed.) (Chicago, IL: University of Chicago Press, 1984). *In*: Shaffer, 1985.
49. Lovejoy, T.E., and Oren, D.C., "The Minimum Critical Size of Ecosystems," *Forest Island Dynamics in Man-Dominated Landscapes*, R.L. Burgess and D.M. Sharp (eds.) (New York: Springer-Verlag, 1981). *In*: Shaffer, 1985.
50. Lyman, T.M., "Progress and Planning for Germplasm Conservation of Major Food Crops," *Plant Genetics Resources Newsletter* 60:3-19, 1984.
51. Lynch, M. and Rey, C., "Diversity of Marine/Coastal Ecosystems," OTA commissioned paper, 1985.
52. MacArthur, R.H., and Wilson, E.O., *The Theory of Island Biogeography* (Princeton, NJ: Princeton University Press, 1967).
53. Margules, C., Higgs, A.J., and Rafe, R.W., "Modern Biogeographic Theory: Are There Any Lessons for Nature Reserve Design?" *Biological Conservation* 24:115-128, 1982.
54. Matthiae, P., Assistant Coordinator of Natural Areas, Wisconsin Department of Natural Resources, personal communication, 1985. *In*: Jordan, et al., 1986.
55. McMahan, J.A., *Nothing Succeeds Like Succession: Ecology and the Human Lot*, 67th Faculty Honor Lecture (Logan, UT: Utah State University Press, 1983). *In*: Jordan, et al., 1986.
56. Morrison, D.G., *Use of Prairie Vegetation on*

Disturbed Sites, Transportation Research Record 822 Landscape and Environmental Design (Washington, DC: National Academy of Sciences, 1981).
57. Myers, N., "Causes of Loss of Biological Diversity," OTA commissioned paper, 1985.
58. Nabhan, G.P., "Native Crop Diversity in Aridoamerica: Conservation of Regional Gene Pools," *Economic Botany* 39:387-399, 1985.
59. National Research Council, *Surface Mining: Soil, Coal, and Society* (Washington, DC: National Academy Press, 1981).
60. Organization of American States (OAS), OAS General Secretariat, *Integrated Regional Development Planning: Guidelines and Case Studies From OAS Experience* (Washington, DC: 1984).
61. Peine, J., *Proceedings of Conference on the Management of Biosphere Reserves*, Nov. 27-29, 1984, Great Smoky Mountains National Park, Gatlinburg, TN.
62. Prescott-Allen, R., "National Conservation Strategies and Biological Diversity," a report to the International Union for the Conservation of Nature and Natural Resources, Conservation for Development Center, draft, February 1986.
63. Prescott-Allen, R., and Prescott-Allen, C., "Park Your Genes: Protected Areas as *In-Situ* Genebanks for the Maintenance of Wild Genetic Resources," *National Parks, Conservation, and Development: The Role of Protected Areas in Sustaining Society*, Proceedings of the World Congress on National Parks, Bali, Indonesia, Oct. 11-22, 1982, J.A. McNeely and K.R. Miller (eds.) (Washington, DC: Smithsonian Institution Press, 1984).
64. Ricklefs, R.E., Naveh, Z., and Turner, R.E., "Conservation of Ecological Processes," *The Environmentalist* 4(8), 1984.
65. Salm, R.V., *Marine and Coastal Protected Areas: A Guide for Planners and Managers* (Gland, Switzerland: International Union for the Conservation of Nature and Natural Resources, 1984).
66. Saunier, R., "Environment and Development . . . A Future Together?" *Development Forum* 11(8), 1983.
67. Saunier, R., and Meganck, R., "Compatibility of Development and the *In-Situ* Maintenance of Biological Diversity in Developing Countries," OTA commissioned paper, 1985.
68. Schlosser, S., "The Use of Nature Reserves for *In Situ* Conservation," *Plant Genetic Resources Newsletter* 61:23-24, 1985.
69. Schonewald-Cox, C.M., and Bayless, J.W., "The Boundary Approach: A Geographic Analysis of Design and Conservation of Nature Reserves," *Biological Conservation* (forthcoming).
70. Schonewald-Cox, C.M., Chambers, S.M., MacBryde, B., and Thomas, W.L., *Genetics and Conservation: A Reference for Managing Wildlife Animal and Plant Populations* (Menlo Park, CA: Benjamin Cummings Co., 1983).
71. Shaffer, M., "The Minimum Viable Population Problem" *Viable Populations for Conservation*, M. Soulé (ed.) (Cambridge, MA: Cambridge University Press, forthcoming).
72. Shaffer, M., "Assessment of Application of Species Viability Theories," OTA commissioned paper, 1985.
73. Shaffer, M., "Determining Minimum Viable Population Sizes for the Grizzly Bear," *International Conference on Bear Research and Management* 5:133-139, 1983.
74. Shaffer, M., "Minimum Population Sizes for Species Conservation," *BioScience* 31:131-134, 1981.
75. Shugart, H., "Planning and Management Techniques for Natural Systems," OTA commissioned paper, 1985.
76. Simberloff, D., "Biogeography: The Unification and Maturation of a Science," *Perspectives in Ornithology*, A.H. Brash and G.A. Clark, Jr. (eds.) (Cambridge, MA: Cambridge University Press, 1983). *In*: Shaffer, 1985.
77. Simon, J., and Wildavsky, A., "On Species Loss, the Absence of Data, and Risks to Humanity," *The Resourceful Earth* J. Simon and H. Khan (eds.) (New York: Basil Blackwell Inc., 1984).
78. Soulé, M.E., and Simberloff, D., "What Do Genetics and Ecology Tell Us About the Design of Nature Reserves?" *Biological Conservation* 35:19-40, 1986.
79. Soulé, M.E., and Wilcox, B.A., *Conservation Biology: An Ecological-Evolutionary Perspective* (Sunderland, MA: Sinauer Associates, 1980).
80. Stenseth, N.C., "Where Have All the Species Gone? On the Nature of Extinction and the Red Queen Hypothesis," *Oikos* 33:196-227, 1979. *In*: Shaffer, 1985.
81. Synge, H. "Status and Trends of Wild Plants," OTA commissioned paper, 1985.
82. Thorsell, J.W., "The Role of Protected Areas in Maintaining Biological Diversity in Tropical Developing Countries," OTA commissioned paper, 1985.
83. U.S. Congress, Office of Technology Assessment, *Technologies To Sustain Tropical Forest Resources*, OTA-F214 (Washington, DC: U.S. Government Printing Office, March, 1984).
84. U.S. Department of Agriculture, Forest Service,

An Annotated Bibliography of Surface-Mined Area Reclamation Research (Washington, DC: 1985).

85. U.S. Department of Agriculture, Forest Service, *Wildlife Habitats in Managed Forests: The Blue Mountains of Oregon and Washington*, USDA Pub. No. 553 1979 (Washington, DC: 1979).
86. U.S. Department of Commerce, National Oceanic and Atmospheric Administration, *National Marine Sanctuary Program: Program Development Plan* (Washington, DC: 1982).
87. U.S. Department of the Interior, Fish and Wildlife Service, "Selkirk Mountain Caribou Management Plan" (Washington, DC: 1985).
88. U.S. Department of the Interior, Bureau of Land Management, "Burro Creek Riparian Management Plan" (Washington, DC: May 1983).
89. Wilcove, D.S., McLellan, C.H., and Dobson, A.P., "Chapter 11: Habitat Fragmentation in the Temperate Zone," *Conservation Biology*, M. Soulé (ed.) (Sunderland, MA: Sinauer Associates, 1986).
90. Wilcox, B.A., "Insular Ecology and Conservation," *Conservation Biology: An Evolutionary-Ecological Perspective,* M.E. Soulé and B.A. Wilcox (eds.) (Sunderland, MA: Sinauer Associates, 1980).
91. Wright, R.M., Houseal, B., and De Leon, C., "Kuna Yala: Indigenous Biosphere Reserves in the Making?" *Parks* 10(3):25-27, 1984.

Chapter 6
Maintaining Animal Diversity Offsite

CONTENTS

Tables

Figures

Boxes

Chapter 6
Maintaining Animal Diversity Offsite

HIGHLIGHTS

- Offsite maintenance of animal diversity includes selective breeding of wild or domestic species and safeguarding genetic diversity through cryopreservation. For wild animals, the programs reinforce rather than replace efforts to maintain diversity onsite. For domestic animals, programs try to maximize usefulness of the animals while preserving their ability to adapt to changing human needs.
- Cryogenic storage could make a considerable contribution to the maintenance of animal diversity. Properly frozen and maintained, sperm and embryos have an expected shelf-life of hundreds of years. Although initial collection and preservation costs are relatively high, subsequent storage costs and space requirements are low.
- The number of individual animals required to start a captive population or a cryogenic store depends on a host of factors. Retaining 99 percent of a source population's genetic diversity for 1,000 generations could require up to 50,000 animals, far too many to be practical under captive management. At a minimum, however, several hundred individual animals are required for captive breeding programs.
- Breeding programs require the international transfer of animals, which risks spreading pests and diseases. For most wild species, regulatory controls are virtually nonexistent. Stringent controls are in place, however, for importing domestic animals. Advances in diagnostic procedures and germplasm transfer technologies are expected to facilitate the international movement of animals.
- No organized program exists, either in the United States or internationally, to sample, evaluate, maintain, and use available sources of animal germplasm. Such a program is needed, in addition to programs to understand the reproductive processes of wild animals, to develop local expertise in reproductive biology and quantitative genetics, and to increase the number of captive maintenance and breeding facilities.

OVERVIEW

Objectives of Offsite Maintenance Efforts

Offsite maintenance of animal diversity is defined as propagation or preservation of animals outside their natural habitat. The programs involve control by humans of the animals chosen to constitute a population and of the mating choices made within that population. The extent of control can vary considerably, but the decision to remove individual animals from a natural habitat implies a major increase in human involvement in propagation of a population.

Captive maintenance of wild species has become progressively more important as increasing numbers of species are threatened or endangered in their natural habitats. These programs can be considered holding actions designed to reinforce rather than replace wild populations. If a natural population is decimated or lost, captive maintenance programs provide a reservoir of individuals to allow re-introduction.

The genetic diversity of the original population must not be lost or seriously reduced during captive maintenance if animals are expected to be able to readapt to life in the wild. Likewise, genetic changes that may be induced during captive maintenance must be minimized. Reciprocal transfers of individuals between wild and captive populations can help reduce genetic pressures. Such exchanges, however, involve the capture of wild animals, and they risk the accidental death of some of them. Therefore, risks should be evaluated carefully before beginning a program of genetic exchanges between wild and captive populations.

For domestic species, all populations are by definition maintained offsite. Most of these animals have existed in association with humans for centuries, and their current genetic diversity is a reflection of this long interaction. Their genes have been manipulated through generations of selective breeding to meet the diverse needs of humans, and this manipulation has led to a wealth of specialized breeds (boxes 6-A and 6-B). Some wild progenitors of domestic species still differ so much from domestic populations that they exist as a reservoir of genetic diversity, but these natural populations are unlikely to contribute much to current commercial stocks through traditional breeding methods.

The aim of programs to maintain genetic diversity in livestock differs somewhat from that for wild animals. In domestic populations, the challenge is to maximize current utility and preserve sufficient diversity to ensure livestock's continued adaptability to changing—and often unforeseen—human needs. In fact, efforts to raise current rates of food production may constitute the greatest threat to future flexibility by concentrating unduly on short-term production goals with attendant losses in genetic diversity that may be important to future generations.

Box 6-A.—Breeds

In its most restrictive form, the designation "breed" is reserved for subpopulations that have both distinctive morphological characteristics and a pedigree-recording system to document the ancestry of individuals within the breed. In practice, however, a breed is often taken to be any differentiated, identifiable domestic subpopulation.

Breeds that are widespread may often be further subdivided into national or regional populations that retain the major characteristics of the parent breed but have become partially differentiated in response to local conditions and selection pressures. Thus, identifiable regional populations exist for Holstein-Friesian dairy cattle in North America and most of the European countries, for Shorthorn beef cattle in nations previously colonized by the British, and for Merino sheep worldwide. Some degree of interbreeding often occurs among such populations, but in most cases they persist and are referred to as "strains" or "stocks" within the breed.

For more industrialized species such as poultry, a few ancestral breeds such as the Leghorn chicken have been used in concert with crossbreeding and selection to develop a wide array of partially differentiated stocks maintained by commercial breeding firms. These stocks are open to genetic manipulation at the discretion of the firms, but they represent another source of differentiated genetic material.

In this assessment, the term breed will be used to describe any differentiated domestic subpopulation and will include both regional strains of major international breeds and identifiable commercial stocks.

Breeding Programs v. Long-Term Cryogenic Storage

Using captive breeding programs to retain a considerable proportion of the genetic diversity of endangered species or rare breeds for

Box 6-B.—Replacement and Genetic Diversity

Traditionally, breed replacement has proceeded on an evolutionary time scale, with gradual changes in breed composition and provision for maintenance of a wealth of local populations. Recently, however, the pace of breed replacement has accelerated, with the emergence of multinational breeding companies in poultry and swine, the widespread use of artificial insemination and intensified sire selection in dairy cattle, and enhanced opportunities for dissemination of germplasm throughout the world. Also, greater standardization of production, marketing, and recording procedures for poultry, swine, and dairy cattle in industrial countries has increasingly promoted replacement of local breeds.

In domestic species, the greatest threat to genetic diversity involves extensive and sometimes indiscriminate crossing of indigenous stocks in developing countries with breeds from North America and Western Europe (3). This crossing stems from needs to increase world food production and from a belief that this goal is best met using stocks with the highest possible genetic merit for individual traits (such as milk or egg production). But breeds developed in temperate-zone industrial countries are often not suited to the more restrictive nutritional, management, and disease conditions of developing countries and may be less efficient than indigenous stocks in using available resources. Only recently has the need for comprehensive evaluation of the total performance of imported breeds begun to be recognized in developing countries. Unfortunately, serious dilution of original breeds may have already occurred. Thus, it is not the process of breed replacement per se that is a problem but the rate of replacement and the danger that useful breeds may be discarded before they can be fairly evaluated.

Regional strains of established breeds are especially vulnerable to loss through intercrossing with more popular strains. Extensive use of Holstein bulls from North America in European Friesian populations threatens serious dilution of the genetic material of these strains. The percentage of Holstein genes in young Friesian bulls entering European artificial insemination programs in 1982 ranged from 8 percent in Ireland to 91 percent in Switzerland and averaged 54 percent for 10 European countries (4).

Genetic diversity can sometimes be reduced in commercial stocks even if population numbers remain large. These losses can occur when selection is intense and control of breeding stock is concentrated in a few large breeding farms (as in the commercial poultry industries) or when artificial insemination allows extensive use of a few selected sires throughout the population (as in the dairy industry). In both cases, the result is increased genetic uniformity within the stock despite the large numbers. Several studies (3,25) have concluded that important losses may be occurring in commercial poultry breeds. Comparable losses have apparently not yet happened in dairy cattle populations. No imminent losses of genetic diversity within major commercial breeds are foreseen for swine, sheep, goats, or beef cattle. Several populations of chickens are currently being maintained without selection and at sufficient population sizes to substantially retard losses in genetic diversity (42). And some artificial insemination organizations retain semen from bulls that have been removed from service. However, these programs do not represent industry or public policy, and parallel programs do not exist for other domestic species.

The loss of endangered species and rare breeds is of particular concern in light of likely future advances in molecular biology and genetics. The ability to extract desirable genes from different species or less productive types and insert them into domestic animals could have important implications for designing superior animals for specific environmental conditions. Unfortunately, knowledge of the genetic material of most wild and domestic animals is rudimentary. For instance, it is unclear if adaptive factors such as heat tolerance and disease resistance are controlled by many or a few genes, and no basis yet exists to assess potential single-gene contributions of local breeds and endangered species.

a substantial period of time requires relatively large numbers of animals. Under the most favorable assumptions, maintenance of 90 to 95 percent of the genetic diversity within a population for 100 to 200 generations would require a captive population of at least several hundred individuals sampled from throughout the range of the species (15,27).

Until relatively recently, zoos have not been concerned with keeping representative levels of genetic diversity within their exhibition stock. Problems in fertility and juvenile survival that often accompany exhaustion of genetic diversity were simply accommodated by obtaining new specimens from the wild. As this became difficult, and in some cases impossible, zoos began to reevaluate their role. The result has been establishment of programs to maintain pedigree information on zoo animals through the International Species Inventory System (ISIS) and to facilitate transfer of individuals among zoos. These efforts help maintain genetic diversity, but existing zoos can support at most 1,000 kinds of terrestrial vertebrates at a minimum population of 250 (2), whereas an estimated 1,500 to 2,000 kinds will be in danger of extinction by the year 2050 (43). The magnitude of the problem will thus outrun currently available facilities for captive breeding.

Recent advances in reproductive biology and cryopreservation may facilitate efforts to preserve genetic diversity. Cryopreservation refers to storage below −130° C: water is absent, molecular kinetic energy is low, and diffusion is virtually nil. Thus, storage potential is expected to be extremely long. Storage in liquid nitrogen (−196° C) or in the vapor above it (ca. −150° C) is a useful technique: Liquid nitrogen is relatively inexpensive, inert, and safer than comparable refrigerants (e.g., liquid hydrogen, liquid oxygen, or freon).

Storage and eventual production of live offspring from frozen semen or embryos have become common for cattle, sheep, goat, buffalo, and horse. The semen of pigs can also be frozen. In 1982, an estimated 10.5 million cattle were produced through artificial insemination with frozen semen. Similarly, bovine embryo transfer has become commercially viable and increasingly involves the use of frozen embryos. Commercial use of frozen semen and embryos is less common in other livestock species, but acceptable results can be achieved. Frozen semen is also regularly used with poultry and with some species of fish (18).

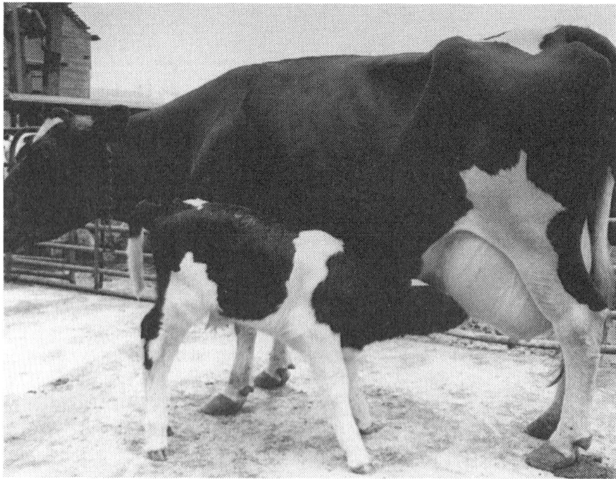

Photo credit: American Breeders Service

This calf, born in 1984, was conceived with semen that had been frozen for 30 years.

Cryopreservation of sperm and embryos of wild species has been much more limited. To date, blackbuck, giant panda, fox, wolf, chimpanzee, and gorilla have been produced from frozen semen (7,9,38); baboon (37) and eland (8), as well as mice, rats, and rabbits, have been produced from frozen embryos. Procedures differ among species, but in theory, semen and embryos from a range of mammalian species can now be successfully frozen.

The contribution of cryopreservation to the maintenance of animal diversity could be tremendous. Properly frozen and maintained, sperm and embryos have an expected shelf-life of hundreds, if not thousands, of years. Although initial collection and preservation costs may be relatively high, subsequent storage costs and space requirements are low, allowing for long-term maintenance of large numbers of individuals and gametes.

These individuals represent a frozen snapshot of the population at the time of collection. If the initial sampling of individuals is done

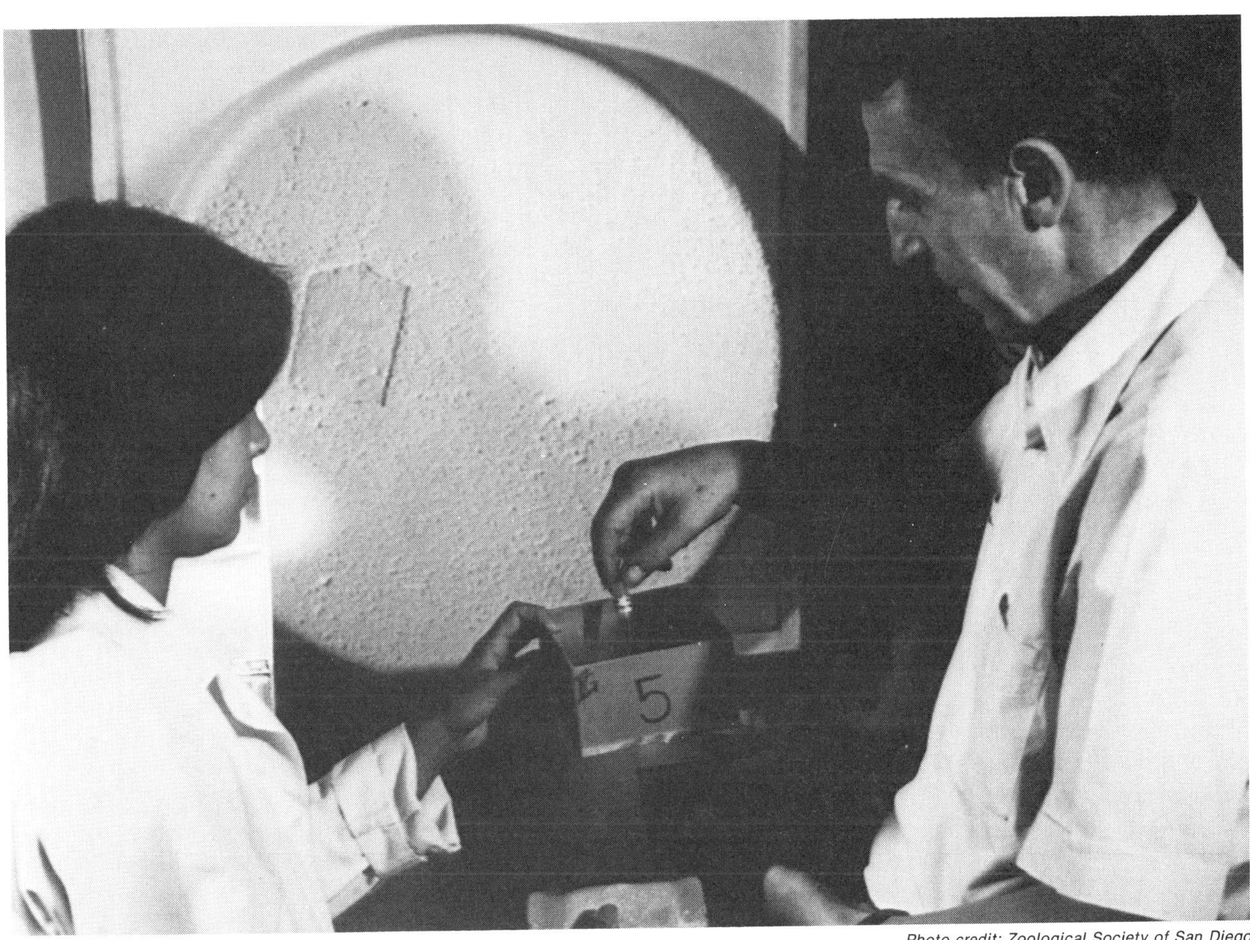

Photo credit: Zoological Society of San Diego

The frozen zoo: Cryogenic storage of cell strains, gametes, and embryos is being undertaken as part of the conservation activities of zoos.

properly, the procedures should allow regeneration of the original population without the genetic changes inherent in the maintenance of captive breeding populations.

The long-term genetic stability of frozen embryos and sperm is a matter of some concern. Freezing and thawing does not appear to increase the mutation rate in these tissues, but long-term exposure to low levels of radiation could be a problem, especially because DNA repair mechanisms would be inoperative at −196° C (1). Mouse embryos and semen have been kept frozen for at most 10 and 30 years, respectively. However, frozen mouse embryos also have been exposed to augmented levels of radiation equivalent to that experienced in 2,000 years of normal storage without apparent ill effects (16). Normal progeny were produced. Thus, risks of genetic damage from background radiation appear negligible.

Just as captive breeding programs reinforce rather than replace natural populations, cryopreservation efforts reinforce rather than replace captive breeding programs. In wild species, females must still be available to gestate frozen embryos or to provide female gametes in matings involving frozen semen. In domestic species, breeding populations selected for biologically or economically important traits may still be required, but cryogenic storage of individuals from the original population provides a valuable measure of insurance. Periodic sampling and preservation of gametes or embryos from rare breeds allow a repository of genetic diversity to be maintained.

Two caveats must be kept in mind regarding the role of cryopreservation of gametes and embryos. First, considerable development work is required to extend the techniques to cover the full range of endangered populations. For wild animals, reliable procedures for collecting, freezing, and using semen and embryos have to be developed further and validated for each species or group of species to ensure that sufficient levels of genetic diversity can be regenerated from the frozen store. Preservation technologies for embryos are well developed only in certain domestic mammals. Similar techniques are needed for birds, reptiles, amphibians, fish, and invertebrates.

Second, cryopreservation of gametes and embryos should not be an 11th-hour effort to protect seriously endangered species. Restraining wild animals to collect semen or embryos is risky. Some animals die, which entails an unacceptable risk if the species is already rare. Therefore, research and the collection of gametes and embryos from many sources should begin before the populations become endangered.

SAMPLING STRATEGIES

Efficient programs for offsite maintenance of animal genetic diversity require a mechanism for monitoring existing populations—to identify when and if intervention is required—and procedures for sampling threatened populations in a way that ensures desired levels of genetic diversity within the conserved population.

Identification of Candidates for Conservation

Three criteria are generally considered when selecting wild species for captive propagation or preservation (35):

1. **Endangerment in the Wild:** Information on the status of wild animals is probably best obtained from the Species Conservation Monitoring Unit (SCMU) of the International Union for the Conservation of Nature and Natural Resources at Cambridge University in England. Funding constraints tend to limit the scope and timeliness of SCMU information, however. Local and regional organizations may also provide useful information, but their effectiveness varies widely.
2. **Feasibility in captivity:** Lack of facilities and expertise may preclude captive breeding or cryogenic storage of some species. The blue whale is an example of a species that cannot be maintained in captivity.
3. **Uniqueness:** Given limited facilities for captive propagation, programs must try to represent as much available taxonomic diversity as possible. Thus, endangered species that are the only representative of their genus, family, or order would receive high priority.

Subspecies present a special problem. Most wild species have several distinct forms or races, analogous to the breeds found in domestic animals. These subspecies usually cannot all be maintained as discrete breeding populations. Instead, captive propagation programs need to concentrate on one or two representative subspecies or amalgamate several of them into a single interbreeding population. Cryogenic preservation of semen or embryos would facilitate conservation of these identifiable subspecies.

Table 6-1 provides some general guidelines for monitoring and intervention to conserve a natural population. Such an approach has three important advantages:

1. a sample of the source population can be obtained before substantial loss of genetic diversity has occurred;
2. conflict over capture and restraint of rare

Table 6-1.—General Guidelines for Intervention To Conserve Natural Populations

Likelihood of extinction	Number of animals	Action
Possible	fewer than 100,000	At least, serious surveillance of status and trends should be initiated
Probable	fewer than 10,000	Well-managed captive propagation programs should be established; reproductive technology research should be vigorously conducted; and germinal tissues should be collected for storage while there are an adequate number of animals to use as founders, subjects, and donors
Certain	fewer than 1,000	Offsite programs should be intensified while onsite efforts are fortified for a "last stand"; offsite programs are imperative
Imminent	fewer than 500	Offsite programs become as important as onsite efforts

SOURCE: D.R. Notter and T.J. Foose, "Concepts and Strategies To Maintain Domestic and Wild Animal Germ Plasm," OTA commissioned paper, 1985.

individuals, e.g., the California condor, can be avoided by taking action before extinction is imminent; and
3. if techniques for semen and embryo preservation are not well developed, material can be made available for experimentation.

For the rare breed of a domestic species, identifying candidates for conservation involves assessment of uniqueness, potential economic contribution, and degree of endangerment. Monitoring the status of domestic animal breeds used for food and fiber production is somewhat coordinated by the Food and Agriculture Organisation of the United Nations, under the auspices of the United Nations Environment Programme. Regional efforts are directed by the European Association for Animal Production, the Society for the Advancement of Breeding Researchers in Asia and Oceania, the InterAfrican Bureau for Animal Resources, the International Livestock Centre for Africa, and the Asociacion Latinoamericana de Produccion Animal (12). Comparable efforts in North America have been less comprehensive and limited to private organizations such as the American Minor Breeds Conservancy.

At least 700 unique strains of cattle, sheep, pigs, and horses have been identified in Europe alone, and 241 of these are considered endangered, under the criteria detailed in table 6-2 (30). Public support for maintenance of all these breeds is not feasible, and choices will have to be made. Two considerations have been suggested for choosing among competing domestic breeds (39):

1. the breed exists as a closed population, and a similar population does not exist elsewhere; or
2. the breed exhibits a specific genetic value, such as superiority in some production trait, the existence of a major gene (i.e., a gene with a known effect on some physiological characteristic), or the expression of a unique characteristic of potential importance.

In the selection of threatened breeds, characterization and evaluation are critical first steps (35). Ideally, breeds would be assessed in their native environments and would be evaluated as both pure breeds and as crosses with other indigenous and improved breeds. This evalua-

Table 6-2.—Criteria for Classifying Domestic Breeds as Endangered

Species	Number of active males	Number of active females[a]	
		Stable population	Decreasing population
Cattle	fewer than 20	fewer than 1,000	1,000 to 5,000
Sheep	fewer than 20	fewer than 500	500 to 1,000
Goats	fewer than 20	fewer than 500	500 to 1,000
Pigs	fewer than 20	fewer than 200	200 to 500

[a]The risks associated with a decreasing population were deemed to be greater than those associated with a stable population. Therefore, larger numbers were suggested for a decreasing population.

SOURCE: Adapted from K. Maijala, A.V. Cherekaez, J.M. Devillard, Z. Reklewski, G. Rognoni, D.L. Simon, and D.E. Steane, "Conservation of Animal Genetic Resources in Europe, Final Report of an E.A.A.P. Working Party," *Livestock Production Science* 11:3-22, 1984.

tion, often lacking for threatened breeds within developing countries, can be extremely important. In the absence of a formal evaluation, bibliographic databases may provide some needed information (12). Following the evaluation, breeds can be put in one of four categories:

1. *Useful under current economic conditions.* Such stocks should be integrated into the production system in a way that uses their genetic material in pure lines, crosses, or selected gene pools. Pure lines should be maintained with selection for net merit in production systems that are characteristic of commercial production within the country of origin or preserved cryogenically if maintenance as a pure line is impossible.
2. *Viable under current economic conditions in relation to other indigenous types, but inferior (in pure lines or in crosses) to improved types; no obvious biological extreme or major gene.* Germplasm preservation in such populations could have two rationales: preservation of frozen semen or embryos to prevent total loss of the germplasm and as insurance during a period of breed replacement with the improved types, or maintenance as pure lines for their cultural-historical value at the option of local governments and producers. A dual philosophy exists here—a unique population should not be discarded until its inferiority is documented, but preservation should not hinder use of improved breeds.
3. *Not competitive under current economic conditions; possesses an extreme phenotype for one or more traits or carries a major gene.* Such breeds should be conserved cryogenically or as pure lines. Research use should be encouraged, and selection to intensify the extreme phenotype should be considered.
4. *Not competitive with existing adapted types; not a biological extreme; no major genes for production traits.* No particular efforts should be made to conserve such breeds unless they can be documented as unique in their genetic origin. Stocks could move from the second category to this one as more productive breeds prove themselves.

Preservation and Collection Considerations

The number of individual animals required to initiate a captive population or a cryogenic store will depend on the nature and extent of the genetic diversity to be maintained, on the population structure in nature, and on the rate at which the captive population reproduces.

The Nature and Extent of the Genetic Diversity

Both natural and artificial selection reflect different fitness or reproductive success for individuals carrying different genes and lead to changes in the frequencies of those genes in a population. The diversity of genes in a large, interbreeding population may be quite extensive, with different individuals possessing a somewhat different genetic composition. It is this diversity that enables populations to adapt to environmental changes. Indeed, preserving the evolutionary potential of the species requires the maintenance of these possibly useful genes.

The objective in sampling a source population, then, should be to obtain a group that represents the bulk of its genetic diversity. Fewer animals are required to obtain an adequate initial sample of a population's diversity than are required to ensure continued maintenance of that diversity over time. Thus, 20 to 30 founder animals should provide an adequate sample of the genetic diversity in most interbreeding populations (6,43), but much larger subsequent population sizes are required to prevent erosion of this diversity over time.

In terms of cryopreservation, enough frozen semen to produce 10 live offspring from each of 25 sires, which would require 50 to 100 units of semen per sire, would constitute a good sample of an interbreeding source population (40). The Council for Agricultural Science and Technology recommends production of 40 to 80 offspring from frozen embryos representing 20 or more unrelated parents (3). Assuming a pregnancy rate of 30 percent and a subsequent survival rate of 80 percent, 167 to 333 frozen embryos would be required for each breed.

For particularly rare breeds, too few individuals may be available to comply with these recommendations. Although a viable population can be established with as few as 4 to 10 animals, such a population may differ considerably in genetic composition from the unendangered source population, and it may have an impaired ability to respond to future changes in environment. Initiation of a captive population with only a few founders could be justified if a reasonable likelihood exists of obtaining additional individuals from the wild at some future point.

Population Structure in Nature

Most domestic and wild populations exist as groups of semi-isolated subpopulations. The extent of this subdivision differs among species and influences the sampling process in developing a captive population. If the population is strongly subdivided, genes present in one subpopulation may be absent in others, and sampling must attempt to include individuals from all major subgroups. According to one calculation, the recommended 20 to 30 founder animals can be decreased by about one-third if the population exists as a small number (2 to 10) of very distinct subpopulations, but it should be increased by about one-third if 50 to 100 distinct subgroups exist (35).

Current assessment of genetic diversity among subpopulations must be based on biochemical, historical, morphological, and ecological criteria. For genes that produce an identifiable protein molecule, genetic differences can be identified by the behavior of the proteins on an electrically charged (electrophoretic) gel. Electrophoretic testing procedures help identify the existence and distribution of various genes in different subpopulations. Rapid advances in molecular biology also hold promise of DNA probes that would directly assess the similarity of DNA molecules among subpopulations. In domestic animals, however, differential selection pressures may result in considerable genetic variation among breeds with similar evolutionary origins.

Reproductive Rate

Reductions in diversity are cumulative over generations in small populations, so the losses associated with a single sampling event are much lower than those that would accumulate over time if the population size remained at the founder number. As soon as a captive breeding population is started, therefore, it should be expanded to a size consistent with continued maintenance of the available genetic diversity. If the reproductive rate is high, maintenance can be achieved rapidly and with only a few founders. If the reproductive rate is low, several intervening generations at limited population size will be required to reach eventual target numbers, and more founders will be needed to assure retention of genetic diversity during this period. Sample sizes for cattle have been suggested to be twice those required for pigs, sheep, and goats, for example (3). One advantage of cryogenic preservation would be that the period of population expansion can be deferred until appropriate facilities and habitat are available.

THE MOVEMENT OF GERMPLASM—DISEASE AND QUARANTINE ISSUES

For many reasons, effective programs for conservation of endangered populations will require extensive international transfer of germplasm. First, facilities, funds, and institutional stability in developing countries may be insufficient to allow endangered species to be conserved onsite, and animals may have to be transferred to countries better equipped to support captive breeding programs. Second, with wild animals, effective maintenance of genetic diversity within captive populations will require the international transfer of animals for breeding purposes. And third, the optimum use of domestic animal germplasm for food production depends on the international movement of desirable breeds and strains to countries where they may be useful.

International transport of animal germplasm is accompanied, however, by the risk of introducing and spreading disease agents and vectors, many of which could have an enormous impact on animal productivity. Indeed, transporting animal germplasm without appropriate safeguards could jeopardize the conservation programs for which the germplasm is required. Thus, technologies to facilitate germplasm transfer must also limit the risk of disease transmission.

Many infectious diseases are caused by organisms that do not naturally occur in the United States, and their introduction could have serious effects on U.S. animals. Those causing most concern are foot-and-mouth disease, African swine fever, rinderpest, foreign bluetongue strains, scrapie, fowl plague, velogenic viscerotropic Newcastle disease, and Venezuelan equine encephalomyelitis (20). Current programs to exclude entry of pathogenic organisms vary with the species, disease, and country of origin.

For most wild species (including nonungulate mammals, most birds, reptiles, amphibians, and most fish), regulatory controls to prevent introduction and transfer of hazardous diseases are virtually nonexistent. Except for inspection at the time of entry, movement of such individuals is not restricted. In contrast, entry requirements for domesticated livestock species are quite stringent, especially for those coming from countries that harbor foot-and-mouth disease, rinderpest, scrapie, or velogenic viscerotropic Newcastle disease.

All imported domestic animals are subjected to a variety of diagnostic tests and to varying periods of quarantine in both the country of origin and the United States. Greater control reflects the wide potential dissemination of these animals throughout the livestock industry. Wild ungulates (hoofed mammals) can carry diseases transmissible to livestock and have importation requirements similar to those of domesticated livestock, but also must remain in permanent post-entry quarantine in U.S. Department of Agriculture (USDA)-approved facilities. They can be moved from one USDA-approved zoo to another, however, and their offspring can be transferred to nonregulated facilities.

Current efforts to control introduction of foreign diseases center on combined strategies of blood (serological) testing and quarantine. Some tests are designed to detect antibodies to specific disease organisms and can thereby identify individuals that have been exposed to the disease at some time; other tests may be used to detect the presence of a specific pathogen. Periods of quarantine support these procedures by allowing an incubation period for animals that may have been infected recently. The tests are conservative, because individuals that have been exposed to a disease but no longer retain the organism still carry antibodies and react positively. However, the procedures also facilitate identification of asymptomatic carriers of the various diseases.

Some serological procedures, such as the complement fixation and the viral neutralization tests, are at times unable to adequately discriminate between pathogenic and nonpathogenic organisms. These limitations have made it very difficult to obtain negative test results for some diseases. Recent advances in diagnostic procedures have yielded tests with much greater accuracy and specificity. Three of the most important are the following:

1. **Indirect Immunofluorescence:** This procedure can provide very rapid screening of samples for a variety of infectious agents. Although it lacks specificity for some diseases, it greatly facilitates the initial screening process.
2. **Enzyme-Linked Immunosorbent Assay (ELISA):** The compounds that are produced by a disease organism and that elicit the production of antibodies by the infected individual are called antigens. This test uses carefully selected and purified anti-

gens unique to a given strain of an infectious agent to identify circulating antibodies. It is rapid and can be highly specific. Continued developments in selection and purification of limited amounts of specific antigens using recombinant DNA technology may ultimately make ELISA the preferred serologic testing method for most infectious agents.

3. **Complementary DNA Probes:** These probes are derived from cloned DNA or RNA of specific infectious agents and can confirm the existence of the infectious agent in tissue samples. The tests would distinguish between animals carrying only antibodies and those that actually carry the infectious agent. The tests would also be of great value in identifying asymptomatic carriers of infectious agents that infect circulating white blood cells without eliciting antibody formation.

In addition to movement of entire animals, increased interest in the international transfer of semen and embryos has produced both opportunities and concerns about disease control. For semen, the risk of disease transmission is usually equated to that associated with the male that produced the semen. When semen is being moved, it undergoes the same tests the donor would undergo if he were being moved. In addition, samples of the semen are usually subjected to various diagnostic tests (44,45).

The risk via either fresh or frozen embryos is less clear. In many cases, infectious agents are attached to the surface of the embryo or found in the associated uterine fluids. Although standard methods of embryo-washing free the embryo of most such organisms (19), it does not remove all of them (e.g., African swine fever) (11). Research is thus needed on the feasibility of purging embryos of undesirable disease agents. Even if the disease organism cannot be disassociated from the embryo, the contaminated germplasm may be rendered noninfectious by highly specific monoclonal antibodies, new antiviral agents, chemical detergents, or immunization of surrogate mothers.

To date, the suitability of embryos for international movement has been equated to the suitability of both parents for such movement. But considerable interest exists in developing procedures that would allow the status of the embryo to be evaluated independently. Such an assessment is likely to become feasible in the future. Indeed, transfer of embryos of wild and domestic animals may ultimately provide the safest means of exchanging germplasm.

Advances in diagnostic procedures and transfer technology should facilitate the international movement of germplasm. For domestic species and wild ungulates, these developments should make foreign breeds more accessible without increasing the risk of introducing disease. Improved serological testing may allow relaxation of the permanent post-entry quarantine now imposed on wild ungulates. For unregulated species, a mechanism for monitoring disease status is needed and should be facilitated by new technologies. These efforts will be particularly important as captive breeding programs enlarge, thereby increasing contact between exotic and indigenous species. Returning individuals from zoos to the wild will also place a premium on ensuring the health status of released individuals.

For improved diagnostic and transfer technologies to be most effective, they must be applied both in the United States and in the countries of origin. Currently, USDA-approved quarantine facilities do not exist in Asia and have only recently been developed in Latin America. To set up such facilities and equip them requires capital inputs—costs that are likely to be borne largely by industrial countries. This approach is reasonable in terms of the ultimate benefits that are expected from global maintenance of animal diversity. The costs of importing animals and semen are currently absorbed by the U.S. importer. Yet this approach ignores societal benefits that accrue from access to foreign domestic animal germplasm and from maintenance of animal diversity as a whole, which argue for a greater U.S. Government role. If widespread maintenance of genetic diversity is the goal, then increased public support for importation, conservation, and use of foreign germplasm is essential.

MAINTENANCE TECHNOLOGIES

Storage Technologies

Cryogenic storage of gametes and embryos introduces a new level of complexity to the procedures already discussed, but it also holds the promise of greatly facilitating conservation of genetic diversity. For both semen and embryos, a critical element for cryopreservation involves development of media to protect cells when they are frozen in liquid nitrogen at −196° C. Likewise, procedures must be developed to regulate the rate of freezing and thawing of this material in a way that will maintain the integrity of the cells.

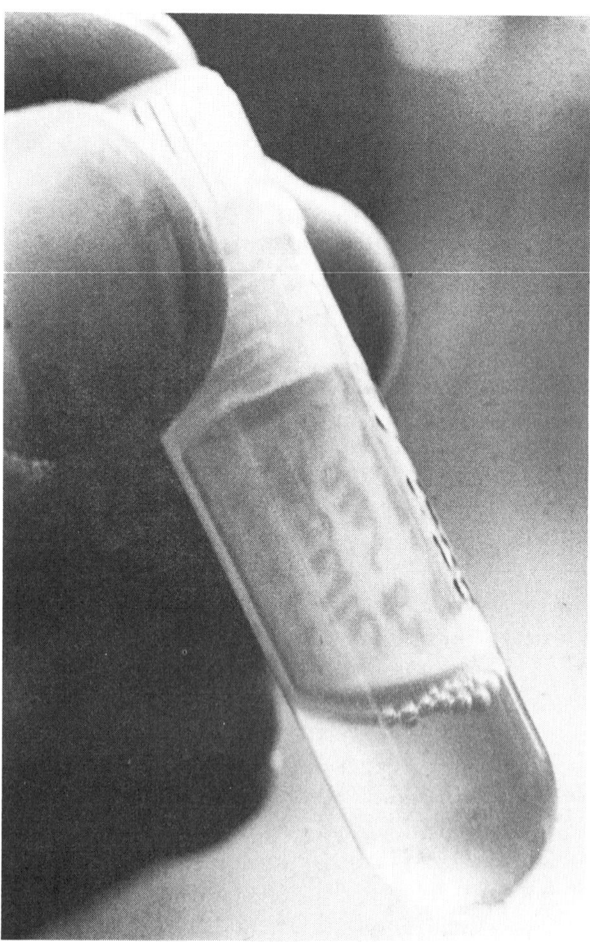

Photo credit: *Zoological Society of San Diego*

In a vial, frozen cells may be stored in suspended animation and later resuscitated. Technologies for storing sperm, ova, and embryos are being developed for domestic and nondomestic species.

The ability to freeze semen successfully resulted from the accidental discovery in 1949 of the cryoprotective action of glycerol. To date, semen has been frozen from at least 200 different species, but little has actually been thawed and tested. Commercial use of artificial insemination with frozen semen is a reality today only for domestic species. Current media for freezing of semen usually include buffering agents, a cryoprotectant such as glycerol, antibiotics, and either egg yolk or milk. Many variations of these media exist, and a somewhat different mix usually must be developed for different species.

The first successful freezing of mammalian embryos with a subsequent live birth was reported in 1972 with mice (50,51). Since then, embryos of 10 mammalian species have been successfully frozen, and the procedure has become routine with the mouse, cow, and rabbit. As with semen, a variety of freezing media and of freezing and thawing procedures are available and are being evaluated. Rapid increases in efficiency have occurred in the bovine embryo transfer industry, and frozen embryos can now be transferred in a manner analogous to artificial insemination. As in the freezing of semen, specific procedures and media appear to be required for each species. Yet the procedure in general rests on a firm mechanistic understanding of the processes responsible for cell injury during freezing, thawing, and dilution. Previous detailed work with mice and primates can act as a model for extension of these techniques to other mammals. Thus, given appropriate research, cryogenic storage of embryos could be developed for a range of species.

Cryopreservation is probably the most promising area of reproduction research today. The potential exists to hold a well-constructed sample of the genetic diversity of a population in suspended animation indefinitely. In practice, frozen semen or embryo storage would probably be used with living populations for a number of reasons: to augment the genetic variation within breeding populations, to allow periodic comparisons between original and current populations, and to validate the viability

Photo credit: American Breeders Service

Liquid nitrogen storage vessels (above) contain enough frozen bull semen to inseminate 4.5 million cows. Liquid nitrogen maintains the temperature at −196° C (−320° F).

of the frozen material. Samples from current populations would likewise periodically be added to the frozen store to retain new variants produced by natural selection or mutation. This process would be particularly important in domestic populations, in which selection could make preserved individuals economically obsolete.

Breeding Technologies

The goals of a propagation program can be defined in terms of how much genetic diversity is to be maintained and for how long. Table 6-3 shows the number of animals required to ensure retention of various proportions of genetic diversity for subsequent generations. Ideally, all of the genetic diversity present in the source population would be maintained indefinitely in the captive population. Table 6-3 suggests that this goal (i.e., retention of 99 percent for 1,000 generations) would require up to 50,000 animals. These numbers are consistent with the guidelines in table 6-1, which suggests natural populations of fewer than 100,000 should be carefully monitored.

Table 6-3.—Captive Animals Required for a Fixed Proportion of Genetic Diversity Over a Number of Generations

Percentage of genetic diversity maintained	Number of generations			
	50	100	200	1,000
50	36	72	145	722
75	87	174	348	1,738
90	238	475	949	4,746
95	488	975	1,950	9,748
99	2,488	4,975	9,950	49,750

SOURCE: Adapted from D.R. Notter and T.J. Foose, "Concepts and Strategies To Maintain Domestic and Wild Animal Germ Plasm," OTA commissioned paper, 1985.

Box 6-C.—Genetic Drift and Inbreeding

Once the decision has been made to propagate a population in a controlled setting, a plan to ensure retention of desired levels of genetic diversity must be developed. In large populations with random mating (i.e., without preferential reproduction by certain individuals), frequencies of the various genes at each locus are essentially constant and genetic change is minimal. Likewise, in large random-mating populations, the mating of close relatives is unlikely. If the population contains only a few individuals, however, the sample of genes passed to the next generation may differ considerably from that found in the parent generation. Rare genes may be lost and others may, by chance, be disproportionately replicated. This process is called genetic drift and is cumulative; once a gene is lost, it can only be re-created through mutation—an exceedingly rare event. Genetic drift can lead to a captive population that differs considerably from the original source population and is much less genetically diverse.

If population size is limited, the likelihood of mating by close relatives is also greater. Because related individuals tend to carry similar genes, mating of relatives can increase genetic uniformity in the offspring. This increased uniformity is known as inbreeding; in most species, rapid increases in inbreeding lead to decreases in fertility and postnatal survival within the population. Indeed, if the level of inbreeding is too high, the overall viability of the population can be seriously compromised.

Both inbreeding and genetic drift can be characterized in terms of decreases in genetic diversity and corresponding increases in genetic uniformity. The losses in diversity are directly related to population size. If several rare genes are present in the original population, the rate of loss of these genes can be considerably greater than the average rate of loss of genetic diversity in the population (6). Likewise, if gene interactions are important sources of diversity (49), specific gene combinations will be lost much more rapidly than will individual genes. Thus, larger populations will be required to ensure retention of rare genes.

In this assessment, genetic diversity is discussed primarily as effects of independent genes. This approach appears acceptable in terms of the probable fitness and viability of the population and of likely future responses to selection (35). However, results and conclusions regarding required population sizes will be conservative for preservation of rare genes and gene combinations.

Genetic drift accumulates generation by generation, not year by year, and animal species differ considerably in this regard (box 6-C). For example, 200 years covers perhaps 100 generations of chickens but only 7 to 8 generations of elephants. Yet in most cases, breeding populations should not experience inbreeding rates in excess of 1 percent per generation; to meet this constraint, at least 50 breeding individuals per generation are required.

Manipulation of the breeding structure of a population can have a significant impact on its genetic characteristics. For example, an appropriate level of subdivision of a population can retard the overall rate of genetic loss in the population as a whole: Subdivision increases the rate of loss within each subgroup, but the specific genes that are lost through random drift vary among subpopulations. And the number of genes that can be maintained within the subdivided population will exceed the number that could be maintained in a random-mating population of comparable size. In practice, the minimum size of the subpopulation will depend on the tolerance of the species to the inbreeding rates. Still, some degree of subdivision should be practiced, and the movement of individuals among subpopulations should be fewer than one per generation unless effects of inbreeding become evident. Subdivision also protects the population against disease outbreaks or other disasters that might annihilate any one subpopulation.

The loss of genes from a captive population can also be retarded by controlling mating. In contrast to selection, which presupposes a differential contribution of different individuals to the next generation, programs that attempt to equalize the contribution of each individual can greatly lower initial rates of genetic loss (21). Likewise, if pedigrees of available individuals are known, matings can be planned in an attempt to equalize the contribution of different lineages. This approach has been used to stabilize founder contributions in a captive population of Speke's gazelle (46). Thus, efforts to record and publish pedigrees of individuals in endangered species (such as the records of ISIS) assume great importance.

For domesticated animals, several population structures and mating systems can be used in conservation programs that are generally not appropriate for wild animals. In many domestic species, the large number of existing breeds (30) precludes conservation of all endangered breeds as pure breeds. One possibility in such cases is to preserve a single breed representative of a group of similar breeds. A better strategy, however, may be to amalgamate into a gene pool individuals from related breeds or from several breeds that excel in a certain characteristic.

Gene-pool populations are designed to conserve genes rather than individual breeds. Thus, several breeds noted for a certain characteristic such as heat tolerance or proliferation might be interbred to provide a single large reservoir of genes for this trait. Although the identity of individual breeds is lost, many genes present in the breeds are retained. Selection to intensify the trait may be appropriate, depending on the potential or current economic importance of the population. Maintenance of a single interbreeding gene pool is less desirable than of a subdivided population for long-term gene conservation. For domestic species, however, the larger population sizes that are possible in a single gene-pool population are expected to facilitate selection for economically important characters within the population. Simmental cattle representing at least five regional or national strains from Europe were imported into North America in the 1970s, and the current American Simmental population represents a gene pool constituted from these breeds. A gene-pool population of pigs was developed in the early 1970s in Nebraska and used in efforts to increase ovulation rate (53).

A program to not only maintain but also generate genetic diversity in domestic breeds has been suggested (26). In this effort, populations would be selected to generate extreme levels of performance in specific traits. These populations could serve as reservoirs of genetic variation and their characteristics would be well known.

Efficient maintenance of captive populations requires a thorough understanding of the reproductive processes of the species. Optimal use of breeding stock is often facilitated by an ability to manipulate and control these processes. In domestic animals, control of the estrous cycle and ovulation through administration of exogenous hormones has become commonplace, greatly assisting programs of controlled mating, artificial insemination, and embryo transfer. In wild species, however, knowledge of basic reproduction remains limited. Efforts to expand knowledge in this area are largely funded by the private sector and are insufficient.

Infertility is a major problem in many species of zoo animals. It reduces the effective breeding size of captive populations and exacerbates genetic losses. Infertility can often be traced to environmental factors such as light, temperature, nutrition, disease, or social influences. Such problems would be more easily overcome if more were known about basic reproductive processes in wild animals.

General principles underlying control of reproduction are relatively uniform across species, yet the particular hormone levels observed and the release patterns of these hormones are species-specific. A practical, reasonably simple, relatively inexpensive kit for monitoring urine hormone levels has recently been developed (29). This test helps confirm ovulation, predict optimal times for insemination, diagnose reproductive dysfunction, and detect pregnancy. Because the test is based on urine sam-

Photo credit: Zoological Society of San Diego

New technologies in reproductive physiology offer possibilities of producing large numbers of offspring from many vertebrate species. Above, an osmotic pump filled with gonadotropin-releasing hormone is prepared for insertion under the skin of a female iguana. The iguana subsequently entered estrus and ovulated.

ples, it also avoids problems restraining and anesthetizing rare animals.

Although reproductive problems in wild animals can often be solved through management changes, hormone therapy can also be used for infertility arising from age or unknown environmental factors. In particular, gonadotropin-releasing hormone has been used to initiate and maintain estrous cycles and ovulation in monkeys, sheep, and cattle. It is administered through a small osmotic pump implanted beneath the skin and appears to have facilitated the birth of two cubs to a previously subfertile cheetah (28). Similarly, a human fertility drug, clomid, is being considered to support ovulation in female gorillas (9).

Growing pressure for the international movement of animal germplasm will also place an increasing premium on knowledge of reproductive biology. In terms of animal safety, convenience, and disease control, movement of semen and embryos (either fresh or frozen) would be preferable to the movement of animals. Although the techniques to allow collection, preservation, transport, and use of these tissues are relatively well developed in domestic animals, comparable methods do not exist for wild species.

Artificial insemination (A.I.) is the introduction of semen into the female reproductive tract by artificial means. It requires technologies to allow collection of semen from the male, storage of semen until it can be used, identification of females in the proper stage of the estrous cycle, and deposition of semen at the appropriate location in the female reproductive tract. Collection of semen from wild species is usually accomplished by electroejaculation, which involves stimulation of ejaculation by application of a mild, pulsating electrical current through a lubricated rectal probe. The process requires restraint and anesthesia of the male, and semen obtained with this procedure is often less fertile than that obtained in a natural ejaculate. Use of A.I. likewise requires the ability to assess the reproductive status of the female quite accurately, and insemination procedures must be developed that are consistent with the biochemical and physical characteristics of the female reproductive tract.

Although artificial insemination has been attempted in many species of wild animals, it has only been successful in a limited number and, in most cases, with one animal in most species. A.I. with frozen semen has been successful with even fewer wild species (see table 6-4) such as the wolf, gorilla, chimpanzee, and giant panda.

Effective use of embryo transfer requires even greater control of an animal's reproductive processes (box 6-D and figure 6-1). Fertilized ova and early embryos are recovered from the

Box 6-D.—Embryo Transfer

Embryo transfer is a well-established practice in the beef and dairy cattle industries. More than 200,000 transfers are performed annually throughout the world, mainly in the United States and Canada. Although the technique was first used with beef cattle, half the transfers are now in dairy cattle. The objective is to increase the number of offspring of cows with valuable genetic traits, such as rapid rates of growth and high levels of milk production. Using this procedure, one valuable cow can produce on average 12 offspring a year.

The procedure involves inducing superovulation in a donor cow using gonadotrophic hormones, so that she will produce six to eight eggs rather than one. The cow is artificially inseminated with semen from a valuable, high-performance bull, and the embryos are collected by nonsurgically flushing the uterus after 6 to 8 days. Embryos that appear viable and healthy by microscopic examination are transferred to recipient cows that are also at the sixth to eighth day of their estrous cycle. Normally one embryo is transferred to each recipient.

Several new technologies hold promise of making the process more efficient and increasing its usefulness to animal agriculture. Among these is the ability to freeze bovine embryos. This procedure is currently used by most embryo transfer companies, and 25 percent of the transfers in the United States are with frozen embryos. Survival of the embryos is not perfect, however: Transfer of unfrozen embryos average a 60-percent pregnancy rate, while frozen and thawed embryos can be expected to yield pregnancy rates of 40 to 50 percent.

Another interesting development in this industry involves cloning bovine embryos. Once developed, this technique would allow the multiplication of large numbers of calves from one valuable embryo. The cloned embryos could be frozen while other embryos from some clonal lines are tested to determine if the line is of high value; valuable ones could be replicated using the frozen clones, providing a powerful tool for livestock improvement. Several research stations are also experimenting with inserting genes for specific productivity traits, such as growth, into embryos before transfer. The application of these new biotechnologies is expected to expand the size and usefulness of the cattle embryo transfer industry.

Although embryo transfer could also be a useful tool in swine production, much of the technology and the industry are not yet well developed. In swine, embryos must be collected and transferred surgically. And the embryos do not survive freezing with present techniques. This procedure therefore has received little use in the swine industry. In addition, the cost of surgically recovering embryos is likely to preclude wide-scale use of this technology in the near future.

Based on the use of embryo transfer in cattle, research on the applicability of this technology for wild species was begun in 1981. Although the nonsurgical collection techniques are similar, working with exotic species entails several unique problems, such as the need to administer drugs by dart or pole syringe and the need for anesthesia to perform even the simplest procedures. The ultimate goal was to develop methods for using a common wild species (e.g., the eland antelope) as a surrogate mother for a less common species (e.g., the bongo antelope).

In 1983, an eland calf was born to a surrogate eland mother, becoming the first nondomestic issue of a nonsurgical embryo transfer. A transfer involving a frozen embryo was accomplished soon thereafter. These successes were followed by attempts at interspecies transfer (i.e., a donor and surrogate of different species). Initial efforts for an eland-to-cow transfer were unsuccessful. The eland, however, proved to be a suitable surrogate mother for an embryo collected from a bongo. This first documented nonsurgical embryo transfer between two different species of wild animals indicates that embryos can be gestated by surrogates of different species, offering hope for the future of endangered wildlife (figure 6-1).

SOURCE: Adapted from materials provided by Dr. Neal First, University of Wisconsin and Dr. Betsy Dresser, Cincinnati Wildlife Research Federation.

Figure 6-1.—Transcontinental Embryo Transfer

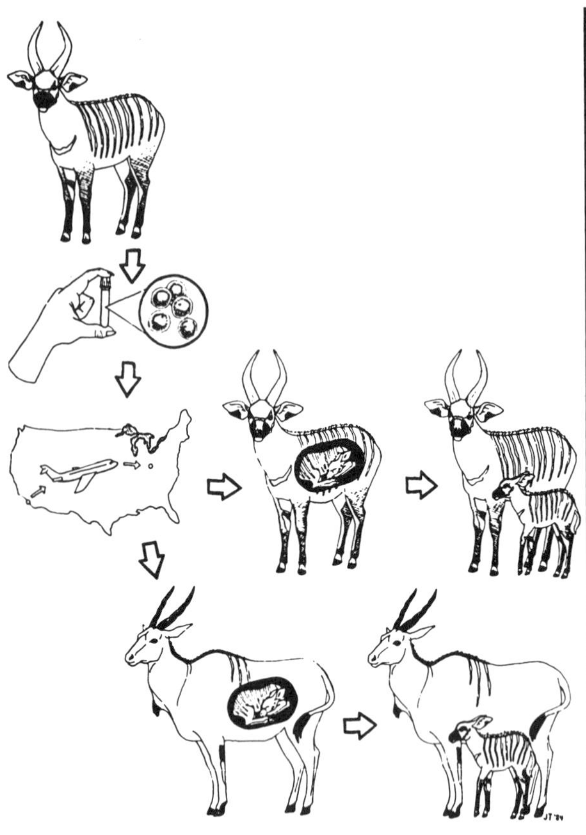

Bongo embryos were collected from animals at the Los Angeles Zoo by the research team of the Cincinnati Wildlife Research Federation and immediately transported by air to the Cincinnati Zoo. The embryos were transferred into a bongo and into a common eland. These embryo transfers resulted in the successful production of offspring from both bongo and eland surrogates.

SOURCE: Betsy Dresser, Director of Research, Cincinnati Wildlife Research Federation, 1986.

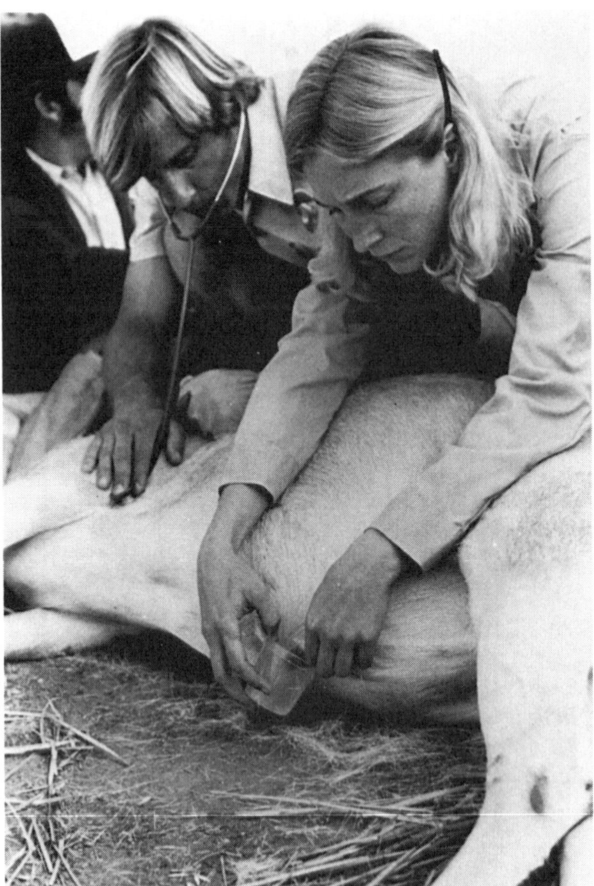

Photo credit: Zoological Society of San Diego

Collection of sperm samples for artificial insemination and cryogenic storage from nondomestic species is part of many offsite conservation programs. Above, an African antelope, the scimitar-horned oryx (*Oryx gazella daimmah*) is tranquilized and undergoing semen collection by electroejaculation.

Table 6-4.—Successful Artificial Insemination in Nondomestic Mammals

Guanaco	Gorilla
Llama	Ferret
Black buck	Fox
Bighorn sheep	Wolf
Brown brocket deer	Persian leopard
Reindeer	Puma
Red deer	Macaca monkey
Speke's gazelle	Papio baboon
Giant panda	Squirrel monkey
Chimpanzee	

SOURCE: B.L. Dresser, Cincinnati Wildlife Research Federation, personal communications, September 1986.

reproductive tract of a donor female (the genetic mother) and transferred into the tract of a recipient female (the foster mother), in whom the embryos develop into full-term individuals. Successful embryo transfer requires synchronization of the estrous cycles of donor and recipient animals (figure 6-2). In domestic animals this synchrony is usually achieved through exogenous hormone treatment. Donors are induced to produce an excess of eggs (superovulated) by injection of fertility hormones. Superovulation has been fairly successful with hoofed mammals, although the results vary considerably. Optimal drugs and dosages have yet

Figure 6-2.—Embryo Transfer Flowchart

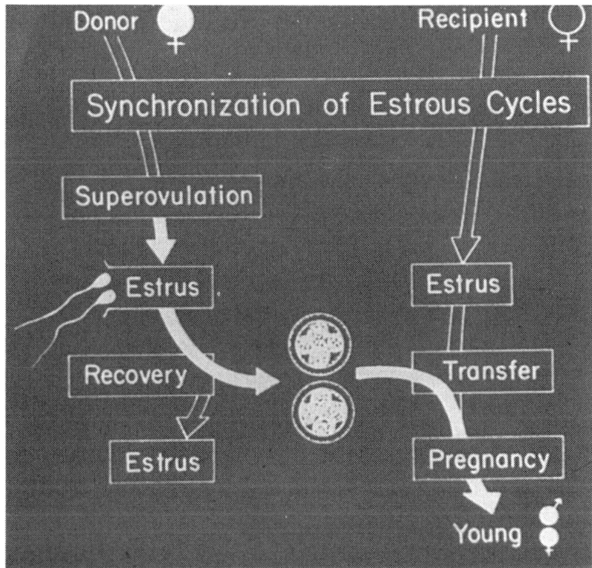

Necessary steps in preparing donor and recipient animals for embryo collection and transfer.

SOURCE: Betsy Dresser, Director of Research, Cincinnati Wildlife Research Federation, 1986.

to be identified in most other species. Donors are mated naturally or by artificial insemination, and fertilized eggs are collected from the female tract (surgically and nonsurgically) and transferred (surgically or nonsurgically) to the recipient female.

Development of embryo transfer techniques is important to maintenance of genetic diversity within captive populations, given the considerations of transfer and disease control previously discussed. In addition, surrogate mothers confer passive immunity to offspring developed from transferred embryos. Thus, animals moved into new environments or reintroduced to the wild may benefit from being carried by mothers acclimated to the new environment.

Several more-advanced techniques, studied primarily in domestic animals, hold considerable potential for all species:

- **Embryo Culture:** This technique involves maintenance of fertilized eggs outside the body during the early stages of embryonic development. The appropriate culture media for development differ among species, but reliable techniques to culture embryos for up to 24 hours exist for cattle, rabbits, mice, sheep, and humans. Successful embryo culture is usually prerequisite to more sophisticated *in vitro* embryo manipulation.
- **Embryo Storage:** This technology involves holding embryos in arrested development for up to several days. Again, specific storage media must be developed for each species. Embryo storage procedures can greatly facilitate transfer of embryos over long distances and *in vitro* embryo manipulation.
- ***In Vitro* Egg Maturation:** This technique involves the culturing of immature eggs to maturity. Coupled with *in vitro* fertilization, this technique could dramatically increase the number of offspring that a given female might produce. The reproductive lifetime of the female is also lengthened because ova suitable for culturing can be obtained prior to sexual maturity as well as after a female is no longer able to conceive naturally.
- ***In Vitro* Fertilization:** In a few species, it is possible to remove unfertilized ova from a female, mix them with semen *in vitro*, and produce fertilized ova that will develop normally when transferred back into a female. In cases of unexpected death of genetically valuable animals, ova can even be collected from ovaries shortly after death.
- **Embryo Splitting:** A single embryo can, under the proper conditions, be split into two or four, and each part can subsequently develop into a live offspring. Although the offspring are genetically identical, this process allows a much larger number of offspring to be produced from each embryo collection.
- **Interspecific Embryo Transfer:** This involves transfer of embryos between related species. Thus, embryos of a rare species could be carried to term by a female of a more common species. This technology has enjoyed some success, but much more research is needed. To date, successful in-

Box 6-E.—Captive Breeding and Przewalski's Horse

Offsite conservation through captive breeding has prevented the extinction of such animal species as the European bison, Peré David's deer, Arabian oryx, and the wild species of horse most closely related to all domestic horse breeds. This latter species, known as Przewalski's horse, has never been domesticated and is known historically to have come from Mongolia and China. Although it is not certain whether it is the direct ancestor of the oldest breeds of domestic horse, this species differs from all domestic horses in possessing a different number of chromosomes, as well as having other genetic variations. First brought to the attention of European scientists in the late 1800s by the Polish explorer N.M. Przewalski, who explored and mapped Mongolia for the Russian czar, the species has a bulky head, erect mane, lack of forelock, and buckskin-dun coloration; it sheds its coat in a manner different from domestic breeds.

This species, *Equus przewalskii*, was already in decline at the time Colonel Przewalski obtained a killed specimen and forwarded the skeleton and skin to the Imperial Zoological Museum at St. Petersburg. The horses depicted on the famous cave paintings at Lascaux in southern France have the same morphological features as Przewalski's horse.

The publication in 1881 of the description of a new species of horse attracted considerable attention from zoologists and horse breeders. By 1902, 52 animals had been captured and were eventually distributed in the Ukraine, Europe, and North America. In Mongolia, young foals were captured shortly after they were born, because attempts to capture adult or even juvenile horses were fruitless. The captured foals were nursed by domestic mares bred to produce their own foals coincident to the capture of the young Przewalski's horses. The Przewalski's horse is thought now to be extinct in its natural habitat. None have been seen since 1967, in spite of annual search efforts by the Joint Soviet-Mongolia Biological Expedition. More than 550 now live in some 70 zoological institutions on all continents except Africa and South America. An international studbook contains the pedigrees of all animals and traces back to 13 individual animals.

Przewalski's horses can interbreed with their domestic relative, and the hybrids produced by such crosses were allegedly incorporated into the Russian czar's cavalry stud.

The disappearance of the species was probably due to hunting and, more important, to restricted access to limited water supplies that were monopolized by nomads and their herds of domestic animals. The horses' habitat remains, and plans exist to restore the species to its former range in Mongolia. This restoration effort involves the cooperative activities of zoological parks in the United States, the Soviet Union, the United Kingdom, and West and East Germany. Animals will be provided by zoological parks from their expanding populations.

The offsite conservation of species and preservation of genetic diversity that these efforts represent may play an increasing role in strategies to prevent extinction. The interaction through movement of individuals from offsite to onsite conservation facilities reduces the chance of extinction and simultaneously provides access to rare species for educational and research purposes.

SOURCE: Dr. Oliver Ryder, Research Department, San Diego Zoo.

terspecific embryo transfers have occurred from mouflon (wild sheep) to domestic sheep, gaur to cattle, bongo to eland, zebra to horse, and Przewalski's horse to pony (see box 6-E) (9).

DEVELOPMENT AND UTILIZATION TECHNOLOGIES

The objectives of developing and using genetic diversity differ between wild and domestic animals. For domestic animals, the potential contribution of rare breeds to food and fiber production on an international scale is of paramount importance. In this context, the most pertinent technologies are those that facilitate the international movement and evaluation of these breeds. Thus, the previously discussed technologies of disease control, artificial insemination, embryo transfer, and cryopreservation of embryos and gametes are extremely important. In particular, aggressive application of state-of-the-art technologies for the control of disease transmission would greatly facilitate use of foreign germplasm.

Equally important, however, is the fact that no organized program exists, either in the United States or elsewhere, to sample, evaluate, preserve, and use available sources of germplasm (3). Current research organizations do not have the resources to evaluate the many unique breeds that exist worldwide. Evaluations of animal germplasm could, however, focus on the present and foreseeable U.S. and world animal production and marketing environments and on the breeds that seem to have the greatest potential for improving animal food and fiber production systems (3).

For wild species, programs of development and utilization are much less clear. The rationale for preservation of such species largely reflects the need to maintain the Earth's ecological structure and, to many individuals, utilization of wild species is inconsistent with this goal. Yet products and processes observed in wild species have been and will continue to be of value to society. Armadillos, for example, provide a unique model of human leprosy. As the understanding of molecular genetics and cellular biology expands, the unique physiological and metabolic processes found in many wild animals are likely to have progressively more important research and development applications.

The domestication of wild animals is an emotional issue. It implies imposition of human control of the mating and husbandry of a previously wild species. To many people, this step is also inconsistent with the preservation of ecological diversity. However, the potential gains from developing adapted populations of previously wild animals to produce food and fiber in harsh or severely restricted environments may be too great to ignore. Thus, populations of red deer in Europe and New Zealand are rapidly becoming domesticated (10), and different species of deer are being crossed to improve production characteristics (32). Eland and oryx in Africa (47), capybara in South America (17), and crocodiles and butterflies in Papua New Guinea (33,34) are also being harvested in semicontrolled programs that may entail domestication of segments of these populations. In such a situation, domestication should not be avoided. Instead, great care must be taken to ensure that protected, viable wild populations are also maintained free of contamination from domesticated subpopulations. Such an approach, though difficult, is necessary to meet the joint goals of food production and maintenance of genetic diversity.

NEEDS AND OPPORTUNITIES

Needs and opportunities for maintaining animal diversity offsite involve both application of available technologies and development of new technologies. Needs differ considerably between wild and domestic animals, and these two groups will be considered separately. For wild animals, many of the needs involve adaptation of techniques that are currently available for domestic animals. In some cases, these adaptations are straightforward. In others, considerable basic research will be required. In domestic animals, efforts to assess and evaluate global genetic resources and facilitate their movement will probably assist in maintaining diversity. Mechanisms to monitor genetic diversity in domestic populations are also badly needed.

Wild Animals

Expertise in Relevant Areas

Maintenance of captive breeding populations of wild animals requires that breeding programs be based on principles of quantitative genetic management to avoid losses in genetic diversity. Likewise, a knowledge of the reproductive biology of the species is required to ensure efficient propagation of the animals in captivity. The need for expertise in these areas has increased dramatically as offsite programs have become more common and more complex. Efficient use of animals for genetic purposes requires extensive movement of germplasm among institutions. These efforts are likely to increasingly rely on transfer of semen or embryos, especially at the international level, placing a premium on scientific expertise.

To date, development of expertise in the application of reproductive biology and quantitative genetics management has largely occurred through the initiatives of individual students within traditional reproductive physiology or quantitative genetics programs. In reproductive physiology, programs are usually directed primarily toward domestic animals; efforts to obtain skills applicable to wild species may be met at best with tolerance or at worst with active discouragement. Still, substantial interest in the reproductive biology of wild animals has been noted, and students of this field are increasingly tolerated. In quantitative genetics, training programs tend to emphasize either the theoretical aspects of quantitative genetics in natural populations or the applied aspects of breeding domestic animals. More opportunities to tailor courses to study of wild populations exist in this area than in reproductive physiology, however.

Fellowships and traineeships in areas that support maintenance of wild animal genetic diversity could be provided on a competitive basis to students in reproductive biology, cryobiology, population genetics, and animal behavior for studies applicable to the genetic and reproductive management of captive populations of wild animals. The program could be administered by the National Science Foundation (NSF). Emphasis would be placed on applying knowledge and theory to managed populations. One advantage of such a program would be sensitization of faculty members to the needs and opportunities in this area.

A grants program to allow selected educational institutions to expand their expertise in supporting maintenance of genetic diversity could be initiated. Grants could be awarded on a competitive basis and could support extension of applied programs to captive wild species. Such a program would be relatively expensive, however, and would tend to concentrate expertise instead of encouraging broad access to needed training.

Facilities for Offsite Maintenance

In recent years, zoo administrators and others have become aware of the need for well-planned breeding programs to ensure maintenance of genetic diversity within captive populations. Substantial theoretical work has gone into developing plans for existing or likely future facilities. The results suggest that today's facilities will not be sufficient to maintain desired levels of diversity. However, zoo personnel appear to have developed mechanisms to make choices (albeit not unanimous choices) among competing possibilities. Still, without additional facilities, losses of diversity appear likely.

Development of captive maintenance and breeding facilities could benefit from additional funding. Such a program would enhance capabilities to preserve biological diversity offsite. Modest levels of funding could have a considerable impact, although substantial funds would be required to address the total problem. Funds could be channeled through the National Zoo in Washington, DC, or through competitive grants to nonprofit zoological parks. Emphasis could be given to species that have limited captive facilities.

Reproductive Biology and Cryopreservation

The reproductive processes of most wild animals are not sufficiently understood to allow optimum rates of reproduction under captive management. This lack of information becomes especially acute in light of increasing interest in artificial insemination and embryo transfer because these technologies require much greater control of the reproductive process. Although the critical elements that control reproduction and cryopreservation in wild species are analogous to those in domestic animals, important differences exist. Thus, extending available knowledge about domestic animals to wild animals will require accumulation of information unique to each species or group of species. Optimum use of available individuals in programs of captive breeding or cryo preservation will depend on collecting this unique information.

Progress is being made in understanding the reproductive processes of wild animals, but not as quickly as it is needed or as it could be used. Without additional research, many available captive animals will continue to experience suboptimal fertility, and fewer total individuals of all species will be maintained at acceptable population sizes in available facilities. Semen from a number of wild species has already been frozen and exhibits near-normal motility and morphology when thawed, but its ability to result in conception is largely untested. Likewise, successful use of frozen embryos has occurred in only a few species.

A program of competitive grants to support research on the reproductive biology and cryopreservation of wild animals could be initiated. This program could be administered through NSF and would channel funds to both basic studies on the reproductive biology and cryobiology of wild animals and to applied studies of control of reproduction, artificial insemination and embryo transfer. Preference could be given to existing programs that emphasize the integration of programs for wild and domestic animals.

Another approach could be establishing a few centers for study of the reproductive biology of wild animals. These centers could serve as focuses for programs of basic and applied research. They should be sufficiently well funded to allow broad programs of research onsite as well as extramural research with cooperating institutions. The centers could likewise serve as repositories for frozen gametes and embryos from endangered populations as techniques are perfected.

Basic Research in Population Biology and Genetics

Much of the basic theory of population genetics was derived in the first half of the 20th century and was adapted to applications in domestic animal breeding in the 1940s and 1950s. Current interest in developing breeding programs to maintain representative levels of genetic diversity within populations of minimum size has introduced several new program-design questions. These questions relate to such things as the amount and nature of genetic diversity that can be lost without compromising the long-term evolutionary potential of the species, the importance to evolutionary potential of rare genes (which are easily lost by genetic drift), the long-term importance of mutation to maintenance of diversity (22), and the importance of genetic diversity (both among and within species) to maintenance of the integrity of entire ecosystems. In many cases, these questions deal with validation of long-term quantitative genetics theory; answering them will require imaginative syntheses of the disciplines of genetics and ecology.

Some of the needed research is currently being done or has been planned. Without direction, however, it will occur in a piecemeal way, with no assurance that issues of the highest priority will be addressed. A program of competitive grants to support development, extension, and validation of quantitative genetic theory related to questions of maintaining biological diversity could be developed. This program could be administered through NSF and would require less funding (because of fewer equipment needs) than programs in reproductive biology or cryopreservation. Such a program could provide a focus for needed efforts in this area and a mechanism for screening competing proposals to identify those that address areas of highest priority.

Domestic Animals

Objective Assessment of Global Genetic Resources

The potential contributions of indigenous stocks of animal agriculture both in their country of origin and internationally needs to be assessed. Experience with the prolific Finnish Landrace and Booroola Merino sheep (31,36) and with Sahiwal cattle (24,48) has shown that local, specialized stocks can often have wide utility outside their country of origin. Likewise, comprehensive performance evaluations of crosses of indigenous and imported breeds suggest that local animals may make important contributions to final performance of the crossbreed (5). The use in West Africa of native cattle resistant to trypanosomiasis (23) is an important example. To assess the contributions of such breeds, objective information must be available to potential users. In many cases, some details exist but they are fragmented and difficult to locate and gain access to. In other cases, only anecdotal information is available.

Considerable international awareness of the need for such assessments exists. Efforts to at least list and broadly categorize breed resources have been initiated in Europe (30), Latin America (13), and Eastern Asia and Oceania (41). These efforts have been coordinated by the Food and Agriculture Organization (FAO) of the United Nations (14). Efforts in most of the developing countries have, however, been hampered by insufficient funding to develop electronic databases and library reference facilities. On balance, efforts to date deserve credit and have achieved some successes but are still insufficient.

No comparable assessment of breed resources has been undertaken for North America yet, so commissioning one would indicate support for efforts elsewhere and represent a minimal contribution by North American countries to a global accounting. The assessment could be coordinated by the National Academy of Sciences (NAS) or the U.S. Department of Agriculture (USDA) with technical support from relevant professional societies (American Society of Animal Science, American Dairy Science Association, Poultry Science Association, and Canadian Society of Animal Science) and private agencies (e.g., American Minor Breeds Conservancy). A recently initiated NAS project on global genetic resources could address domestic animal genetic resources and develop options for improving the present efforts.

Limited additional financial and technical support for development of databases and library reference facilities in existing foreign centers could be provided (14). In many cases, funds for microcomputers, software, and reference materials could provide a major improvement in the capabilities of existing institutions at limited cost. Necessary funds and consulting personnel could be channeled through USDA, FAO, or the U.S. Agency for International Development (AID).

Another approach could be the development of an international center for animal genetic resources that would be charged with maintenance of a comprehensive base of information on domestic animal germplasm resources. The center could maintain and update files on the status, trends, and characteristics of domestic breeds worldwide and provide information to potential users of this germplasm. Charges would be made to clients requesting information, but to function properly, considerable public subsidization would probably be required. The center could be a branch of USDA or a part of the National Agricultural Library. This plan has the potential disadvantage of moving responsibility for maintenance of the necessary databases out of national and regional institutions, or at least deemphasizing the roles of such institutions. Such an approach would tend to reduce the emphasis on breed evaluation and preservation at the grassroots level in the countries of origin.

Major new funding to support breed evaluation and characterization efforts could be provided. Even though considerable information already exists on many foreign breeds, the material is often fragmented and limited to only descriptive characteristics. The initiation and support of several major projects to objectively evaluate and compare indigenous breeds to potential imported breeds for the full array of productive traits in the country or region of origin would be a tremendous asset in terms of knowledge of global genetic resources. Funding could be channeled through USDA or AID.

Such a project would require major new funding, including support for development of necessary foreign facilities.

Facilitation of International Movement of Germplasm

Effective use of global germplasm requires that mechanisms exist to facilitate the movement of such resources. This is especially important for specialized breeds in developing countries, such as prolific Chinese pigs (52), which may have utility in crossbreeding programs in industrial countries. The international movement of germplasm is often difficult because of different countries' health-related import-export requirements. This area involves both technologies for actual movement of germplasm (embryo transfer, semen and embryo collection, etc.) and technologies for prevention of disease transmission.

The United States currently maintains facilities for quarantine and disease-testing at Plum Island, NY, and Flemming Key, FL. These stations provide U.S. breeders access to foreign breeds. The approach taken has usually been to provide use of these facilities to importers in the private sector and to require that the cost of importation be borne completely by the importer. Importation of some breeds of sheep and swine has been supported by public (USDA) funds, but these cases are the exceptions. Assessing private sector importers for importation costs does allow the expense to be borne by those likely to receive economic benefit from the sale of imported animals, but it ignores the public benefits likely to accrue from access to foreign germplasm. When the decision to import a breed lies solely within the private sector, preference will be given to more traditional breeds judged to have the most speculative potential while unique breeds of undocumented value will usually be ignored.

USDA and the Animal and Plant Health Inspection Service (APHIS) could be directed to pursue an aggressive program of screening, importation, and evaluation of promising foreign breeds. Such a program would involve both a redirection of existing funds and appropriation of modest new funds. Such a program would recognize the existence of promising foreign breeds and likewise acknowledge that the procurement of these breeds is a matter of public interest. A considerable improvement in U.S. access to foreign germplasm could be accomplished through such a program with existing technology.

New funding for research and development on the diagnosis and neutralization of foreign diseases could be provided to APHIS and other research laboratories through a system of competitive grants. This new funding could be accompanied by a mandate to aggressively pursue importation of promising foreign germplasm into the United States. Objectives of the program would be, first, to validate and apply recently developed technologies for disease diagnosis (ELISA, DNA probes, etc.) and, second, to improve on and extend these technologies. Such a program should be able to accelerate access to foreign germplasm.

The training of foreign professionals in areas that support germplasm transfer could be supported. These areas would include veterinary pathology, reproductive biology, with emphasis on techniques for gamete and embryo collection and transfer, and cryobiology. In many cases, germplasm transfer is limited by insufficient expertise and facilities in the country of origin. An expanded training program for foreign students and professionals would increase the chances that the needed expertise existed onsite. Considerable opportunities for foreign professionals to receive this kind of training already exist, however. A major problem is that students receive sophisticated training in highly technical areas but have insufficient facilities and equipment to put their training to use when they return home.

The development and improvement of foreign centers for transfer of germplasm could be supported. This improvement would require new funding to allow development of centers in major geographical areas of the world. These centers could serve as focuses for a full range of considerations relating to maintenance of biological diversity. In particular, equipment and expertise for collection, preparation for shipment, and preservation of gametes and embryos could be concentrated in such institutions. Facilities for quarantine and diagnostic testing using advanced technologies would greatly facilitate germplasm transfer. To be effective, these centers would have to be well funded and equipped on a continuing basis. Ideally, they would address a range of biological diversity issues, for wild as well as domestic animals, including maintenance of information centers and repositories for cryopreservation of frozen semen and embryos of rare native breeds.

Losses of Genetic Diversity Among and Within Breeds

Indiscriminate crossbreeding of so-called improved breeds from industrial nations coupled with increasing intensification within the poultry, swine, and dairy industries have resulted in reductions in global breed diversity and may lead to substantial losses of rare breeds. Within some of the major commercial breeds of livestock, losses in genetic diversity may also be occurring because of narrow selection goals and intensified use of individual sires and their sons through artificial insemination.

A National Board for Domestic Animal Resources could be established, composed of representatives from USDA, universities, private foundations, and industry. The board could provide a mechanism to coordinate animal germplasm conservation activities. The program could be established through a directive to a lead agency such as USDA and would not require additional legislation. Such a board would identify potential sources of foreign germplasm for import and monitor the status of genetic diversity within commercial breeds. It could also monitor the status of rare breeds within the United States and make recommendations for their preservation and use. The board could act as a liaison with institutions in other countries and show a U.S. commitment to maintenance of domestic animal biological diversity. It would be primarily advisory in nature but should possess some funding to implement its recommendations to function effectively.

An International Board on Domestic Animal Resources could also be established. This board could provide international coordination of programs, set standards and coordinate the exchange and storage of germplasm, and provide funds to support activities in developing countries, probably at the regional level. Some efforts have already been made in this direction, and the United States could support and expand these efforts.

A program to identify, conserve, and use endangered breeds of potential value worldwide could be developed. It could identify rare breeds of potential value worldwide, with subsequent negotiation of procedures to protect and maintain the genetic integrity of these populations within the country of origin. If maintenance of such populations within the country of origin could not be assured, the United States could support collection and cryogenic storage of gametes and embryos. Semen of all mammalian livestock can be successfully frozen, as can embryos of all mammalian livestock species except the pig. Such storage could be located in this country to ensure maximum safety of the preserved material, and it would include material that could not be imported as live animals under current animal health regulations. Efforts such as these would require close cooperation with the countries of origin of the various breeds to avoid the perception of exploitation of foreign resources for the sole benefit of the United States.

A program like this could also monitor the status of genetic diversity within commercial populations in the United States. This monitoring would involve interacting with industry to ensure maintenance of genetically diverse poultry control strains, retaining semen from a wide sample of dairy bulls as a reservoir of genetic diversity, and monitoring the status of other species.

CHAPTER 6 REFERENCES

1. Ashwood-Smith, M.J., and Grant, E., "Genetic Stability in Cellular Systems Stored in the Frozen State," *The Freezing of Mammalian Embryos*, K. Elliott and J. Whelan (eds.) (Amsterdam: Elsevier, 1977).
2. Conway, W.G., "The Practical Difficulties and Financial Implications of Endangered Species Breeding Programs," *International Zoo Yearbook*, 1986.
3. Council for Agricultural Science and Technology, "Animal Germ Plasm Preservation and Utilization in Agriculture," Council for Agricultural Science and Technology Report No. 101, 1984.
4. Cunningham, E.P., "European Friesians—The Canadian and American Invasion," *Animal Genetic Resources Information,* January 1983, pp. 21-23.
5. DeAlba, J., and Kennedy, B.W., "Criollo and Temperate Dairy Cattle and Their Crosses in a Humid Tropical Environment," Animal Genetic Resources Conservation by Management, Data Banks and Training, FAO Animal Production and Health Paper 44/1, 1984, pp. 102-104.
6. Denniston, C., "Small Population Size and Genetic Diversity Implications for Endangered Species," *Endangered Birds*, S. Temple (ed.) (Madison, WI: University of Wisconsin Press, 1977), pp. 281-289.
7. Douglass, E.M., and Gould, K.G., "Artificial Insemination in Lowland Gorilla (*Gorilla gorilla*)," *Proceedings of the American Association of Zoo Veterinarians,* 1981, pp. 128-130.
8. Dresser, B.L., Kramer, L., Dahlhausen, R.D., Pope, C.E., and Baker, R.D., "Cryopreservation Followed by Successful Transfer of African Eland Antelope (*Tragelaphus oryx*) Embryos," *Proceedings of the 10th International Congress on Animal Reproduction and Artificial Insemination,* 1984, pp. 191-193.
9. Dresser, B.L., and Leibo, S.P., "Technologies To Maintain Animal Germ Plasm in Domestic and Wild Species," OTA commissioned paper, 1986.
10. Drew, K.R. (ed.), *Advances in Deer Farming* (Wellington, New Zealand: Editorial Services, Ltd., 1978).
11. Eaglesome, M.D., Hare, W.C.D., and Singh, E.L., "Embryo Transfer: A Discussion of Its Potential for Infectious Disease Control Based on a Review of Studies on Infection of Gametes and Early Embryos by Various Agents," *Canadian Veterinary Journal* 21:106-112, 1980.
12. Fitzhugh, H.A., Getz, W., and Baker, F.H., "Biological Diversity: Status and Trends for Agricultural Domesticated Animals," OTA commissioned paper, 1985.
13. Food and Agriculture Organisation of the United Nations, "Recursos Geneticos Animales en Americana Latina," FAO Animal Production and Health Paper 22, 1981.
14. Food and Agriculture Organisation of the United Nations, "Animal Genetic Resources Conservation by Management, Data Banks and Training," FAO Animal Production and Health Paper 44/1, 1984.
15. Franklin, I.R., "Evolutionary Change in Small Population," *Conservation Biology*, M.E. Soulé and B.A. Wilcox (eds.) (Sunderland, MA: Sinauer Associates), 1980, pp. 135-149.
16. Glenister, P.H., Whittingham, D.G., and Lyon, M.F., "Further Studies on the Effect of Radiation During the Storage of Frozen 8-Cell Mouse Embryos at −196° C," *Journal of Reproduction and Fertility* 70:229-234, 1984.
17. Gonzalez-Jeminez, E., "The Capybara: An Indigenous Source of Meat in Tropical America," *World Animal Review* 21:24-30, 1977.
18. Graham, E.F., Schmehl, M.R., and Deyo, R.C.M., "Cryopreservation and Fertility of Fish, Poultry and Mammalian Spermatozoa," *Proceedings of the 10th Technical Conference on Artificial Insemination and Reproduction,* National Association of Animal Breeders, 1984, pp. 4-29.
19. Hare, W.C.D., and Singh, E.L., "Control of Infectious Agents in Bovine Embryos," *Proceedings of an International Symposium on Microbiological Tests for the International Exchange of Animal Genetic Material,* 1984, pp. 80-85.
20. Heuschele, W. P., "Management of Animal Disease Agents and Vectors Potentially Hazardous for Animal Germ Plasm Resources," OTA commissioned paper, 1985.
21. Hill, W.G., "Estimation of Genetic Change, I: General Theory and Design of Control Populations," *Animal Breeding Abstracts* 40:1-15, 1972.
22. Hill, W.G., "Predictions of Response to Artificial Selection From New Mutations," *Genetical Research* 40:255-278, 1982.
23. International Livestock Centre for Africa, "Trypanotolerant Livestock in West and Central Africa, Volume I: General Study," International Livestock Centre for Africa Monograph 2, Addis Ababa, Ethiopia, 1979.
24. International Livestock Centre for Africa, "Sa-

hiwal Cattle: An Evaluation of Their Potential Contribution to Milk and Beef Production in Africa," International Livestock Centre for Africa Monograph 3, Addis Ababa, Ethiopia, 1981.
25. King, J.W.B., "Genetic Exhaustion in Single-Purpose Breeds," *Proceedings of the FAO/UNEP Technical Consultation on Animal Genetic Resources Conservation and Management,* FAO Animal Production and Health Paper 24, 1981, pp. 230-242.
26. Land, R.B., "An Alternate Philosophy for Livestock Breeding," *Livestock Production Science* 8:95-99, 1981.
27. Lande, R., and Barrowclough, G.F., "Effective Population Size, Genetic Variation and Their Use in Population Management," *Viable Populations,* M.E. Soulé (ed.) (Oxford: Blackwell Publishing Co., 1986).
28. Lasley, B.L., and Wing, A., "Stimulating Ovarian Function in Exotic Carnivores With Pulses of GnRH," *Proceedings of the American Association of Zoo Veterinarians,* 1983, p. 14.
29. Loskutoff, N.M., Ott, J.E., and Lasley, B.L., "Monitoring the Reproductive Status of the Okapi (*Okapia johnstoni*)," *Proceedings of the American Association of Zoo Veterinarians,* 1981, p. 149.
30. Maijala, K., Cherekaev, A.V., Devillard, J.M., Reklewski, Z., Rognoni, G., Simon, D.L., and Steane, D.E., "Conservation of Animal Genetic Resources in Europe, Final Report of an E.A.A.P. Working Party," *Livestock Production Science* 11:3-22, 1984.
31. Maijala, K., and Osterberg, S., "Productivity of Pure Finnsheep in Finland and Abroad," *Livestock Production Science* 4:355-377, 1977.
32. Moore, G.H., "Deer Imports," *New Zealand Agricultural Science* 18:19-24, 1984.
33. National Research Council, "Butterfly Farming in Papua New Guinea" (Washington, DC: National Academy Press, 1983).
34. National Research Council, "Crocodiles as a Resource for the Tropics" (Washington, DC: National Academy Press, 1983).
35. Notter, D.R., and Foose, T.J., "Concepts and Strategies To Maintain Domestic and Wild Animal Germplasm," OTA commissioned paper, 1985.
36. Piper, L.R., Bindon, B.M., and Nethery, R.D. (eds.), *The Booroola Merino* (Melbourne, Australia: Commonwealth Scientific and Industrial Research Organization, 1980).
37. Pope, C.E., Pope, V.Z., and Beck, L.R., "Live Birth Following Cryopreservation and Transfer of a Baboon Embryo," *Fertility and Sterility* 42:143-145, 1984.
38. Seiger, S.W.J., "A Review of Artificial Methods of Breeding in Captive Wild Species," *The Dodo, Journal of the Jersey Wildlife Preservation Trust* 18:79-93, 1981.
39. Simon, D.L., "Conservation of Animal Genetic Resources—A Review," *Livestock Production Science* 11:23-36, 1984.
40. Smith, C., "Genetic Aspects of Conservation in Farm Livestock," *Livestock Production Science* 11:3-48, 1984.
41. Society for the Advancement of Breeding Researchers in Asia and Oceania, "Evaluation of Animal Genetic Resources in Asia and Oceania," Kuala Lumpur, Malaysia, 1981.
42. Somes, R.G., "International Registry of Poultry Genetic Stocks," Storrs Agricultural Experiment Station Bulletin 469, 1984.
43. Soulé, M.E., Gilpin, M., Conway, W., and Foose, T., "The Millennium Ark: How Long the Voyage, How Many Staterooms, How Many Passengers?" *Zoobiology* 5:101-114, 1986.
44. Stalheim, O.H.V. (ed.), "Proceedings of an International Symposium on Microbiological Tests for the International Exchange of Animal Genetic Material" (Madison, WI: American Association of Veterinary Laboratory Diagnosticians, 1984).
45. Stalheim, O.H.V., Bartlett, D.E., Carbrey, E.H., Knutson, W.W., Landford, E.V., and Seigfried, L., "Recommended Procedures for the Microbiologic Examination of Semen" (Madison, WI: American Association of Veterinary Laboratory Diagnosticians, 1979).
46. Templeton, A.R., and Read, B., "The Elimination of Inbreeding Depression in a Captive Herd of Speke's Gazelle," *Genetics and Conservation,* C.M. Schonewald-Cox, S.M. Chambers, B. MacBryde, and L. Thomas (eds.) (Menlo Park, CA: Benjamin/Cummings, 1983).
47. Thresher, P., "The Economics of Domesticated Oryx Compared With That of Cattle," *World Animal Review* 36:37-43, 1980.
48. Trail, J.C.M., "Cattle Breed Evaluation Studies by the International Livestock Centre for Africa," *Animal Genetic Resources Information,* January 1983, pp. 17-20.
49. Wallace, B., and Vetukhiv, M., "Adaptive Organization of the Gene Pools of *Drosophila* Populations," *Gold Spring Harbor Symposium on Quantitative Biology* 20:303-309, 1955.
50. Whittingham, D.G., Leibo, S.Y., and Mazar, P.,

"Survival of Mouse Embryos Frozen to −196° and −260° C," *Science* 178:411-414, 1972.
51. Wilmut, I., "Effect of Cooling Rate, Warming Rate, Cryoprotection Agent, and Stage of Development on Survival of Mouse Embryos During Cooling and Thawing," *Life Science II*, Part 2:1071-1079, 1972.
52. Zhang, W., Wu, J.S., and Rempel, W.E., "Some Performance Characteristics of Prolific Breeds of Pigs in China," *Livestock Production Science* 10:59-68, 1983.
53. Zimmerman, D.R. and Cunningham, P.J., "Selection for Ovulation Rate in Swine: Population, Procedures and Ovulation Response," *Journal of Animal Science* 40:61-69, 1975.

Chapter 7
Maintaining Plant Diversity Offsite

CONTENTS

Tables

Figures

Boxes

Chapter 7

Maintaining Plant Diversity Offsite

HIGHLIGHTS

- Seed storage techniques are being used to conserve the genetic diversity of cereals, legumes, and many other important crop species. Some plants, however, do not produce seeds that can be stored. Eventually, this problem may be resolved with techniques for *in vitro* storage of plant tissue, from which whole plants can be regenerated. At present, the main alternative to seed storage is to grow entire specimens in the field.
- New technologies, including cryogenic storage of seeds and clones, use of biochemical methods to characterize accessions, and improved methods to detect pathogens in plant materials transferred internationally, have the potential to increase the cost-effectiveness of maintaining plant diversity offsite. Progress, however, is constrained by a lack of fundamental research on plant physiology, reproductive processes, and the mechanisms of genetic and cellular change.
- Priorities and protocols for collecting and maintaining germplasm of major crop plants are internationally coordinated. However, they are not well-organized for minor crops or for wild plants that are endangered or have economic potential.
- Long-term public and private support for germplasm storage facilities depends on whether the stored materials prove to be valuable. The use of offsite collections can be improved by characterizing the genetic diversity contained in the collections and evaluating collections for important traits.
- Major breakthroughs in biotechnologies might eventually lead to fundamental changes in how biological diversity is maintained. Even so, a large portion of public resources for technology development should be used in improving the application of existing technologies, such as cryogenic storage of germplasm, that are important to society but are not attractive to the private sector to support.

OVERVIEW

One approach to maintaining plant diversity involves collecting samples of agricultural and wild species and storing them in offsite collections. Such collections can assemble agriculturally, geographically, and ecologically diverse plants for use in crop improvement, genetic research, or plant conservation. This chapter assesses technologies of collecting, storing, and using plants offsite.

Objectives of Offsite Collections

Offsite collections of agricultural crops bring together varieties and related species from widely dispersed areas (box 7-A). These collections conserve plant genetic resources threatened with loss or extinction. They also serve as a convenient source of new genes for public and private plant improvement. The highly suc-

Box 7-A.—Definitions Relevant to Offsite Plant Collections

Most offsite plant collections are devoted to assembling a diversity of agricultural species, although collections exist for wild species (69,112). Offsite collections of agricultural crops can be classified according to their principal goals.

- **Base collections** focus on long-term preservation of genetic diversity to ensure against loss of valuable plant germplasm. Base collections often include crops and their wild relatives, typically maintained as seeds in long-term, subfreezing storage. The base collections important to agriculture in the United States are held at the National Seed Storage Laboratory in Fort Collins, Colorado, a part of the National Plant Germplasm System (NPGS) (see ch. 9).
- **Active collections** provide public and private plant breeders and researchers access for use in crop improvement, pharmaceutical studies, taxonomic investigations, or genetic research. Long-term security for plants in active collections is provided by duplicating their holdings in base collections. Within the NPGS, the Regional Plant Introduction Stations maintain active collections for numerous crops.
- **Genetic stock collections** constitute an array of plants with one or more unique genetic characteristics. These collections provide references for gene nomenclature and genetic mapping studies. Leading research institutions maintain many of these collections.
- **Working collections** are held by public and private plant breeders and contain many interrelated lines that serve crop improvement activities. Working collections are commonly obtained from active collections when new characteristics (e.g., resistance to a new disease) are needed for crop improvement. The genetic diversity in working collections is frequently limited to the specific goals of an individual breeding or research program.

Many of these collections, regardless of their specific goals, include a wide variety of accessions. Accessions in offsite plant collections can be classed as wild species, landraces, breeding lines, agricultural varieties, and engineered lines (18,46,52,103,115,116):

- **Wild species** related to agricultural crops are of particular interest because they may provide new and valuable gene complexes. Others may prove to be valuable sources for industrial, forest, and pharmaceutical products. But the bulk of wild species have no immediately obvious use. However, offsite maintenance becomes important when such species are endangered in their natural environment.
- **Landraces** have developed through generations by farmers who selected plants with the characteristics they desired. They represent populations that can be genetically diverse and are often specifically adapted to local environments and are considered to be valuable sources for the genetic traits that have enabled them to survive. However, a landrace may be productive in one region, but because it is so highly adapted, it may not grow well in another.
- **Breeding lines** are intermediate forms produced in the process of developing new agricultural varieties. Although saving all such lines would be impractical, some have permanent value to breeders (46). Probably most valuable of the breeding lines are those from which modern agricultural hybrids are produced.
- **Agricultural varieties** (cultivated varieties or cultivars) can include those under cultivation as well as those no longer in popular use. These generally are the products of breeding and development programs. If still widely being cultivated, agricultural varieties are often not collected or stored, because they are considered easily available.
- **Engineered lines** are the products of modern biotechnology. These are plants produced by methods of cell or tissue selection in culture, culture-induced mutation, or genetic engineering. Such materials are generally of more concern to scientists than breeders. The promise of transferring traits, such as nitrogen fixation or improved protein quality, may make such varieties important. At present, though, engineered lines are of minor concern to offsite collections.

cessful rice variety IR36 developed by the International Rice Research Institute resulted from the crossbreeding of 13 accessions in their collections of rice from the United States and several Asian countries (79). Wild plant collections help to preserve endangered species, supply materials to restore degraded lands, and provide material for genetic improvement of crops.

Botanic gardens and arboretums are the primary repositories for wild plant species (69, 112). Arboretums have been particularly important for maintaining individual trees and shrubs that may have little commercial significance (69). These facilities may also have commitments to public education and display that can result in selecting plants with special or unusual characteristics rather than those representing the genetic diversity within the species. However, many such institutions are now giving greater attention to the potential contributions they can make to maintaining plant diversity (12,34).

Considerations in Selecting Technologies

No single technology is appropriate for all the plants stored in offsite collections. Several considerations affect the selection of appropriate technologies such as biological limitations of the species, reliability of the technology, and cost.

Biological Limitations

Seeds are the most commonly and easily stored propagules of plants. When placed in conditions that reduce their moisture content to approximately 5 to 6 percent, seeds of many species will remain viable for years. Lowered temperatures can further extend storage life. Seeds able to withstand reduction in moisture and temperature are called ***orthodox*** seeds. Most of the major food crops (e.g., cereals and legumes) have orthodox seeds, and many, when properly dried, withstand cooling to -196° C, the temperature used in cryopreservation (58, 85,89,100,108).

Seeds that cannot survive a reduction in moisture content are called ***recalcitrant*** seeds. Recalcitrant seeds are found in many important tropical species, a few temperate tree species, and some aquatic plants (8,42,83,100,108,118).

Reducing the water content of recalcitrant seeds severely shortens their life span. Thus, they cannot be stored like orthodox seeds: cooling to subfreezing temperatures would lead to

Photo credit: International Board for Plant Genetic Resources

Many temperate and tropical species such as coffee and oil palm have seeds that cannot be stored for long periods; for this reason, collection of woody cuttings is preferred.

formation of ice, resulting in damaged cells and death of seeds. Instead, plants with recalcitrant seeds are commonly stored as field collections. Research on the physiology of recalcitrant seeds may lead to methods for long-term seed maintenance of these species. *In vitro* plantlet or embryo culture, coupled with cryogenic storage, may also eventually become useful for these species.

Two other biological limitations may restrict the maintenance of some plants. First, the qualities that distinguish a particular variety (e.g., flower color and shape in roses; or the color, flavor, and texture of a peach) may not be preserved in plants grown from seeds. For many fruit and nut varieties, retention of these specific qualities is only possible through clonal propagation, which entails producing plants from cuttings or by grafting. Second, some plants do not produce seeds readily because of inappropriate environmental conditions, physiological barriers, or genetic inabilities. In some cases, such as many varieties of banana or the tropical yams (Dioscorea), plants are sterile and seeds cannot be obtained. Basic studies of physiology are needed to improve understanding of the processes controlling flowering and seed production for both cultivated and wild plant species.

Reliability

The reliability of technology refers to both the potential for loss (by natural causes, accident, or equipment failure) and the likelihood of genetic alteration during storage.

The potential for loss by natural causes is higher in field collections than in seed collections. Pests, diseases, and environmental conditions can decimate field collections. Therefore, collections should be duplicated in different locations to ensure against loss. Greenhouses or other controlled environments may reduce the potential for loss by reducing environmental exposure. Research of *in vitro* culture may lead to alternatives to field collections that are free from disease and environmental uncertainties.

Collections are also subject to equipment failure. As a backup measure, mechanical refrigeration compressors should have alternative power systems to prevent warming, which may adversely affect the viability of stored seeds (100). Cryogenic techniques do not rely on external power sources or mechanical cooling systems. And though containers may develop leaks, the risks to security are considered much less than for mechanical refrigeration (100,101).

Some novel approaches to reducing dependence on mechanical cooling systems are being tested. The Nordic Gene Bank in Sweden recently established a long-term storage facility in old mines dug into the permafrost (125). The Polish Government proposed establishing a world seed collection in Antarctic ice caves, but this approach raises questions about storage temperatures and about ease of access and political control (55). An approach being developed in Argentina is to use the cold nights of mountain environments to cool a specially constructed storage vault (55). These options, while interesting, are suitable for only certain countries and are still experimental.

Genetic stability of plants can be affected in several ways. Mutations in orthodox seeds may increase as samples lose their ability to germinate (84,108). Growing out samples subjects them to conditions that will select against certain individuals and thus reduce genetic diversity in the sample (88). Cryogenic storage can improve genetic stability by slowing viability loss and lengthening the time needed between regenerations. For *in vitro* cultures, genetic mutation becomes a concern when the tissues are growing as unorganized calli. Cryogenic storage of such cultures could suppress mutation by arresting growth, but more research on plant development and cryopreservation of *in vitro* cultures is still needed (90,108).

Costs

A final consideration in the selection of technologies is cost. In seed storage facilities, expenses are incurred with monitoring viability and regenerating samples. Mechanical refrigeration systems can be expensive because of continuous energy costs. Cryogenic storage can lower long-term costs by reducing the frequency of viability monitoring and regeneration. However, cryogenic storage is economical for only certain plant species; larger seeds, such as beans, may be stored more economically under mechanical refrigeration (102).

In vitro cultures may be more economical than extensive field collections, particularly if the cultures are stored cryogenically. However, further research and development is needed on both *in vitro* culture and cryogenic storage.

COLLECTING SAMPLES

Collecting samples involves the development of strategies as well as the actual collection of plants. Developing strategies can be facilitated by analyzing plants already in storage. Ideally, a strategy would provide for the collection of all genetic variants of a species without redundancy (37) (box 7-B).

Strategies

Considerations of economic or esthetic importance, rarity, degree of endangerment, access, genetic diversity, and similarity to plants already stored offsite can all influence the setting of strategies.

For the major agricultural species, collection strategies have been established by considering the data on plants already collected, geographic distribution of crop species, particular needs of breeders, and the economic or social importance of the crop (117). The International Board for Plant Genetic Resources has established a system of priority ratings to guide collectors: Priority 1 crops are those with most urgent global collection needs; Priorities 2, 3, and 4 indicate descending orders of urgency (51,73).

Strategies are less clearly established for the collection of most wild species (66). In general, those threatened in their natural environment or of display value have received greater attention (69,112). A formal system for coordinating conservation activities has only recently been established (34,112). Botanic gardens, commercial institutions, and private collections have been growing and propagating rare plants for years but without concern for obtaining a range of genetic diversity (69). Organizations such as the Botanic Gardens Conservation Coordinating Body of the International Union for the Conservation of Nature and Natural Resources (IUCN) and the Center for Plant Conservation at the Arnold Arboretum of Harvard University are beginning to focus the expertise and resources of botanic gardens, arboretums, and private collectors to improve offsite maintenance of wild plants.

Selecting Sites for Collecting

Studies of the geographic distribution of plants, the experiences of scientists and plant collectors, computer-based models, and re-

Photo credit: M. O'Grady

Collecting natural rubber (*Hevea*) in the Amazon forest. Natural rubber still has many industrial uses for which synthetic rubber cannot be employed.

Box 7-B.—The History of Plant Collection

Humans have collected plants and maintained them in collections for centuries (69,79,124). One of the earliest documented collecting expeditions was sent by Egyptian Pharaoh Hatshepsut in 1500 B.C. to obtain samples of the cedar tree. In 1492, Christopher Columbus returned to Europe with seeds of a new crop: corn. Later he took European wheat to the West Indies. As a result of such activities, the center of production of the world's agricultural crops today may be very distant from the places where the crops originated. Wheat, for example, is now a major U.S. crop for some European customers.

Botanic gardens were early repositories for the plants collected by explorers and grew in importance in the Middle Ages with the proliferation of pleasure gardens in Europe and the Middle East. During the colonial period of the 16th to 19th centuries, botanic gardens were established throughout the tropics. The palm, an important source of oil in Asia, was introduced in the early 19th century to Indonesia and Malaysia through botanic gardens in Java and Singapore. The Dutch East India Co. supplied trading ships with a variety of fresh fruits and vegetables from a botanic garden in its Capetown colony at the end of the 17th century.

The first botanic garden in the New World was established in 1766 by the British to acclimatize South Sea Island crops on St. Vincent (79). Captain William Bligh sought to introduce varieties of the starchy Tahitian staple, breadfruit, to British colonies. His first attempt ended in a well-known mutiny, but a later voyage in 1793 succeeded, and he brought six varieties of breadfruit to the botanic garden in St. Vincent. Though first rejected by the local population, breadfruit eventually became a regular part of Carribean diets. The garden at St. Vincent was instrumental in the introduction of many other important crops and spices to the Americas.

Medicine provided an important impetus to developing collections at many botanic gardens (79). These gardens were essential units within medical schools and similar institutions. The chemicals in many of these plants formed the basis for the pharmaceuticals industry, and some remain important sources for modern medicines (110).

Unlike many gardens of the 18th century, the Royal Botanic Gardens at Kew, England maintained a variety of plants, not just those of medical importance (79). Scientists at Kew developed a collection that today exceeds 50,000 species from throughout the world. Kew has been involved in the introduction of many agricultural crops important to Europe and North America, such as the tomato, potato, and rubber tree.

Introducing new crops became an official activity of the U.S. Government in 1819, when the Secretary of the Treasury enlisted the help of foreign diplomats and U.S. Navy personnel to collect plants from abroad. Prompted initially by the desire to introduce new plants and later by concern over loss of crop germplasm, the Federal Government instituted various national systems to collect, describe, maintain, evaluate, and distribute plant germplasm. These activities evolved to what is today the National Plant Germplasm System (115).

search on the origins of plants have enabled scientists to locate regions rich in diversity of crop species. The regions where most of the major crop species were originally domesticated and developed are known as centers of diversity, after a scheme first proposed by botanist N.I. Vavilov of the Soviet Union (45,57). At least 12 centers of diversity are now recognized (figure 7-1). Primitive varieties and related wild species that are able to survive in diverse habitats and resist a variety of crop-specific diseases can be located in these areas.

Some guidelines are available for collecting a genetically diverse sample (12,47). For instance, plants growing in areas of different soil, water, or light conditions may represent types that have genetic adaptations and could prove to be valuable sources of particular genetic traits. But guidelines for collecting seeds or

Photo credit: Centro Internacional de la Papa (CIP)

Collecting wild potato species in South America, a center of potato diversity.

other propagation materials (cuttings, tubers, etc.) are only general and may be altered by specific conditions in the field. For example, the Central America and Mexico Coniferous Resources Cooperative (CAMCORE) has established guidelines for environmental and geographic factors to consider, depending on the number of trees to collect from, and the amount of seed to be collected (25,26). But if collectors encounter small populations of tree species, collection is made from any tree with seed (26). Thus, the guidelines for a collecting expedition depend heavily on the expertise and judgment of the individual collector.

Scientists and experienced collectors also provide helpful information on collecting sites. Several of the large so-called genetic stock collections of crop germplasm in the United States, such as the one for tomato at the University of California-Davis, are overseen by a few individuals with special interests in that crop. The knowledge these people have about origins and distribution of a crop is frequently the result of extensive observations and field collecting experiences.

Computer-based modeling is another potentially useful tool for predicting the location of sites appropriate for collecting. Data from a few key locations may provide information on the distribution of a particular crop trait, such as drought tolerance. A map can then be constructed by computer-based extrapolation of neighboring regions. Areas likely to contain plants with similar characteristics could be selected (2). However, this technology has its limitations. This kind of analysis requires precise data on latitude, longitude, and elevation for collected plants, for example—information that is not currently available in most crop databases (2). And because overall geographic information comes from satellite imagery, it can be prohibitively expensive (2). Political or other (e.g., geographic) restrictions on collecting in some areas may also make acquisition of plant samples difficult. Finally, data for the initial profile are obtained from sites chosen by statistical analysis, and plant distribution may not parallel these mathematically chosen sites. Refinements in existing databases, collection information, and artificial intelligence systems may someday allow such models to assist in collecting. However, it seems the importance of using existing data for such tasks has been overlooked (46).

Quarantine

After samples have been collected, complications may arise in transporting them. Movement of plants from one region to another always carries some risk that pests (nematodes, snails, insects, etc.) or pathogens (viruses, bacteria, or fungi) will also be transported (14,59,

Figure 7-1.—Regions of t'-• World Where Major Food Crops Were Domesticated

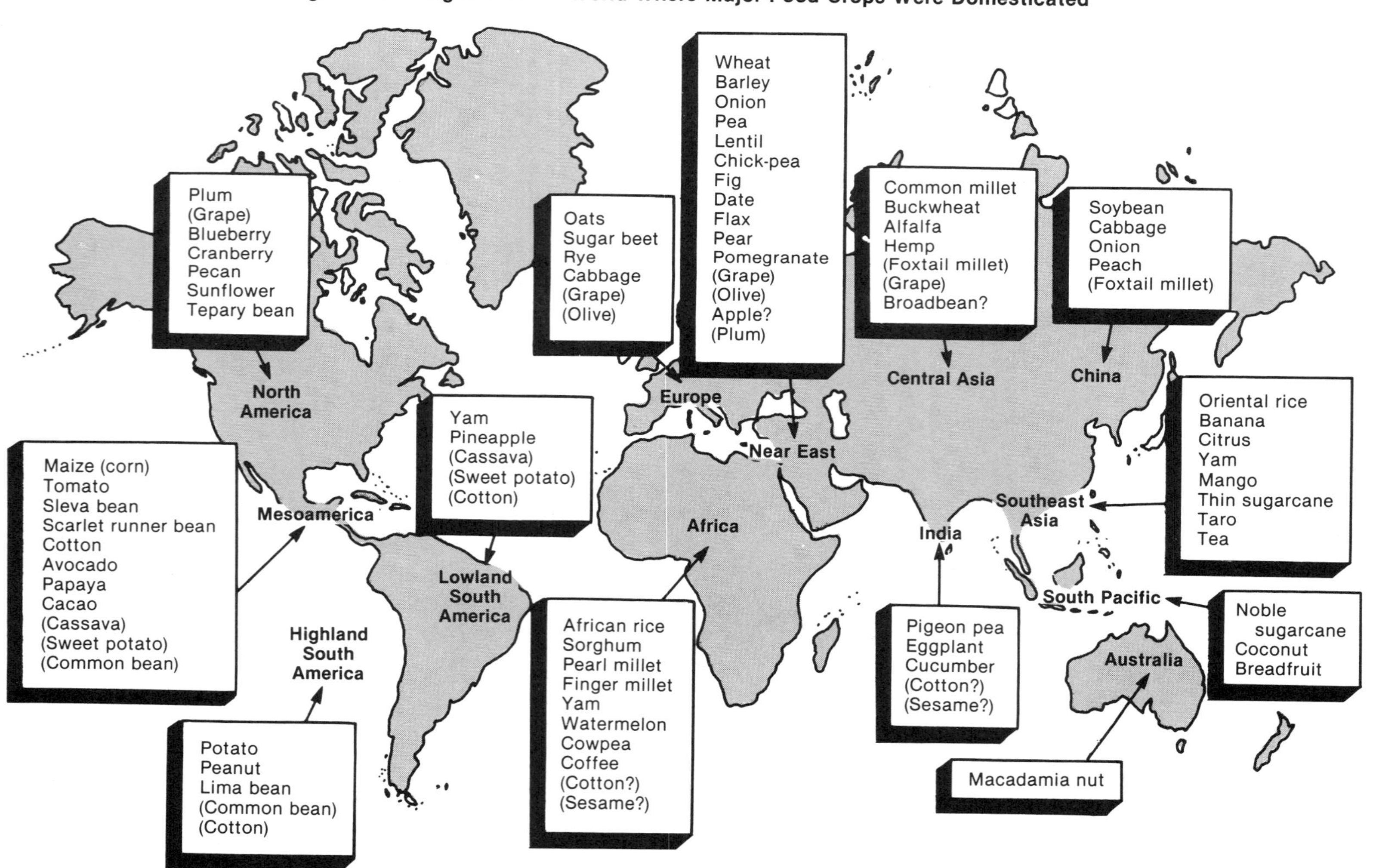

Crops that apparently originated in several different areas are shown in parentheses. A question mark after the name indicates doubt about the location of origin.

SOURCE: J.R. Harlan, "The Plants and Animals That Nourish Man," *Scientific American* 235(3):88-97, September 1986.

60). Plants destined for offsite collections, particularly those from centers of diversity, can present particular quarantine concerns. These centers possess not only considerable crop diversity but also widely adapted crop pests and pathogens. An area where coffee and the disease coffee-rust coexist, for example, would be a likely place to obtain plants with rust resistance genes, but it could also have pathogens that have adapted to coffee plants (60).

The presence of most exotic pests can be determined by inspection or by treatment of plants upon entry, but some imported plants may be detained while tested for obscure pathogens. Such testing requires well-equipped laboratories and personnel as well as considerable time. This last constraint—5 or more years for the detection of certain viruses and virus-like organisms in woody plants—can profoundly affect importation of some plants (60).

Quarantine and Plant Importation

Establishing quarantine policies and practices for a particular plant species depends on both knowledge of its risk of carrying pests or pathogens and availability of technologies to detect such pathogens (table 7-1). Most plant species, when imported according to regulations (e.g., clean and free of soil, and subject to inspection at an authorized port of entry), are considered unlikely to be carrying harmful organisms and thus to be of low risk (60).

Some agricultural crops, such as rice, sorghum, or sugarcane and their related wild species, require greater attention because they might contain pathogens not easily detected by current technologies.

Certain agricultural plants or plant parts used for vegetative propagation (e.g., sugarcane stems or potato tubers) represent the greatest risk to agriculture because they may be infected with undetected pathogens (60). The U.S. Department of Agriculture (USDA) allows small quantities of these plants to be imported for scientific use only. Permits typically require that plants be grown under the supervision of a knowledgeable specialist and may require diagnostic testing for pathogens as well as specialized growing practices (60). Once plants have cleared safeguard restrictions, they may be distributed to the general public.

In the United States, some plants (e.g., apples, pears, and potatoes) are held at one of the Agricultural Research Service's Plant Protection and Quarantine facilities until they are considered free of any pests or pathogens (60). Plants in this group face the most constraints because the hazards associated with importing them are highest (60). These plants may be held for several years after their original importation.

In developing countries, plant quarantine systems may lack scientific expertise, facilities, or appropriate governmental infrastructures to support them (14). Therefore, they depend heavily on such regulatory constraints as import refusal, lengthy quarantine, or treatment for pests or pathogens. These restrictions can result in considerable delay in importation of plants to facilities in these areas.

Safeguards for Reducing Risk in Imported Germplasm

A number of actions and regulations, either at the place of origin or at the port of entry, reduce the risks associated with plant importation.

Inspection, certification, testing, or treatment of plants before export can reduce potential quarantine delays (60). Most plants require little more than inspection to move quickly through quarantine. *In vitro* plantlet cultures can, in some cases, be imported with fewer restrictions than the plants from which they originate. Such cultures, though, are not considered free of disease without diagnostic testing (60).

Upon entry, plants likely to contain pathogens can be tested with a variety of technologies (table 7-2) (59,60). However, procedures vary in reliability and in the resources they require. Indexing, which uses highly sensitive indicator plants, is the most reliable and widely used method, but it requires considerable greenhouse space to maintain the plants necessary for this test (60). Serologic methods that use an-

Table 7-1.—Crops and Trees Commonly Prohibited Entry by Quarantine Regulations in 125 Countries

Crops and trees	Number of countries in which crops/genera are prohibited	Percentages of countries that name one or more pests or pathogens	Percentages of countries prohibiting: Plants only[a]	Plants and seeds	Seeds only
Forest crops:					
Maple	14	43	100	0	0
Chestnut	34	23	76	24	0
Conifers[b]	27	26	100	0	0
Hawthorn	14	86	100	0	0
Walnut	21	48	100	0	0
Poplar	27	44	93	7	0
Oak	25	47	92	8	0
Willow	22	45	100	0	0
Ash	24	58	96	4	0
Elm	32	47	94	16	0
Fruit crops:					
Citrus	62	45	55	45	0
Coconut	28	32	29	64	7
Strawberry	20	55	65	35	0
Banana	39	39	54	46	0
Pome fruits[c]	37	68	85	15	0
Prunus (cherry, plum, etc.)	37	68	85	15	0
Currant	16	38	69	31	0
Grapevine	41	41	90	10	0
Vegetable crops:					
Sweet potato	23	35	61	39	0
Potato	41	41	90	10	0
Other crops:					
Coffee	49	31	24	57	18
Cotton	52	23	25	61	14
Sunflower	15	40	20	80	0
Rubber	28	50	29	71	0
Tobacco	26	31	35	58	7
Oil palm	16	38	56	44	0
Rice	33	42	21	61	18
Rose	22	41	100	0	0
Cacao	43	42	19	79	2
Tea	20	45	45	55	0
Sugarcane	40	10	63	37	0

[a]Includes plants as well as any parts for vegetative propagation.
[b]Specifically, the genera *Picea, Larix, Pinus,* and *Abies.*
[c]Includes the genera *Chaenomeles, Cydonia, Malus,* and *Pyrus.*

SOURCE: R.P. Kahn, "Technologies To Maintain Biological Diversity: Assessment of Plant Quarantine Practices," OTA commissioned paper, 1985.

tibodies to pathogens provide rapid results but may not detect all forms of a particular pathogen (60). Molecular techniques to detect the genetic material of pathogens are available, but these may require better-equipped laboratories and greater expertise than is available in many quarantine programs. Identifying the presence of a pathogen can thus be difficult because a negative result may be due to inadequate technology. It is essential, therefore, that the limits of any technology be understood. Basic research on the biology of pathogenic organisms and the technologies used to detect them is needed to improve testing procedures, develop them for other pests and pathogens, and understand the limits.

STORING SAMPLES

Storage technologies aim to preserve an adequate amount of plant germplasm, sustain its viability, and preserve its original genetic constitution (81). Plants can be maintained offsite in a number of forms and with a number of technologies (table 7-3). They may be maintained

Table 7-2.—Technologies and Practices To Detect Pests and Diseases of Quarantine Significance and the Pathogens To Which They Are Most Frequently Applied

Technology/practice: Description	Range[a]
Physical examination:	
Physical manifestations of disease-producing agents	A
Seed health testing:	
Germinating seed in culture conditions that allow growth of fungi or bacteria. Microscopic examination of seed for pathogens	B,F
Grow-out testing:	
Growing plants under controlled conditions until diseases are no longer detected	B,F,V,O
Indexing:	
Attempted transfer of pathogens from a plant under examination to another species that is highly sensitive to infection by them. Can involve transfer by mechanical abrasion with extracts or by grafting	B,F,V
Electron microscopy:	
Examination of extracts or tissues for the presence of pathogens or their spores	B,V
Inclusion bodies:	
Light microscopic examination of tissues for structures characteristic of pathogen infection	B,F,V
Serologic testing:	
An array of procedures utilizing the ability of test animals to produce antibodies that are highly specific for a particular pathogen. Important variations include enzyme-linked immunosorbent assay, radioimmunosorbent assay, and immunosorbent electron microscopy	B,V,O
Polyacrylamide gel electrophoresis:	
Detects ribonucleic acid (RNA) of pathogens in small amounts of tissue	B,V,O
Nucleic acid hybridization:	
New technology that uses recombinant DNA procedures to locate the genetic material of a pathogen in DNA extracted from tissue samples	B,V,O

[a]A = Most pests and pathogens; B = bacteria; F = fungi; V = viruses; O = others (e.g., mycoplasmas, viroids).

SOURCE: Based on data from R.P. Kahn, "Technologies To Maintain Biological Diversity: Assessment of Plant Quarantine Practices," OTA commissioned paper, 1985.

Table 7-3.—Storage Technologies for Germplasm of Different Plants

		Storage technology[a]				
Plant group	Storage form[b]	Field collections	*In vitro* culture	Cool temperature	Liquid nitrogen	Collection[c]
Cereals and grain legumes	seeds			X	R	B,A
(wheat, corn, barley, rice, soybean)						
Forage legumes and grasses	seeds			X	R	B,A
(alfalfa, orchardgrass, bromegrass, clover)	plants	X	X,R	X		A
Vegetables	seeds			X	R	B,A
(tomato, bean, onion, carrot, lettuce)	plants	X	R	R	R	A
Forest trees	seeds			X	R	B,A
(pines, firs, hardwoods)	plants	X	R	R		B,A
Roots and tubers	seeds			X		B,A
(potato, sweet potato, tropical yam, aroids)	plants	X	X,R	R	R	A
Temperate fruit and nuts	seeds			X	R	B(?),A
(apple, grape, peach, strawberry, raspberry)	plants	X	X,R	R	R	A
Tropical fruit and nuts	seeds			X	R	A
(avocado, banana, date, citrus, papaya, cashew)	plants	X	R	R	R	A
Ornamentals	seeds			X	R	A
(carnation, zinnia, lilac, rhododendron)	plants	X	X,R			A
Oilseeds	seeds			X	R	B,A
(soybean, sunflower, peanut, oilpalm, rape)	plants	X	R	R		A
New crops	seeds			X	R	B,A
(jojoba, amaranth, guayule)	plants	X	X,R			A

[a]X = currently in use; R = under research and development.
[b]Refers to source of materials for storage (e.g., plants are the source of materials for initiating tissue cultures.
[c]B = base collections available; A = active collections available.

SOURCE: Adapted from L. Towill, E. Roos, and P.C. Stanwood, "Plant Germplasm Storage Technologies," OTA commissioned paper, 1985.

Figure 7-2.—Maintenance Process of a Plant Seed Bank

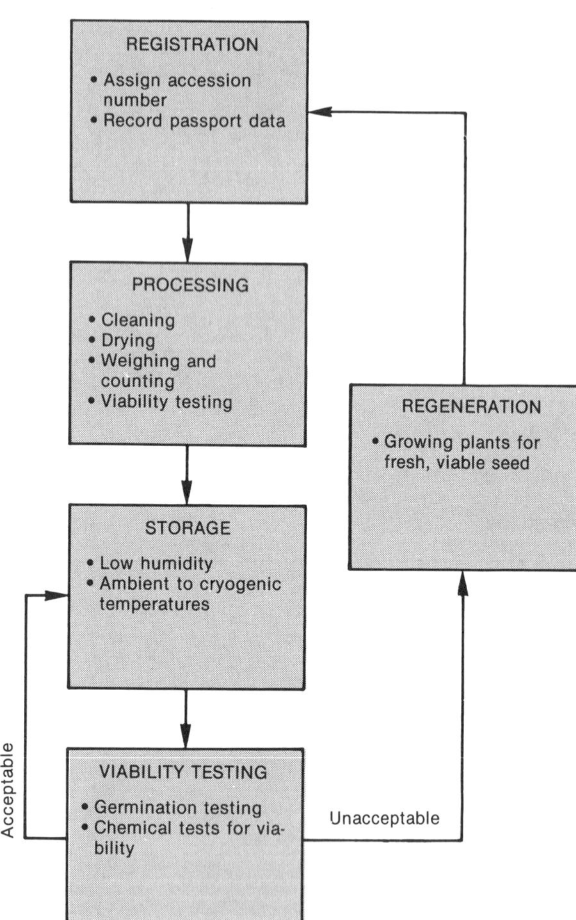

SOURCE: Office of Technology Assessment, 1986.

as seeds, in fields, or in greenhouse collections. Pollen storage and *in vitro* plantlet cultures may supplement storage of many of these species. Finally, cryogenic storage (in liquid nitrogen) and emerging DNA technologies may hold potential for improving maintenance of plants.

Conventional Seed Storage

Most agricultural crops held by the National Plant Germplasm System (NPGS), international centers, national programs, and private collections are maintained as seeds. The process of conventional seed storage can be divided into several steps: registration, processing, storage, viability testing, and regeneration (figure 7-2).

Registration

When seeds arrive at a storage facility, passport information must be recorded and a number assigned to facilitate recordkeeping. Passport data may include information on the origin of the sample, its source (if acquired from another facility), and any pertinent physiological details that would aid storage. Data of interest to potential users, such as disease resistance, also may be included.

The information accumulated may reflect the focus of a particular collection. The Royal Botanic Gardens at Kew, England, requires details on the location and habitat in which a wild plant species was collected, an estimate of the total number of plants represented by the sample, a taxonomic classification, and the location of a reference herbarium specimen. The more detailed this preliminary information is, the more useful the accession is for crop development or conservation.

Collections may receive the accessions of another collection, thus duplicating materials. Although such duplication does provide security against loss, the number of accessions held by all collections does not reflect duplicates. In barley, for example, the total of more than 280,000 accessions in storage is considerably greater than the estimated 50,000 distinct accessions worldwide (70).

Processing

Once registered, other data such as estimates of the number of seeds received, viability in terms of percentage of germination, and taxonomic identification must be obtained. In addition, seeds may require preparation, like cleaning and drying for storage.

Seeds are tested for germinating ability to determine if the sample is of high viability or whether it must be planted to produce fresh seed before storing (29,30,31,43).

To reduce moisture in seeds, procedures using chemical desiccants have been developed and are widely applied (111). Facilities with large amounts of seeds to process, such as the U.S. Plant Introduction Stations or the National

Seed Storage Laboratory (NSSL), use dehumidified rooms to reduce moisture.

Storage

Four factors affect seed storage: 1) moisture content, 2) storage temperature, 3) storage atmosphere, and 4) genetic composition of the sample (4,87,88,89). Reduced moisture content is considered the most crucial to maintaining viability. In general, each 1-percent decrease in seed moisture between the 5- and 14-percent range will double the lifespan of a seed sample. Reduction of storage temperature also increases seed longevity. A 5° decrease in temperature between 0° and 50° C doubles longevity (89). Control of storage atmosphere generally does not provide significant advantages over manipulation of moisture and temperature, particularly when the latter is below freezing. Genetics relate to differences between individual accessions or between individuals in a mixed sample and cannot be altered to increase longevity.

Viability Testing

Seeds must be tested periodically for viability. This information helps determine when an accession needs to be grown-out to produce a fresh sample of seeds. The most obvious test is to germinate a portion of the seeds to estimate the viable percentage.

Viability testing involves placing seeds in appropriate conditions (damp blotter paper, agar medium, etc.) and counting the number of seeds that germinate over a period of time. However, if seeds are dormant, obtaining viability estimates can be difficult. Citrus species, for example, were thought to have died when prepared for conventional storage but were shown instead to be dormant (82). Heating, cooling, lighting, and treatments (e.g., removal or cracking of the seed coat) may be required to overcome dormancy in some species.

Typically, 200 to 400 seeds are required for viability testing (43). However, a sequential approach reduces the number needed for testing. Forty seeds can be tested to determine whether the accession should be regenerated, stored, or whether another 40 seeds are needed (30,42). But a small sample of seeds may need to undergo numerous tests before an answer is reached, which may take longer than testing a single large sample.

Other tests—involving dyes, physiological tests, or biochemical assays—have been developed to determine seed viability (89). The validity of such tests relies on the ability to demonstrate a correlation with actual germination rates. These tests can be useful to determine if dormancy or inappropriate storage conditions are producing misleading results (30). Some, such as the tetrazolium dye test for a range of seeds, or X-ray contrast with heavy metals in tree seeds, have been widely used (30,89). Others, such as enzyme tests, provide information useful for the study of seed physiology but are more expensive and difficult to perform than standard germination tests.

Preserving the genetic variability in seed accessions is a major concern in offsite collections. Many accessions, particularly those of primitive landraces and wild species, are genetically diverse populations and display considerable genetic variation between individuals in a sample. Genetic differences in storage lifespan can mean that the genetics of a population could be altered by decline in viability (86,87,88,108).

Although seeds generally should be regenerated when germination drops 15 percent, practical considerations of labor, space, and time can delay this step (32,43,83,108). One recent report stated that NSSL does not regenerate samples until viability has dropped 40 percent (113). This practice, however, may be based more on lack of resources than on scientific considerations.

Regeneration

Variations in growth requirements for species and even for varieties within a species complicate the growing-out of seed. In beans, for example, different accessions may require different conditions, e.g., daylength, for growing out. Thus, both subtropical and temperate sites must be used to grow-out a range of bean

Photo credit: OTA staff

Conventional seed storage. Seeds are stored in airtight containers (top photo), then placed in drawers (bottom left) in refrigerated rooms (bottom right). Many storage facilities increasingly use laminated foil envelopes instead of cans.

varieties. Other factors such as control of pollination can also be important for regenerating certain crops (95,108). Wind-pollinated accessions can readily cross with others, and thus, individual accessions must be grown in widely separated fields to ensure that they are not genetically mixed.

Genetic loss by natural selection during grow-out may be undetected in regenerated seed. At NSSL, the designated grower is responsible for ensuring that the sample returned is from plants grown under conditions that would minimize genetic loss. No testing beyond visual examination and a viability test of the returned sample is conducted.

Grow-outs are expensive in terms of facilities and personnel, and they subject stored materials to damage from pests, pathogens, and environmental conditions, which may reduce genetic diversity in an accession. But the most effective and least expensive way to maintain diversity is to reduce the frequency of regeneration through technologies that extend storage.

Cryogenic Storage of Seeds

A critical factor in cryogenic storage is the amount of water in the tissue to be frozen. Most orthodox seeds can be easily stored at cryogenic temperatures because their water content is low enough to avoid damage associated with freezing (99,100,102,108).

Cryogenic technologies may be able to extend the storage life of orthodox seeds to more than a century, which would greatly reduce the need for viability testing and regeneration (100,102, 121,122).

However, limitations on cryogenic storage exist for some species depending on a plant's seed coat, oil or moisture content, and seed size (108). Some plants, such as plums and coffee, have orthodox seeds that tolerate low moisture levels but are sensitive to cooling below −40° C (100). If cooled or warmed incorrectly, many seeds can crack (100). Seeds as big as cotton seeds (about eight seeds per gram) are appropriate for cryogenic storage (99,100,102). Larger seeds, such as beans, can also be frozen, but increased costs, due in part to the greater amount of space required, may reduce or eliminate potential cost-savings over mechanical refrigeration (102).

Photo credit: OTA staff

Cryogenic storage, applicable to seeds of some plant species, can reduce dependency on mechanical systems and greatly increase storage times.

This method could hold considerable cost advantages with regard to operating a seed bank and regenerating seed (table 7-4) (102). Cryogenic storage facilities will cost about the same to establish, but operation over time would be cheaper, in part due to reduced need for viability testing and grow-out. Investment in a facility to produce liquid nitrogen might be necessary in some areas, but the operational savings in the seed bank could allow recovery of costs in 6 to 14 years (99).

Major obstacles to this technology are the lack of appropriate facilities and scientific exper-

Table 7-4.—Estimated Costs[a] of Conventional and Cryogenic Storage

Source	Storage for 100 years	
	Conventional[b]	Cryogenic
Storage:		
Includes equipment, supplies, and replacement of equipment	$ 5.30	$ 5.00
Operations:		
Includes utilities, equipment maintenance, liquid nitrogen coolant, and monitoring of viability (every 5 years for mechanical; every 50 years for cryogenic	60.00	11.80
Seed replacement:		
From regenerating when viability or sample size declines (four times for mechanical; one time for cryogenic)	100.00	25.00
Total 100-year cost	$165.30	$41.80
Average yearly cost	$ 1.65	$ 0.42

[a]Costs for accession of onion (a species that survives poorly under conventional storage). Savings for other crop species—particularly those with large seeds—may be less dramatic or nonexistent.
[b]Assumes storage conditions of −18° C and seed moisture of 4 to 7 percent, under which storage life of onion seed is approximately 25 years.

SOURCE: P.C. Stanwood and L.N. Bass, "Seed Germplasm Preservation Using Liquid Nitrogen," *Seed Science and Technology* 9:423-437, 1981.

tise at many locations, particularly in developing countries, and the lack of scientific data on genetic stability of seeds stored cryogenically. Certainly the capacity to use cryogenic technologies should be part of any newly constructed facility for seed storage.

Field Maintenance and Controlled Environments

Accessions may also be stored as vegetative plants in field collections or controlled environments e.g., greenhouses (108). This approach may be necessitated by physiological restrictions on storing seed, by the need to preserve particular combinations of characters or by inabilities to obtain satisfactory seed samples (81,108). Field collections can preserve the genetic diversity of many aquatic plants; tropical species (e.g., coconut, cacao, mango, or rubber trees); tropical forest trees; and some temperate trees (e.g., oaks)—which all have recalcitrant seeds (28,42,63,64,84).

Botanic gardens and arboretums maintain diverse field collections, though many institutes focus on a narrow taxonomic group, as mentioned earlier. Arboretums conserve limited samples of tree and shrub species with very small natural gene pools that are under pressure of destruction, or plants with distinctive characteristics (34,69,80). In the United States, establishing a network among botanic gardens and arboretums, facilitated by the newly formed Center for Plant Conservation at the Arnold Arboretum at Harvard University, could allow a division of labor and sharing of expertise that would enable more species and more genetic diversity within species to be maintained (34, 106).

Trees in field collections, however, may have been selected for economically important traits and thus may only represent a narrow range of the total diversity available for a species. CAMCORE, for example, collects seeds only from coniferous trees with commercially valuable trunk characteristics (i.e., tall and straight) (25). Trees in field collections, nonetheless, can be useful sources of seeds for restoration and reforestation projects (8).

Many clonally propagated crops are maintained in field collections. Clonally propagated crops include fruit and nut species; many ornamentals, such as roses; and some root and tuber crops important to developing countries (e.g., sweet potato, cassava, and taro). Seeds may be available for many varieties. However, most of these crops are genetically heterogeneous, and clones grown from their seeds may not retain the particular qualities of the parent plants (e.g., the seeds of a Macintosh apple do not produce Macintosh apple trees, but rather a range of trees that result from recombination of the genes in the Macintosh apple). The Centro Internacional de la Papa (CIP) in Peru, for instance, maintains an active field collection of potato landraces but also has a base collection of seeds from these accessions (50).

Pollen Storage

Pollen is not a conventional form of germplasm storage, but information is available on preserving pollen for breeding purposes for

Photo credit: International Board for Plant Genetic Resources

Current practice for conserving short-lived, clonally propagated materials is to grow individual accessions as field collections every year as with these potatoes. However, this approach is expensive in terms of space and labor, and it has the disadvantage of possible losses due to pests and diseases and to genetic change.

many species, particularly for crossing materials that flower at different times (107,108). A population of pollen grains collected from genetically different individuals would contain the nuclear genes; cytoplasmic (nonnuclear) genetic factors would not be transmitted, however, because these are not inherited through the pollen (108).

Pollen can be separated into types that are tolerant or intolerant of drying (81,107,108). Tolerant types store best when dried and maintained at low temperatures, much like orthodox seeds. But the pollen of many species does not survive low moisture or freezing temperatures. Some intolerant types, notably maize, have been successfully preserved in liquid nitrogen, but data on success are sparse (3,107, 108).

Considerable information is still needed on stability and longevity of storing pollen, however, before its use in storage will be possible (108). Pollen is undesirable as the sole propagule for base collection storage because whole plants cannot generally be obtained from it (81,108). In addition, pollen storage does not circumvent potential plant health problems because some pathogens are pollen-borne.

Biotechnology

Biotechnology provides additional opportunities to improve offsite maintenance of plants. Of particular relevance are *in vitro* cultures of plants that are now maintained in field collections. And developments in genetic engineering may make the storage of isolated DNA practical in the future.

In vitro Culture

In vitro cultures of plants have been advocated for a variety of species, especially those that are clonally propagated (20,53,56). Although this technology can be adapted to many species (16,33,93), it is generally unnecessary for those that have orthodox seeds. However, the methods may be necessary if there is a need to maintain specific genetic types, if seed progeny are highly variable, if plants have long juvenile stages (e.g., many tree species), or if seeds are lacking (e.g., clonal crops such as banana, taro, and sugarcane) (20).

In vitro maintenance is defined as the growing of cells, tissues, organs, or plantlets in glass or plastic vessels under sterile conditions (108). When plants originate from intact isolated meristems, the cultures may be free of pathogens (108). The media for growing *in vitro* cultures may vary among species and among individuals within a species. By altering the balance of nutrients and growth regulators in the media, *in vitro* cultures can be made to develop unorganized growth (termed callus), produce multiple shoots or plantlets, form structures similar to the embryo in a seed, or develop roots to enable transfer to field conditions. Not all plants, however, are amenable to growth or manipulation by *in vitro* culture (108).

One aspect of *in vitro* technology of particular concern for plant germplasm conservation is the occurrence of genetic modification (somaclonal variation) in plants derived from callus cultures (67,90). Such variation is considered useful in the development of new varietal characteristics but is unacceptable when preservation of specific genotypes is the objective. Although it is known that certain types of cultures and conditions, such as callus cultures, can produce higher frequencies of somaclonal variation, the cellular processes that produce them are not well understood (90,120). Furthermore, growing cultured plants to maturity remains the only satisfactory way to examine the consequences of such changes. Consequently, each method of culture must be carefully evaluated before it is applied. For germplasm preservation, *in vitro* plants directly derived from buds or shoot-tips are considered most suitable.

Photo credit: International Board for Plant Genetic Resources

In vitro culture could become an important method of long-term maintenance for plants with seeds that cannot be stored under dry, cold conditions and for plants that can only be maintained in field collections.

No *in vitro* long-term base collections of agricultural crops exist at present, although some active collections are being developed: potato at CIP, cassava at the International Center for Tropical Agriculture (CIAT) in Colombia, and yam and sweet potato at the International Institute for Tropical Agriculture in Nigeria (122, 123). The NPGS Clonal Repositories are investigating *in vitro* cultures as backup to field collections of some crops (108).

In vitro cultures can be stored under normal growth conditions, in reduced temperatures or in a medium that inhibits growth (53,56,119, 120,122). Cultures can thus be maintained for weeks to months without subculture (i.e., transfer to fresh medium). However, all treatments that retard growth put additional stress on the culture, which may increase the potential for somaclonal variants.

In vitro culture techniques could be important for the long-term maintenance of plants with recalcitrant seeds. One recent proposal has been to excise the embryo from the seed and store it under cryogenic conditions. The embryo could be thawed and then grown and multiplied *in vitro* (40). Research in cooperation with the Royal Botanic Gardens at Kew, England, has demonstrated the feasibility of this procedure for two tree species (*Araucaria husteinii* and *Quercus robur*). Further research is needed to apply it to other plants with recalcitrant seeds (40).

Cryogenic storage may help avoid the stresses of continuous *in vitro* culture (49,62,90,121,122, 123). Considerably greater attention would be needed in preparation, freezing, storing, thawing, and subsequent culturing than is the case for orthodox seeds. Although some generalizations can be made, methods acceptable to one species or variety may not be satisfactory for others. However, research has demonstrated that *in vitro*-cultured shoot-tips from some herbaceous plants (e.g., potato, carnation, and cassava); berries (e.g., strawberry, raspberry, and blueberry); and buds of some woody species (e.g., apple) can survive cryogenic storage (108).

Many questions remain before cryogenic storage of *in vitro* cultures is widely applied. Among these is whether a single procedure can be developed that works well for an array of plants. Further, it is not yet understood how the process of freezing and thawing affects regeneration of cultures or their genetic constitution (108,121). Additional investigation for individual crops is needed, and current technologies have not yet been adapted for handling the large numbers of specimens that might be expected in an offsite facility.

DNA Storage

Future storage technologies may include, as a supplemental strategy, the preservation of the isolated genetic information (DNA and RNA) of plants. Existing technologies can locate, excise, and reinsert genes. In some cases, these genes retain nearly normal function (75,108). A much better understanding of gene structure, function, and regulation is needed, however, before isolated DNA can be used for germplasm storage (76,108).

Management of Stored Materials

Offsite collections of plants must be well managed to guard against loss of materials and to use financial and technological resources most efficiently. Some duplication between collections can prevent catastrophic loss, but excessive redundancy can waste resources. Disease organisms that might be brought into a collection by new accessions need to be managed. Finally, information on the accessions must be easily available both to managers and users.

Duplication of Collections

Duplication of collections provides the best insurance against natural catastrophes, pests, diseases, mechanical failures, or abandonment (81,108). CIP protects its collection of landrace potatoes with field plantings at other locations, with seed storage, and with *in vitro* culture (50). At the NPGS Clonal Repositories (see ch. 9), greenhouse collections back-up field-maintained collections of fruit and nut species. Duplication is equally critical for seed banks, where malfunctioning of a mechanical compressor can result in loss of cooling.

Plants in an offsite collection also can be lost if institutional priorities change, or if the person responsible for the species or collection leaves (108). The situation is particularly critical for older varieties of fruits, berries, and vegetables that may be held only by private individuals or groups (112). For wild species, too, a large collection is frequently the result of the interest of one person or a few individuals. When these efforts cease, a valuable collection can rapidly deteriorate. Information on the focus and extent of various collections can aid coordination of duplication and minimize the potential for loss (34,112).

Assessing Diversity in a Collection

The diversity of a collection can be assessed by collecting morphologic, biochemical, or phytochemical information, frequently called characterization data.

Most characterization data can help distinguish one accession from another but not assess potentially useful traits. This is particularly true for assays of proteins or DNA, which give little indication of such traits as crop yield, disease, or stress resistance.

Morphological Assessment.—Assessing the morphology of an accession is the first step to developing accurate characterization. Morphological information for wild species is important for taxonomic identification and forms the essential baseline from which all other data are related (68). The information on agricultural crops can be used to distinguish individual accessions, as well as identify them taxonomically (114). However, data are on gross appearance and do not reflect the full genetic composition of an accession.

Care must be taken to ensure reliable results when plants are grown-out for morphological assessment (10). Spacing of plants must be adequate to ensure that results do not reflect overcrowding, for example. Samples thought to be duplicates are frequently grown side by side for comparison (13). Biological factors, such as the potential for cross-pollination among accessions, must be taken into consideration.

The major constraints to assessing morphology are adequate space, funds, and trained personnel. Though not technically difficult, such assessments require attention to possible environmental effects and may take a significant amount of time to perform, analyze, and record.

Biochemical Analysis.—Analysis of proteins or DNA using electrophoretic techniques is another way to assess diversity (94). Isoenzymes, the protein products of individual genes, can change in number or chemical structure when the genes for them are altered, and these changes can be detected by an electrophoretic assay. Examination of DNA can allow comparison of the entire genetic composition of accessions.

Isoenzyme analysis on either starch or polyacrylamide gels has probably been the most popular technique for assessing genetic diversity. Surveys of isoenzyme polymorphism have been performed for maize, wheat, tomato, pea, and barley (94,114). In addition, surveys have been done on hundreds of other cultivars and wild species (94,104,105,114).

A potential application of this technology is the development of isoenzyme "fingerprints" to permit reliable identification of specific plant varieties to certify breeding materials, to isolate genetically similar cultivars, or to monitor otherwise undetected genetic changes in accessions. Electrophoretic analysis has been used to detect duplication in some offsite collections, such as the CIP collection of potato germplasm (50) and is increasing in application at NPGS facilities (19). However, since the data are generally restricted to a few biochemical characteristics and do not reflect performance data or the full genetic composition of an accession, such analysis has been considered risky (39).

Two-dimensional electrophoresis is used to separate complex protein mixtures such as those found in seed or leaf extracts so that several hundred proteins can be distinguished in a single gel (9,17,114). The results can be difficult to reproduce, however. The technique requires specialized equipment and may be too lengthy for routine use because only one sample can be evaluated at a time. Managers of offsite collections are unlikely to have the time, expertise, or resources to use this technique routinely.

Restriction fragment length polymorphisms (RFLPs) have been used to directly examine

DNA. DNA is "cut" enzymatically (using restriction endonuclease enzymes) into pieces or restriction fragments that can be separated on electrophoretic gels. Because RFLPs represent the whole genetic composition of the sample, comparing individual analyses within a sample or among accessions would indicate the variability that exists. The techniques, however, are expensive and require technical expertise to execute and interpret. Thus, use of RFLPs appears limited at present to appropriately equipped laboratories (114). RFLPs have been useful in developing detailed genetic maps for use in breeding programs (48) but are not routinely applied to characterization of germplasm.

Phytochemical Analysis.—Phytochemical analysis deals with the distribution and chemistry of organic compounds synthesized by plants (114).

Analysis involves three general processes: extraction, isolation, and identification (44,114). Plant materials are homogenized in aqueous alcohol, then purified by evaporation of the alcohol and chemical partitioning to remove contaminating substances and isolate the chemicals of interest (114). Chemicals can then be identified by chromatographic or spectroscopic techniques (114).

During the past 10 years, the major technological advance in separation and purification of organic chemical mixtures has been the development of high-performance liquid chromatography (HPLC). HPLC is more rapid than other chromatographic procedures and can isolate a range of plant chemicals. Developing appropriate HPLC procedures, however, requires considerable investment of funds and time. Some facilities of NPGS, however, are using techniques such as HPLC to help characterize germplasm (19).

Recent advances in microcomputers have provided sophisticated, low-cost spectrophotometers that can identify plant chemicals (114). Other techniques, such as nuclear magnetic resonance spectroscopy and mass spectrometry, also can determine chemical structure but they require considerable technical expertise and expensive instrumentation. These technologies have been used extensively, however, by scientists studying the taxonomy and systematics of plants (114) and by university and industry scientists interested in developing potential uses for wild plants in medicine and industry.

Controlling Pests and Pathogens in Collections

Managing stored samples requires efforts to ensure that seeds or plants are not lost to pests or pathogens. Because a collection may distribute seeds to other regions, precautions must be taken to reduce the possibility of sending pests or pathogens as well (61).

Stored seed can be severely damaged by rodents, insects, or fungus (58). With rodents, the major damage is not from consumption but rather from the pests scattering and mixing up different accessions (58). Many insect, fungal, and bacterial contaminants can be controlled with the use of chemical fumigants, although such treatments might also harm the seeds (58, 59,81). Sanitary storage facilities that obviate the need for such treatment are therefore preferable (81). When dried seeds are kept at subfreezing temperatures, the potential risk is minimal (58,101).

The risk of disseminating pathogens is considerably greater for crops maintained through clonal propagation (59,61). Some facilities with a specific focus may have greater expertise with a crop and its diseases than a national quarantine facility concerned with all potential introductions. Cooperation between scientists and quarantine officials can improve control of pathogens and aid technology development.

Information Management

Offsite collections are repositories not only of germplasm but also of information. This information can aid collection management, can provide more efficient access to specific accessions, and might help develop collection strategies.

The current focus has been on standardization of terminology in order to facilitate exchange between collections and to provide more consistent information to users (117). Development of uniform crop descriptors that include information about the storage history of an accession as well as data on the original collector, collection site, vegetative and reproductive characteristics, disease or pest susceptibility, and biochemical characteristics (e.g., isoenzyme profiles) is important for consistent and accurate information (7,98,117). This task for NPGS has been assigned to crop advisory committees (see ch. 9).

Several data storage and retrieval methods are now used (65). A collection of only a few hundred accessions might use file cards or books. As the collection grows, a computer-based system may be more appropriate. Large collections, such as the Royal Botanic Gardens in England, have developed systems adapted to their specific needs (97). The nature of the data in computer-based information management systems depends on whether the materials being stored are agricultural crops (1,125) or wild species (34,97).

The Germplasm Resources Information Network (GRIN) of NPGS is an example of a large information system designed to coordinate data from multiple collections in the United States. GRIN, once fully established, is expected to provide information on all accessions held by NPGS. Although the capacity of the system is more than adequate, entering information is, after several years, still in the preliminary stages. OTA has found GRIN praised by managers of NPGS facilities for its recordkeeping operations but criticized by potential users because detailed information is unavailable on individual accessions and obtaining results of searches can take considerable time. GRIN at present does not collect information on privately held collections of agricultural plants, such as those coordinated by the Seed Savers Exchange or the North American Fruit Explorers, nor does it hold information on wild species.

USING PLANTS STORED OFFSITE

Collections are used for crop development as well as conservation. Plant breeders and scientists who may depend on the genetic diversity in such collections require specific information about accessions to select appropriate plants. Genes in selected accessions are incorporated into improved crop varieties using traditional plant breeding practices. In addition, biotechnology may provide methods that could enable development of improved crop varieties or more efficient use of genes in plants.

Evaluation

Evaluation of plant germplasm involves the examination of accessions for the presence and quality of particular traits that may be of use to crop breeders.

In general, evaluation examines traits that may be genetically quantitative (i.e., controlled by many genes) and subject to environmental influence, such as drought tolerance or earliness of maturity in a fruiting crop. This assessment can be complicated in a genetically variable accession because all individuals may not express the trait equally (81). Further, changes in conditions (e.g., appearance of a new disease) can require further evaluation for new traits. In addition, new accessions must be evaluated. Thus, evaluation may be considered a never-ending task (81).

Evaluations vary according to the species or trait being examined and may be both lengthy and complex (37). A test for yield potential, for example, would require different growing conditions than a test for genes that enable plants to grow in acid soils. And sufficient space to grow plants to maturity is needed, along with trained personnel to design the tests and to analyze results (81). The time required too can be considerable. Testing the wheat held by the U.S. National Small Grains Collection for resistance to stem and leaf rust, for example, is expected to require more than 10 years (96). Evaluation of some traits may require repeating tests over several years and in different regions (81).

Evaluation has been perceived as a serious deficiency in the overall effort to maintain crop germplasm (23,37,78,108,117). Insufficient information has meant accessions have been underused. However, the situation has been improving (81,108). International Agricultural Research Centers have evaluated many of their accessions for important traits, chiefly disease and pest resistance, yield, and quality factors (13,50,108). In the United States, the four regional plant introduction stations (see ch. 9) have included examination for several agriculturally important traits in their preliminary characterizations. This is not, however, sufficient to meet all the needs of users, and more extensive efforts are necessary (108).

Although the usefulness of a collection may depend on its evaluations, questions remain as to who has responsibility for this task. Collection managers might be able to gather morphological data, but they may not be able to perform the lengthy and detailed trials needed to evaluate traits. Further, it has been argued that such evaluations are best done under the conditions in which they will be used because expression may be altered by environmental differences (36). It would seem, therefore, that evaluation trials for specific traits are best performed by the breeders who require those traits. Duplicating efforts could be minimized by putting results into centralized database systems—a proposed function of the GRIN system in the United States.

Traditional Breeding

Traditional breeding typically involves identifying particular genes or characteristics and incorporating them into existing varieties. Crop development through breeding, a major contributor to modern gains in agricultural production, is a time-consuming process: It may take 10 to 15 years to develop a single new variety (box 7-C).

Traditional breeding has provided as much as 60 percent of the production increases of many agricultural crops (22,24,35,77,124). Nevertheless, the process must continue in order to sustain agricultural yields—pests and diseases adapt to new varieties, and the needs of growers and consumers constantly change (77).

Traditional breeding involves several basic steps: 1) locate a genetically stable trait (e.g., yield, pest resistance, or stress tolerance); 2) isolate plants with the most desired expression of the trait; 3) breed genes into breeding lines of plants similar to those that will be improved to provide more usable material; and 4) cross these plants with other breeding lines to produce plants from which improved crop varieties can be selected (41,81).

The third step, called developmental breeding, is important because the desired trait may be located in a wild species or variety that is difficult to cross with domesticated ones. This is the case, for example, with genetic resistance to some 27 serious diseases of the tomato (81). Wild species or landraces may have different growth requirements that make crossing them with other varieties difficult. Developmental breeding overcomes such differences but may require growing plants at multiple locations. Incorporation of genes from over 500 exotic sorghums, for example, required growth in two locations because the exotics required shorter days to flower than commercial U.S. varieties. A cooperative effort, therefore, was established between the Texas Agricultural Experiment Station and the USDA Federal Station in Mayaguez, Puerto Rico, to perform the crosses and test the progeny (48,81).

The major constraint to traditional breeding is its dependence on the sexual process of plants. Multiple crossings and testing of offspring may take years. Molecular biological techniques to locate and map genes in plants may greatly shorten the time needed for breeding improved varieties (6).

Biotechnological Improvement

Biotechnology provides greater precision and speed in the manipulation of genes by avoiding the sexual reproductive process (24,41,75). Three general areas have potential impact on the use of stored plant diversity: 1) somaclonal

Box 7-C.—Breeding and the Development of Gaines Wheat

The first American soft, white, semi-dwarf, winter wheat variety was released in 1956 and was known by the varietal name "Gaines." It was developed by farmers of the Inland Empire region of the Pacific Northwest—a region of rolling hills and deep soils that included eastern Oregon and Washington and northern Idaho. The flour of Gaines wheat can be used in pastries, cookies, and other soft white wheat products.

Gaines wheat was responsible for major increases in the yields of farmers in the Palouse region in Washington. Fifty years ago, these farms yielded an average of 15 to 17 bushels of wheat per acre. Today, many farms harvest more than 90 bushels per acre. These increases have been the result of a breeding program that dramatically restructured the wheat plant.

High yields of Gaines wheat result from a greater proportion of the plant's energy being channeled into grain production. Moisture and nutrients are more efficiently utilized. Genes that reduce the amount of straw relative to grain produced per plant were incorporated into breeding lines. Short-stature Gaines wheat has not only increased farmers' yields, but it also has served as the main source of genes for the wheat varieties of the Green Revolution for which Dr. Norman Borlaug received the 1970 Nobel Peace Prize.

Wheat is not native to the United States, and thus, the germplasm to develop Gaines had to come from international sources. Many of the necessary breeding stocks were already part of the USDA's National Small Grains Collection, which contains wheats from many countries. Other materials came from individual breeders both in the United States and other countries. The above chart shows the parentage of Gaines wheat and illustrates the numerous crosses and selections that must occur for the development of a crop variety. Today, Gaines has been replaced by improved varieties that were developed from it.

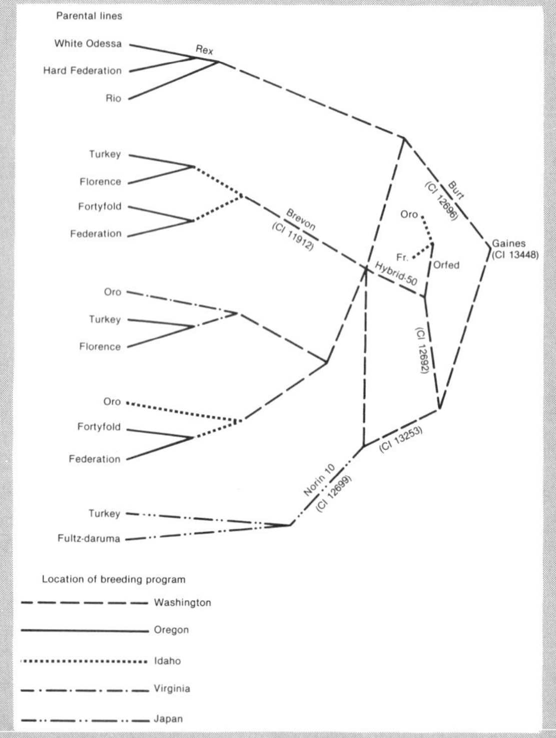

SOURCE: Dr. S. Deitz, USDA-ARS Western Regional Plant, Introduction Station, Pullman, WA.

SOURCE: Adapted from materials provided by Dr. Sam Dietz, Regional Plant Introduction Station, USDA/ARS, Pullman, Washington, 1987.

variation, 2) somatic hybridization, and 3) recombinant DNA technologies.

Somaclonal Variation

During the process of *in vitro* culture of unorganized plant tissues, modifications frequently arise that can be genetically stable and heritable (5,75,90,91). A number of significant somaclonal variants have been isolated from buds produced in unorganized *in vitro* cultures (table 7-5).

However, somaclonal variations may not always persist (74,90). In some cases, variations can be passed on to succeeding generations, but in others they are lost (75,90). In addition to the problem of genetic stability, variant cells

Table 7-5.—Somaclonal Variation in Economically Important Plant Species

Plant	Source of tissue	Characters	Transmission[a]
Oats (*Avena sativa*)	Immature embryo, apical meristem	Plant height, heading date, leaf striping	Seed
Wheat (*Triticum aestivum*)	Immature embryo	Plant height, spike shape, maturity tillering, leaf wax giliadins, amylase	Seed
Rice (*Oryza sativa*)	Seed embryo	Tiller number, panicle size, seed fertility, flowering date, plant height	Seed
Sugarcane (*Saccharum officinale*)	Various	Disease resistance, auricle length, isoenzyme alterations, sugar yield	Vegetable
Corn (*Zea mays*)	Immature embryo	Endosperm and seedling mutants, pathogen toxin resistance, DNA sequence, changes in mitochondria	Seed
Potato (*Solanum tuberuosum*)	Protoplast leaf callus	Tuber shape, yield, maturity date, plant form, stem, leaf, and flower structure, disease resistance	Vegetable
Tobacco (*Nicotiana tabacum*)	Anthers, protoplasts, leaf callus	Plant height, leaf size, yield grade index, alkaloids, reducing sugars, leaf chlorophyll	Seed
Alfalfa (*Medicago sativa*)	Immature ovaries	Leaves, petiole length, plant form and height, dry matter yield	Vegetable
Brassicas (*Brassica* spp.)	Anthers, embryos, meristems	Flowering time, growth form, waxiness, glucosinolates, disease tolerance	Seed

[a]Seed = inherited in seeds of variant plants; vegetable = transmitted to clonally reproduced plants.

SOURCE: W.R. Scowcroft, S.A. Ryan, R.I.S. Brettel, and P.J. Larkin, "Somaclonal Variation: A 'New' Genetic Resource," *Crop Genetic Resources: Conservation and Evaluation,* J.H.W. Holden and J.T. Williams (eds.) (London: George Allen & Unwin, 1984).

and tissues may not regenerate into whole plants or may produce abnormal or sterile plants (75).

Progress in developing somaclonal variation for plant improvement has been promising for a few plant species (5,90,92). However, its general application remains unproven. Further, it is not yet possible to select through *in vitro* culture many valuable traits, such as yield or quality characters. This inability reflects a basic lack of knowledge of the genetic mechanisms controlling many such traits (38).

Somatic Hybridization

A report conducted more than a decade ago on the fusion of leaf protoplasts (cells from which the cell walls have been enzymatically removed) from two species of tobacco heralded exciting possibilities (11). The process, termed somatic hybridization, held promise of bridging many barriers to hybridization. Questions of whether "impossible hybrids" could be obtained were partially answered with reports of a successful protoplast fusion from a tomato and potato (72). Unfortunately, as with a hybrid sexually produced 50 years earlier by crossing radish and cabbage, the resulting plant exhibited the least desirable characteristics of each parent and was sterile (75).

Research on irradiation and protoplast fusion shows promise. By irradiating one set of protoplasts, the genetic material is broken into short sequences, some of which will make its way into the fusion partner. The technique, with considerable development, may eventually enable transfer of genes between sexually incompatible species.

Recent studies show the potential for transferring cellular organelles with their genetic information (chloroplasts and mitochondria) to other species (15,41,75). This technique may be useful in the transfer of genes for the limited number of traits (e.g., photosynthetic efficiency, herbicide tolerance, cytoplasmic male sterility) found in these organelles.

Application of somatic hybridization has been limited to plants from three families: *Solonaceae* (e.g., potato, tomato, tobacco); *Cruciferaceae* (e.g., cabbage, rape); and *Umbelliferaceae* (e.g., carrots) (75). Regeneration of whole plants from protoplasts often remains an obstacle because little is known about the culture conditions needed to cause protoplasts or undifferentiated tissue to regenerate into whole plants (41,75,111).

Recombinant DNA Technologies

The technologies associated with recombinant DNA allow insertion of specific genetic information into plants to produce altered characteristics. Basic principles of the technologies have been discussed in earlier OTA studies (109,111). Current constraints relate largely to inabilities to culture and regenerate isolated cells of most plant species (41,75).

Genetic engineering techniques may allow scientists to develop, by gene transfer, new agricultural varieties (75), but considerable scientific development is needed before such technologies can become routine. Further, genetic engineering technologies face legal, social, and political questions in light of warnings that potential products might cause health, environmental, or economic problems. With continued research, genetic engineering could augment, but not substantially replace, standard breeding practices.

NEEDS AND OPPORTUNITIES

In the past 10 years, new technologies for germplasm collection, maintenance, and use have been developed (108). Improved germplasm maintenance in the United States will require not only the addition of new technologies, but careful planning for facilities and resources to support them. Determining the appropriateness of a particular technology involves consideration of the biology of the species, the reliability of the technology, the effect of the technology on a collection's composition, and costs. This section discusses several areas of offsite maintenance that need attention and the opportunities for doing so.

Develop a Standard Operating Procedure

Studies have only recently begun to systematically address problems of records maintenance, regeneration procedures, seed-drying techniques, storage, liability testing conditions, or improper management (21,30,31,43). This assessment has highlighted numerous appropriate procedures. Implementation of these technologies in the United States and internationally could provide a basis for improving maintenance in offsite collections and developing appropriate avenues for training personnel.

Standard operating procedures for maintaining offsite collections of plants could be developed that include newly developed technologies and incorporate additional procedures as they are developed. Such procedures could be developed by a task force composed of representatives of government, industry, and academia. The task force could specifically consider the use of technologies by the National Plant Germplasm System.

Development of recommendations will not assure improvement of germplasm maintenance in existing U.S. collections. Issues such as the need for additional storage space at NSSL and implementation of better viability testing and regeneration protocols must be addressed by increased funds if necessary. A plan to improve storage and maintenance in NPGS collections should be drawn up, therefore, that would address both the needs for new facilities and support of basic operations. Such a plan could be developed by USDA with or without the suggested task force, or by a separate committee drawn from sectors served by NPGS.

Storage

Cryogenic techniques could greatly extend the storage time of seeds and could reduce costs associated with monitoring seed viability and regenerating samples. USDA funding of research on the effects of cryogenic storage could increase the number of species that can be maintained and allow investigation of concerns about genetic stability.

In vitro plants can be used for a range of species with recalcitrant seeds or for those that must be maintained as clones. However, the techniques are not now used extensively for germplasm storage, and uncertainties about genetic stability in the *in vitro* environment have been noted. Cryogenic technologies could be particularly important, but they require further development.

Funds to develop technologies for maintaining plants in offsite collections are already provided through the Agricultural Research Service (ARS) to NPGS researchers. These efforts could be enhanced by making funds available to researchers outside USDA on a competitive basis. As an alternative, the USDA/Competitive Research Grants Office could develop a program that would focus on germplasm maintenance and the application of technology.

Characterization and Evaluation of Offsite Collections

Characterization and evaluation data are not available for most plants held by NPGS, but the development of descriptors by crop advisory committees (CACs) (see ch. 9) will provide guidelines for preliminary characterizations of many crops. Technologies for biochemical characterization exist, and consideration should be given to ones that are appropriate for particular crops. Further, careful consideration of the agronomic traits to be evaluated will be necessary.

Improving characterization and evaluation data will require additional funding and personnel. A 10-year NPGS program to provide detailed evaluations of the genetic diversity and potentially useful agronomic characters in cultivated species and their relatives might cost $5 million annually. Such a program would probably require increased collaboration between NPGS facilities and scientists to expand the available expertise, develop a computerized file for each accession, and increase involvement of CACs and breeders in determining which agronomic traits to evaluate.

By examining analyses of the roles of CAC, NPGS facilities, and users of NPGS in recording evaluation data, different ways to improve present efforts might be revealed. Such an examination could be performed by an expert committee appointed by USDA. Recommendations could include specific roles for components of NPGS and mechanisms for accomplishing those goals.

Grant funds could be made available through ARS to researchers and breeders screening for particular traits. Such funds could encourage greater use of germplasm collections as well as increase the information about accessions. Data from evaluations could then become part of the permanent GRIN record.

Maintenance of Endangered Wild Species

The efforts of botanic gardens and arboretums to obtain and store seeds or plants of endangered wild species have only recently been coordinated. Additional funding for facilities and personnel to develop and maintain such collections will be needed. Further, each species presents a potentially unique set of requirements for maintenance and regeneration that must be taken into account.

Funds have come in part from the Institute for Museum Services (34). They have been used for daily operations as well as to establish storage facilities. Continued funding could provide for the maintenance of many endangered wild plants. However, it has been estimated that maintaining the 3,000 or so rare and endangered plant taxa will cost at least $1.2 million annually (71).

One possibility is to expand the scope of NPGS activities to include endangered wild species. NPGS personnel have considerable expertise in offsite maintenance of plants, and including endangered wild plants as a responsibility would take advantage of this expertise. However, an enlargement of NPGS's scope would require additional funding for personnel and facilities. And because responsibilities are currently divided among various parts of NPGS on a crop-by-crop basis, an administrative mechanism for assigning responsibility for a particular species would be needed.

As an alternative, an existing private organization, such as the Center for Plant Conservation (CPC), could become the mechanism within NPGS for coordinating maintenance of endangered wild plants. Funds could be designated through USDA/ARS for this purpose, and CPC could be responsible for coordinating efforts and administering funds to cooperating botanic gardens and arboretums.

Improve Movement of Germplasm Through Quarantine

Technologies that identify viruses in imported plants could reduce delays associated with the testing of a few plant species. Although many potentially useful technologies exist, few are applied routinely to quarantine testing, because USDA's Animal and Plant Health Inspection Service (APHIS) lacks sufficient trained personnel, facilities, or operating funds needed to implement a particular technology. Cooperation between APHIS and NPGS facilities could enhance the technical expertise applied to quarantine-testing and other solutions to improve quarantine efforts.

A panel representing APHIS, the research community, and NPGS could be convened to assess the adequacy of facilities and programs relating to quarantine. It could make recommendations for implementing newer technologies, improving present facilities, constructing new facilities, and mechanisms for promoting cooperation with NPGS facilities. The panel could also redirect existing budgets within USDA to address specific problems and, if necessary, develop legislation for increasing USDA appropriations to meet quarantine needs. The panel might also consider mechanisms for incorporating new technologies and the appropriateness of facilities and personnel for performing them.

Promote Basic Research on Maintenance and Use of Plant Germplasm

Although technologies to maintain plants offsite have advanced considerably in recent years, several fundamental questions still need to be addressed.

In the past, storage has essentially referred to orthodox seed storage. It is increasingly apparent that new techniques for storage of nontraditional forms of germplasm (e.g., recalcitrant seeds, pollen, and *in vitro* cultures) are needed. Although cryogenic storage has been used for several years on animals, its use with plants has only recently been investigated. Questions about the nature of genetic control and the mechanisms involved in somaclonal variation are as yet unresolved. These new storage technologies all require better understanding of developmental processes, of cell and seed physiology, and of mechanisms of cellular deterioration and repair.

New methods for storage of naked DNA and RNA and possible recovery of DNA from dead cells could lead to a new concept in germplasm conservation. Caution must be exercised, however, to ensure that limited funds are not disproportionately channeled into this high-technology area. If genetic conservation is a goal, then existing technologies and those showing promise should receive adequate funding before more speculative approaches are pursued.

Improved understanding of the biochemical, genetic, and physiological control of development may lead to techniques for characterizing and evaluating germplasm. The genetic control of most important traits is not yet understood. Additional research on the basic structure and function of genes can also improve the biological knowledge necessary for genetic manipulation of plants.

Funding for research on germplasm has come from several agencies. But research priorities at the National Science Foundation (NSF) or USDA's Competitive Research Grants Office (CRGO), however, generally do not encompass projects that focus on germplasm maintenance. Perhaps a new program within USDA/CRGO or NSF could be created to address research appropriate to germplasm maintenance and use.

CHAPTER 7 REFERENCES

1. Astley, D., "Management Systems at the National Vegetable Research Station Gene Bank," *Documentation of Genetic Resources: Information Handling Systems for Genebank Management*, J. Konopka and J. Hanson (eds.) AGPG: IBPGR/85/76 (Rome: International Board for Plant Genetic Resources, 1985).
2. Atchley, A.A., USDA/ARS Germplasm Resources Laboratory, personal communications, May 23, 1986.
3. Barnabas, B., and Rajiki, E., "Storage of Maize (*Zea mays L.*) Pollen in Liquid Nitrogen," *Euphytica* 25:747-753, 1976.
4. Bass, L., "Seed Viability During Long-Term Storage," *Horticultural Reviews*, J. Janick (ed.) (Westport: Avi Publishing Co., 1980).
5. Berlin, J., and Sasse, "Selection and Screening Techniques for Plant Cell Culture," *Plant Cell Culture* 31:99-131 (Berlin: Springer-Verlag, 1985).
6. Bishop, J.E., "New Genetic Technology Shortens Time Required To Breed Food-Plant Varieties," *Wall Street Journal*, May 17, 1986.
7. Blixt, S., and Williams, J.T. (eds.) *Documentation of Genetic Resources: A Model* (Rome: International Board for Plant Genetic Resources, 1982).
8. Bonner, F.T., "Technologies To Maintain Tree Germplasm Diversity," OTA commissioned paper, 1985.
9. Brown, J.W.S., and Flavell, R.B., "Fractionation of Wheat Giliadin and Glutenin Subunits by Two-Dimensional Electrophoresis and the Role of Group 6 and Group 2 Chromosomes in Giliadin Synthesis," *Theoretical and Applied Genetics* 59:349-359, 1981. *In*: Weeden and Young, 1985.
10. Burton and Davies, "Handling Germplasm of Cross-pollinated Forage," *Crop Genetic Resources: Conservation and Evaluation*, J.H.W. Holden and J.T. Williams (eds.) (London: George Allen & Unwin, 1984).
11. Carlson, P.S., Smith, H.H., and Dearing, R.D., "Parasexual Interspecific Plant Hybridization," *Proceedings, National Academy of Sciences, U.S.A.* 69:2292-2294, 1972. *In*: Orton, 1985.
12. Center for Plant Conservation, *Recommendations for the Collection and Ex Situ Management of Germplasm Resources From Rare Wild Plants* (Boston: CPC, 1986).
13. Chang, T.T., "The Role and Experience of an International Crop-Specific Genetic Resources Center," *Conservation of Crop Germplasm: An International Perspective* (Madison, WI: Crop Science Society of America, 1984).
14. Chiarappa, L., and Karpati, J.F., "Plant Quarantine and Genetic Resources," *Crop Genetic Resources: Conservation and Evaluation*, J.H.W. Holden and J.T. Williams (eds.) (Boston: George Allen & Unwin, 1984).
15. Cocking, E.C., "Use of Protoplasts: Potentials and Progress," *Gene Manipulation in Plant Improvement*, J.P. Gustafson (ed.) (New York: Plenum Press, 1984). *In*: Orton, 1985.
16. Conger, B.V. (ed.), *Cloning Agricultural Plants Via In-Vitro Culture* (Boca Raton, FL: CRC Press, 1981).
17. Cremer, F., and Van de Walle, C., "Method for Extraction of Proteins From Green Plant Tissue for Two-Dimensional Polyacrylamide Gel Electrophoresis," *Analytical Biochemistry* 147:22-26, 1985. *In:* Weeden and Young, 1985.
18. de Bakker, I.G., "The Gene Bank, A New Concept, An Old Cause," *Zaadlelangen* 37:7-25, 1983.
19. Deitz, S.M., director, Western Regional Plant Introduction Station, personal communication, August 1986.
20. de Langhe, E.A.L., "The Role of *In-vitro* Techniques in Germplasm Conservation," *Crop Genetic Resources: Conservation and Evaluation*, J.H.W. Holden and J.T. Williams (eds.) (London: George Allen & Unwin, 1984).
21. Dickie, J.B., Linington, S., and Williams, J.T., *Seed Management Techniques for Genebanks*. AGPG:IBPGR/84/68 (Rome: International Board for Plant Genetic Resources, 1984).
22. Duvick, D.N., "Genetic Rates of Gain in Hybrid Maize Yields During the Past 40 Years," *Maydica* 22:187-196, 1977.
23. Duvick, D.N., "Genetic Diversity in Major Farm Crops on the Farm and in Reserve," *Economic Botany* 38:161-178, 1984.
24. Duvick, D.N., "Plant Breeding: Past Achievements and Expectations for the Future," *Economic Botany* 40(3):289-297, 1986.
25. Dvorak, W.S., "Strategy for the Development of Conservation Banks and Breeding Programs for Coniferous Species from Central America and Mexico," *Proceedings of the Southern Forests Tree Improvement Conference*, 1983.
26. Dvorak, W.S., director, Central America and Mexico Coniferous Resources Cooperative (CAMCORE), personal communication, May 1986.
27. Dvorak, W.S., and Laarman, J.G., "Conserving

the Genes of Tropical Conifers," *Journal of Forestry* 84:43-45, 1983.
28. Ellis, R.H., "Revised Table of Seed Storage Characteristics," *Plant Genetic Resources Newsletter* 58:16-33, 1984.
29. Ellis, R.H., "Information Required Within Genetic Resources Centres To Maintain and Distribute Seed Accessions," *Documentation of Genetic Resources: Information Handling Systems for Genebank Management*, J. Konopka and J. Hanson (eds.) AGPG:IBPGR/85/76 (Rome: International Board for Plant Genetic Resources, 1985).
30. Ellis, R.H., Hong, T.D., and Roberts, E.H., *Handbook of Seed Technology for Gene Banks, Volume I: Principles and Methodology*, Handbooks for Gene Banks No. 2 (Rome: International Board for Plant Genetic Resources, 1985).
31. Ellis, R.H., Hong, T.D., and Roberts, E.H., *Handbook of Seed Technology for Gene Banks, Volume II: Compendium of Specific Germination Information and Test Recommendations*, Handbooks for Gene Banks No. 3 (Rome: International Board for Plant Genetic Resources, 1985).
32. Ellis, R.H., and Roberts, E.H., "Procedures for Monitoring the Viability of Accessions During Storage," *Crop Genetic Resources: Conservation and Evaluation*, J.H.W. Holden and J.T. Williams (eds.) (London: George Allen & Unwin, 1984).
33. Evans, D.A., Sharp, W.R., Ammirato, P.W., Yamada, Y. (eds.), *Handbook of Plant Cell Culture, Volume 1* (New York: Macmillan Press, 1983).
34. Falk, D.A., and Walter, K.S., "Networking To Save Imperiled Plants," *Garden*, January/February 1986.
35. Fehr, W.R. (ed.), "Genetic Contributions To Yield Gains of Five Major Crop Plants," *Crop Science Society of America*, Spec. Pub. No. 7, 1984.
36. Frankel, O.H., and Brown, A.D.H., "Plant Genetic Resources Today: A Critical Appraisal," *Crop Genetic Resources: Conservation and Evaluation* J.H.W. Holden and J.T. Williams (eds.) (London: George Allen & Unwin, 1984).
37. Frey, K.J., Meredith, C.P., and Long, S.R., "Genetic Improvement," *Crop Productivity—Research Imperatives Revisited*, M. Gibbs and C. Carlson (eds.), an international conference held at Boyne Highlands Inn, MI, Oct. 13-18, 1985 and Airlie House, VA, Dec. 11-13, 1985.
38. Gibbs, M., and Carlson, C. (eds.), *Crop Productivity—Research Imperatives Revisited*, an international conference held at Boyne Highlands Inn, MI, Oct. 13-18, 1985 and Airlie House, VA, Dec. 11-13, 1985.
39. Goodman, M.M., Department of Crop Science, University of North Carolina, Raleigh, personal communication, May 1986.
40. Grout, B.W.W., "Embryo Culture and Cryopreservation for the Conservation of Genetic Resources of Species With Recalcitrant Seed," *Plant Tissue Culture and Its Agricultural Applications*, L.A. Withers and P.G. Alderson (eds.) (London: Butterworths, 1986).
41. Hansen, M., Busch, L., Burkhardt, W.B., and Lacy, L.R., "Plant Breeding and Biotechnology," *BioScience* 36:29-39, 1986.
42. Hanson, J., "The Storage of Seeds of Tropical Tree Fruits," *Crop Genetic Resources: Conservation and Evaluation*, J.H.W. Holden and J.T. Williams (eds.) (London: George Allen & Unwin, 1984).
43. Hanson, J., "Practical Manuals for Genebanks: No. 1," *Procedures for Handling Seeds in Genebanks* (Rome: International Board for Plant Genetic Resources, 1985).
44. Harborne, J.B., *Phytochemical Methods* (London: Chapman & Hall, 1973).
45. Hartmann, H.T., Flocker, W.J., and Kofranek, A.M., "Plant Science," *Growth, Development and Utilization of Cultivated Plants* (Englewood Cliffs, NJ: Prentice-Hall, Inc., 1981).
46. Hawkes, J., *Plant Genetic Resources: The Impact of the International Agricultural Research Centers* (Washington, DC: Consultative Group on International Agricultural Research, World Bank, 1985).
47. Hawkes, J.G., *Crop Genetic Resources Field Collection Manual* (Rome: International Board for Plant Genetic Resources and European Association for Research on Plant Breeding, 1980).
48. Helentjaris, T., King, G., Slocum, M., Siedenstrang, C., and Wegman, S., "Restriction Fragment Polymorphisms as Probes for Plant Diversity and Their Development as Tools for Applied Plant Breeding," *Plant Molecular Biology* 5:109-118, 1985.
49. Henshaw, G.G., Keefe, P.D., and O'Hara, J.F., "Cryopreservation of Potato Meristems," *In Vitro Techniques: Propagation and Long Term Storage*, A. Schafer-Menuhr (ed.) (Boston: Martinus Nijoff/ Dr. W. Junk Publication, 1985).
50. Huaman, Z., "The Evaluation of Potato Germ-

plasm at the International Potato Center," *Crop Genetic Resources: Conservation and Evaluation,* J.H.W. Holden and J.T. Williams (eds.) (London: George Allen & Unwin, 1984).
51. International Board for Plant Genetic Resources, *Revised Priorities Among Crops and Regions*, AGP:IBPGR/81/34 (Rome: 1981).
52. International Board for Plant Genetic Resources, *Practical Constraints Affecting the Collection and Exchange of Wild Species and Primitive Cultivars*, AGPG:IBPGR/83/49 (Rome: 1983).
53. International Board for Plant Genetic Resources, "IBPGR Advisory Committee on *In Vitro* Storage," First Meeting, Aug. 16-20, 1982 (Rome: 1983).
54. International Board for Plant Genetic Resources, *IBPGR Advisory Committee on Seed Storage*, Report of the Second Meeting, Sept. 19-20, 1983, AGPG:IBPGR/83/116 (Rome: 1984).
55. International Board for Plant Genetic Resources, *Cost-Effective Long-Term Seed Stores* (Rome: 1985).
56. International Board for Plant Genetic Resources, "IBPGR Advisory Committee on *In-Vitro* Storage," Report of the Second Meeting (Rome: 1985).
57. Janick, J., Schery, R.W., Woods, F.W., and Ruttan, V.W., *Plant Science. An Introduction to World Crops*, 2d ed. (San Francisco: W.H. Freeman & Co., 1974).
58. Justice, O.L., and Bass, L., *Principles and Practices of Seed Storage,* Agriculture Handbook No. 506 (Washington, DC: U.S. Department of Agriculture, 1978).
59. Kahn, R.P., "Plant Quarantine: Principles, Methodology, and Suggested Approaches," *Plant Health and Quarantine in International Transfer of Plant Genetic Resources*, W.B. Hewitt and L. Chiarappa (eds.) (Cleveland, OH: CRC Press, 1977). *In*: Kahn, 1985.
60. Kahn, R.P., "Technologies To Maintain Biological Diversity: Assessment of Plant Quarantine Practices," OTA commissioned paper, 1985.
61. Kaiser, W.J., "Plant Introduction and Related Seed Pathology Research in the United States," *Seed Science and Technology* 11:197-212, 1983.
62. Kartha, K.K., "Meristem Culture and Germplasm Preservation," *Cryopreservation of Plant Cells and Organs*, K.K. Kartha (ed.) (Boca Raton, FL: CRC Press, 1985). *In*: Towill, et al., 1985.
63. King, M.W., and Roberts, E.H., *The Storage of Recalcitrant Seeds—Achievements and Possible Approaches* (Rome: International Board for Plant Genetic Resources, 1979).
64. King, M.W., and Roberts, E.H., "The Desiccation Response of Seeds of *Citrus limon L.*," *Annals of Botany* 45:489-492, 1980. *In*: Towill, et al., 1985.
65. Konopka, J., and Hanson, J. (eds.), *Documentation of Genetic Resources: Information Handling Systems for Genebank Management* (Rome: International Board for Plant Genetic Resources, 1985).
66. Koopowitz, H., Developmental and Cell Biology, University of California, Irvine, personal communication, August 1986.
67. Larkin, P.J., and Scowcroft, W.R., "Somaclonal Variation: A Novel Source of Variability From Cell Cultures for Plant Improvement," *Theroet. Appl. Genet.* 60:197-214, 1981.
68. Lucas, G., Royal Botanical Gardens, Kew, England, personal communication, May 1986.
69. Lucas, G., and Oldfield, S., "The Role of Zoos, Botanical Gardens and Similar Institutions in the Maintenance of Biological Diversity," OTA commissioned paper, 1985.
70. Lyman, J.M., "Progress and Planning for Germplasm Conservation of Major Food Crops," *Plant Genetic Resources Newsletter* 60:3-21, 1984.
71. McMahan, L., Center for Plant Conservation, Boston, personal communication, August 1986.
72. Melchers, G., Saeristan, M., and Holder, A.A., "Somatic Hybrid Plants of Potato and Tomato Regenerated From Fused Protoplasts," *Carlsberg Res. Comm.* 43:203-218, 1978.
73. Oldfield, M.L., *The Value of Conserving Genetic Resources* (Washington, DC: U.S. Department of the Interior, 1984).
74. Orton, T.J., "Somaclonal Variation: Theoretical and Practical Consideration," *Gene Manipulation in Plant Improvement*, J.P. Gustafson (ed.) (New York: Plenum Press, 1984). *In*: Orton, 1985.
75. Orton, T.J., "New Technologies and the Enhancement of Plant Germplasm Diversity," OTA commissioned paper, 1985.
76. Peacock, W.J., "The Impact of Molecular Biology on Genetic Resources," *Crop Genetic Resources: Conservation and Evaluation*, J.H.W. Holden and J.T. Williams (eds.) (London: George Allen & Unwin, 1984).
77. Plucknett, D.L., and Smith, N.H., "Sustaining Agricultural Yields," *BioScience* 36:40-45, 1983.
78. Plucknett, D.L., Smith, N.J.H., Williams, J.T.,

and Anishetty, N.M., "Crop Germplasm Conservation and Developing Countries," *Science* 220:163-169, 1986.

79. Plucknett, D., Smith, N., Williams J.T., and Anishetty, N.M., *Gene Banks and The World's Food* (Princeton, NJ: Princeton University Press, in press).
80. Richardson, S.D., "Gene Pools in Forestry," *Genetic Resources in Plants—Their Exploration and Conservation*, O.H. Frankel and E. Bennet (eds.) IBP Handbook No. 11:353-365 (Oxford: Blackwell Scientific Publications, 1970).
81. Rick, C.M., "Plant Germplasm Resources," *Handbook of Plant Cell Culture,* P.V. Ammirato, D.A. Evans, W.R. Sharp, and Y. Yamada (eds.), vol. 3:9-36 (New York: Macmillan Press, 1984).
82. Roberts, E.H., "Loss of Seed Viability During Storage," *Advances in Research and Technology of Seeds*, Part 8, J.R. Thompson (ed.) (Wageningen: Pudoc, 1983).
83. Roberts, E.H., "Monitoring Seed Viability in Genebanks," *Seed Management Techniques for Genebanks*, J.B. Dickie, S. Linington, and J.T. Williams (eds.) AGPG:IBPGR/84/68 (Rome: International Board for Plant Genetic Resources, 1984).
84. Roberts, E.H., and Ellis, R.H., "The Implications of the Deterioration of Orthodox Seeds During Storage for Genetic Resources Conservation," *Crop Genetic Resources: Conservation and Evaluation*, J.H.W. Holden and J.T. Williams (eds.) (London: George Allen & Unwin, 1984).
85. Roberts, E.H., King, M.W., and Ellis, R.H., "Recalcitrant Seeds: Their Recognition and Storage," *Crop Genetic Resources: Conservation and Evaluation*, J.H.W. Holden and J.T. Williams (eds.) (London: George Allen & Unwin, 1984).
86. Roos, E.E., "Induced Genetic Changes in Seed Germplasm During Storage," *The Physiology and Biochemistry of Seed Development, Dormancy and Germination*, A.A. Kahn (ed.) (New York: Elsevier Biomedical Press, 1982).
87. Roos, E.E., "Genetic Shifts in Mixed Bean Populations, Part I: Storage Effects," *Crop Science* 24:240-244, 1984.
88. Roos, E.E., "Genetic Shifts in Mixed Bean Populations, Part II: Effects of Regeneration," *Crop Science* 24:711-715, 1984.
89. Roos, E.E., "Precepts of Successful Seed Storage," *Physiology of Seed Deterioration*, Spec. Pub. No. 11:1-25 (Madison, WI: Crop Science Society of America, 1986).
90. Scowcroft, W.R., *Genetic Variability in Tissue Culture: Impact on Germplasm Conservation and Utilization*, AGPG:IBPGR/84/152 (Rome: International Board for Plant Genetic Resources, 1984).
91. Scowcroft, W.R., and Larkin, P.J., "Somaclonal Variation, Cell Selection and Genotype Culture," *Comprehensive Biotechnology*, vol. 3, C.W. Robinson and H.J. Howells (eds.) (Sydney: Academic Press, 1984). *In*: Scowcroft, et al., 1984.
92. Scowcroft, W.R., Ryan, S.A., Brettel, R.I.S., and Larkin, P.J., "Somaclonal Variation: A 'New' Genetic Resource," *Crop Genetic Resources: Conservation and Evaluation*, J.H.W. Holden and J.T. Williams (eds.) (London: George Allen & Unwin, 1984).
93. Sharp, W.R., Evans, D.A., Ammirato, P.V., and Yamada, Y. (eds.), *Handbook of Plant Cell Culture*, vol. 2 (New York: Macmillan Press, 1984).
94. Simpson, M.J.A., and Withers, L.A., *Characterization of Plant Genetic Resources Using Isozyme Electrophoresis: A Guide to the Literature* (Rome: International Board for Plant Genetic Resources, 1986).
95. Singh, R.B., and Williams, J.T., "Maintenance and Multiplication of Plant Genetic Resources," *Crop Genetic Resources: Conservation and Evaluation*, J.H.W. Holden and J.T. Williams (eds.) (London: George Allen & Unwin, 1984).
96. Smith, D.H., curator, U.S. National Small Grains Collection, personal communications, Jan. 19, 1986.
97. Smith, R.D., Linington, S.H., and Fox, D.J., "The Role of the Computer in the Day to Day Running of the Kew Seed Bank," *Documentation of Genetic Resources: Information Handling Systems for Genebank Management*, J. Konopka and J. Hanson (eds.), AGPG:IBPGR/85/76 (Rome: International Board for Plant Genetic Resources, 1985).
98. Sprague, G.F., "Germplasm Resources of Plants: Their Preservation and Use," *Annual Review of Phytopathology* 18:147-165, 1980.
99. Stanwood, P.C., "Cryopreservation of Seeds," Report Second Meeting, IBPGR Advisory Committee on Seed Storage, App. III (Rome: International Board for Plant Genetic Resources, 1984).
100. Stanwood, P.C., "Cryopreservation of Seed Germplasm for Genetic Conservation," *Cryopreservation of Plant Cells and Organs*, K.K. Kartha (ed.) (Boca Raton, FL: CRC Press, 1985).
101. Stanwood, P.C., USDA-ARS National Seed Storage Laboratory, Fort Collins, CO, personal communication, May 1986.

102. Stanwood, P.C., and Bass, L.N., "Seed Germplasm Preservation Using Liquid Nitrogen," *Seed Science and Technology* 9:423-437, 1981.
103. Strauss, M.S., "Technology and Plant Gene Banking in Developing Countries," OTA staff paper, 1986.
104. Tanksley, S.D., and Orton, T.J., *Isozymes in Plant Genetics and Breeding, Part A* (New York: Elsevier Biomedical Press, 1983).
105. Tanksley, S.D., and Orton, T.J., *Isozymes in Plant Genetics and Breeding, Part B* (New York: Elsevier Biomedical Press, 1983).
106. Thibodeau, F., and Falk, D., "The Center for Plant Conservation: A New Response to Endangerment," *The Public Garden*, January 1986.
107. Towill, L.E., "Low Temperature and Freeze-Vacuum-Drying Preservation of Pollen," *Cryopreservation of Plant Cells and Organs*, K.K. Kartha (ed.) (Boca Raton, FL: CRC Press, 1985).
108. Towill, L., Roos, E., and Stanwood, P.C., "Plant Germplasm Storage Technologies," OTA commissioned paper, 1985.
109. U.S. Congress, Office of Technology Assessment, *Impacts of Applied Genetics: Micro-Organisms, Plants, and Animals*, OTA-HR-132 (Washington, DC: U.S. Government Printing Office, April 1981).
110. U.S. Congress, Office of Technology Assessment, *Plants: The Potentials for Extracting Protein, Medicines, and Other Useful Chemicals—Workshop Proceedings*, OTA-BP-F-23 (Washington, DC: U.S. Government Printing Office, September 1983).
111. U.S. Congress, Office of Technology Assessment, *Commercial Biotechnology: An International Analysis*, OTA-BA-218 (Springfield, VA: National Technical Information Service, January 1984).
112. U.S. Congress, Office of Technology Assessment, *Grassroots Conservation of Biological Diversity in the United States—Background Paper#1*, OTA-BP-F-38 (Washington, DC: U.S Government Printing Office, February 1986).
113. U.S. Department of Agriculture, Agricultural Research Service, *Research Progress in 1985, A Report of the Agricultural Research Service* (Washington, DC: 1986).
114. Weeden, N., and Young, D.A., "Technologies To Evaluate and Characterize Plant Germplasm," OTA commissioned paper, 1985.
115. Wilkes, G., "Current Status of Crop Plant Germplasm," *CRC Critical Reviews in Plant Science* 1:133-181, 1983.
116. Wilkes, G., "Germplasm Conservation Toward the Year 2000: Potential for New Crops and Enhancement of Present Crops," *Plant Genetic Resources, A Conservation Imperative*, AAAS Selected Symposium 87:131-164, 1984.
117. Williams, J.T., "A Decade of Crop Genetic Resources Research," *Crop Genetic Resources: Conservation and Evaluation*, J.H.W. Holden and J.T. Williams (eds.) (London: George Allen & Unwin, 1984).
118. Williams, J.T., and Damania, A.B., *Directory of Germplasm Collections, 5: Industrial Crops, I: Cacao, Coconut, Pepper, Sugarcane, and Tea* (Rome: International Board for Plant Genetic Resources, 1981).
119. Withers, L.A., "Germplasm Storage in Plant Biotechnology," *Plant Biotechnology*, S.H. Mantell and H. Smith (eds.) (Cambridge, MA: Cambridge University Press, 1983).
120. Withers, L.A., "Germplasm Conservation *In Vitro*: Present State of Research and Its Application," *Crop Genetic Resources: Conservation and Evaluation*, J.H.W. Holden and J.T. Williams (eds.) (London: George Allen & Unwin, 1984).
121. Withers, L.A., "Cryopreservation of Cultured Cells and Meristems," *Cell Culture and Somatic Cell Genetics of Plants*, 2:253-316 (New York: Academic Press, 1985).
122. Withers, L.A., "Cryopreservation and Genebanks," *Plant Cell Culture Technology*," M.M. Yeoman (ed.), Botanical Monographs, 23:96-140 (Oxford, U.K.: Blackwell Scientific Publications, 1986).
123. Withers, L.A., and Williams, J.T., "Research on Long-Term Storage and Exchange of *In Vitro* Plant Germplasm," *Biotechnology in International Agricultural Research* (Manila, Philippines: International Rice Research Institute, 1985).
124. Witt, S., *Briefbook: Biotechnology and Genetic Diversity*, California Agricultural Lands Project, 1985.
125. Yndgaard, F., "Genebank Security Storage in Permafrost," *Plant Genetic Resources Newsletter* 62:2-7, 1985.

Chapter 8

Maintaining Microbial Diversity

CONTENTS

Table

Box

Chapter 8
Maintaining Microbial Diversity

HIGHLIGHTS

- Micro-organisms provide benefits and harbor danger. But safeguarding a diversity of these organisms remains important, for few have been cataloged and the potential contribution to agriculture, industry, and medicine is therefore unknown.
- The most cost-effective way to preserve economically important micro-organisms today is through offsite collections. Micro-organisms that are isolated and identified can be stored from a few days to as long as 30 years.
- Technologies used are freeze-drying (the most common method), ultra-freezing (which costs more in labor and materials), immersion in mineral oil, low-temperature freezing, and desiccation. The last three methods are suitable for short-term storage only.
- A high priority in efforts to maintain a diversity of micro-organisms is the need for an integrated database of current collections. Also needed is research on microbial ecology, to better understand the extent to which plants and animals depend on bacteria and fungi to survive.

OVERVIEW

Micro-organisms constitute a vast, though largely unseen, part of the biotic world. Although most frequently discussed in terms of their harmful effects on humans, they are essential to the proper functioning of ecosystems as well as to the survival of many species of plants and animals (19,25) (see box 8-A). The public is less concerned, however, about potential losses of microbial diversity than about plant, animal, or ecosystem diversity (19).

What Is Microbial Diversity?

The wide range of micro-organisms not typically classed as plants or animals includes bacteria, cyanobacteria (blue-green algae), fungi (including yeasts), protozoa, and viruses (see table 8-1). Although the microscopic, single-celled bacteria are generally considered synonymous with the term micro-organism, the field includes such different things as the large marine algae of ocean kelp beds and the submicroscopic viruses that infect humans, animals, plants, and other micro-organisms. Even smaller than viruses are those infective agents, such as viroids, that have been found to be nothing more than pieces of genetic material, lacking even the typical protein coats of a virus. Thus, the diversity of micro-organisms is immense, with only a relatively small fraction of micro-organisms having been identified (19,25).

The concept of a species, borrowed from animal and plant biology, cannot be easily applied to all micro-organisms (5,19,25,29). Research frequently focuses on populations of microbial cells that share common nutritional, chemical, or biochemical characteristics. Such populations, each typically descended from a single

Box 8-A.—The Importance of Microbial Diversity

Micro-organisms are important to humans both for the benefits they provide and for their harmful effects (7,14,19,26). Micro-organisms are essential parts of the environment, contributing to the maintenance of stable ecosystems. Medicine, agriculture, and industry all benefit significantly from products derived from micro-organisms.

Studies of microbial ecology can provide basic information about the environment (25). Microbial communities provide unique model systems for the study of ecological principles and the organization of natural populations. Because microbial communities can occupy little space and grow rapidly, it can be easier to study the principles of community structure and relationships with such materials than with more complex ecosystems that include plants and animals.

Monitoring of environmental disturbances from pollution or natural causes can be improved by study of micro-organisms in the environment. Species of all types are subject to extinction when the environments in which they live are destroyed. Such loss is generally preceded by more subtle changes in the environment (e.g., pollution or nutrient loss) that can be detected far earlier in the microbial population. Changes in the microbial composition of ecosystems can thus warn of impending environmental destruction resulting from pollution or other environmental disturbance. Micro-organisms can even improve water quality by degrading environmental pollutants and naturally occurring organic matter. Development of new strains of micro-organisms promises to provide important weapons for combating pollution.

World health has been assaulted by micro-organisms, aided when they are found to be the specific agents of disease, and improved through development of vaccines, antibiotics, and chemical agents to combat these organisms. Bacteria from the swamps of southern New Jersey, for example, led to development of a drug that has proven effective against many drug-resistant and hospital-acquired infections (20). The potentials of micro-organisms to aid in the treatment of human disease are still largely unexploited or unrecognized, however.

Farming, too, has benefited from a better understanding of the fungal and bacterial organisms associated with crops. Studies in both developed and developing countries seek to improve the availability of nitrogen to crops through the nitrogen-fixation activities of various micro-organisms. At present, no crop is able to obtain nitrogen until it is converted into a biologically useful form by industrial processes or the action of micro-organisms. The U.S. corn crop alone requires $1 billion per year in industrially produced fertilizers (6). Soil fungi have been found to colonize plant roots and greatly improve nutrient availability and augment the functioning of the plant's own root system (6). Micro-organisms, such as the *Bacillus thuringiensis* commonly used to combat caterpillar infestations, also have been developed as biological control agents for pests of agricultural crops (23).

The industrial applications of micro-organisms are numerous. Microbial fermentation provides a large number of the foods, beverages, and chemicals that are a part of modern society (13). Butyric acid-producing bacteria have been used for centuries for the retting of flax and hemp. The same bacteria also have been applied to production of the important industrial chemicals acetone and butanol. Micro-organisms have been used as biological catalysts to perform what might otherwise be complex chemical conversions in the production of a variety of medical and industrial products (8,13). Microbially produced polysaccharides, such as xanthan gum, are used to thicken, suspend, bind, lubricate, or stabilize materials in numerous food, industrial, and oil field applications (3).

Finally, micro-organisms are a focus of renewed research interest because of technological advances that allow them to be modified with recombinant DNA technologies (19,30). In the field of genetic engineering, micro-organisms are essential, because the production of recombinant nucleic acids, whether destined for use in plants, animals, or micro-organisms, is accomplished in bacteria or other single-celled micro-organisms (6,19,30,31).

Table 8-1.—Summary of the Characteristics, Problems, and Uses of Micro-Organisms

Organisms	Characteristics	Problems	Uses
Bacteria	Single-cells; spherical rod and spiral forms. Most are saprophytes (use dead matter for food), although some bacteria are photosynthetic.	Some cause disease in humans, animals, and plants.	Break down organic matter and assist soil fertility, waste disposal, and biogas production; source of antibiotics and other chemicals.
Fungi	Variety of forms; microscopic molds, mildews, rusts, and smuts; larger mushrooms and puffballs.	Rot textiles, leather, harvested foods, and other products; cause important plant and animal diseases.	Assist in recycling complex plant constituents such as cellulose; mushrooms and yeasts important as foods; many used in chemical and pharmaceutical industries.
Algae	Single cells, colonies, or filaments containing chlorophyll and other characteristic pigments; no true roots, stems, or leaves; aquatic.	Cover pond surfaces, producing scum and unpleasant odor and taste (in drinking water); absorb O_2 from ponds and some produce toxins.	Red and brown seaweeds are important foods in Asia and Polynesia; red algae produce agar; important food source for many ocean fish.
Protozoa	Single cells, or groups of similar cells, found in fresh and sea water, in soil, and as symbionts or parasites in man, animals, and some plants.	Responsible for serious human and animal disease (e.g., malaria, sleeping sickness, dysentery).	Assist in breakdown of organic matter such as cellulose in ruminant nutrition.
Viruses	Infective agents; capable of multiplying only in living cells; composed of proteins and nucleic acids.	Cause variety of human, animal, and plant diseases (e.g., influenza, AIDS, rabies, mosaics).	Important as carriers of genetic information; some infecting pests can be used as biological control agents.

SOURCE: National Academy of Sciences, *Microbial Processes: Promising Technologies for Developing Countries* (Washington, DC: National Academy Press, 1979).

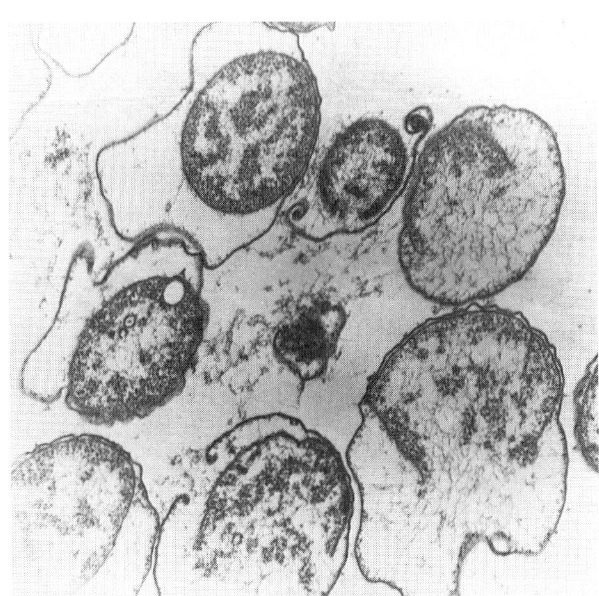

Photo credit: G.E. Pierce and M.K. Mulks

Pseudomonas putida, a bacterium capable of degrading hydrocarbons, is one type of micro-organism.

cell, are termed strains (7). Individually identifiable strains of micro-organisms are thus often regarded as the basic units of microbial diversity.

Micro-organisms are found in nearly all environments (8). Bacteria, for example, have been found living in deep-sea steam vents at temperatures of 350° C (22). Although some micro-organisms are widely distributed, others may be restricted to a narrow ecological range. Most micro-organisms are found as parts of complex microbial communities or as integral parts of larger ecosystems. Many of these micro-organisms cannot be isolated and grown under controlled laboratory conditions (25).

Status of Microbial Diversity Onsite

Assessing the status and changes in microbial diversity onsite can be difficult. As noted earlier, few micro-organisms have been isolated and described or identified (25). Therefore, any changes that occur cannot be determined as temporary or permanent.

For example, the composition of microbial populations within environments can be dramatically altered by pollutants (18,19,32). Studies of microbial populations at a pharmaceutical dump site in the Atlantic Ocean indicate that the survival and growth of certain marine micro-organisms over others in the population occur as a result of pollution (27). Although it is clear that pollution or environmental disturbance can produce quantitative changes, definitive evidence of extinction of micro-organisms, as is seen in plants and animals, is rare. But it would seem likely that where micro-organisms are highly adapted to a specific environment, loss of that environment could result in extinction of the micro-organisms (10).

One group of micro-organisms for which the potential for loss has been a particular concern is the macrofungi—more specifically, the edible wild mushrooms. Morels, chanterelles, and other mushrooms have long been collected by fanciers, particularly in the Northwest United States (33). However, increased collection to serve a growing commercial demand for wild mushrooms has raised fears that the most sought-after species could become rare or extinct (1,33). Scientists currently disagree about whether this is possible, for it is not even known if wild mushrooms can be overharvested (33). In the absence of information, some efforts have been made to limit collection (33). Research on the biology and ecology of these mushrooms and their distribution is needed.

Microbial Diversity Offsite

The principal repositories for those few micro-organisms that have been isolated are offsite collections. Offsite collections of micro-organisms provide easier and quicker access to specific strains than repeatedly returning to onsite sources to obtain them. In addition, it may not always be possible to obtain the same micro-organism from the same place. The fungus from which penicillin was derived, for example, could not again be isolated from air or dust samples in the laboratory where it was first found as a culture contaminant. Offsite collections also are used as reference standards for taxonomic and comparative studies. In microbiology, a "type" strain of a micro-organism is maintained for use as a reference and as a source for subsequent studies (17). It would be impossible to isolate type-strains from natural environments each time comparative studies were initiated (19,21).

Current offsite collections of micro-organisms are actively used as resources by industry and by the scientific community. Yet such collections are rarely established or maintained for preserving microbial diversity (26).

MAINTENANCE TECHNOLOGIES

Onsite maintenance is the only feasible long-term method for maintaining the major portion of microbial diversity, because most of the micro-organisms in any single environment have yet to be identified (19). But existing programs to maintain animal and plant diversity will likely cover all but a few specialized environments (e.g., deep-sea steam vents), so establishing reserves specifically for maintaining micro-organisms should not be necessary.

The most cost-effective approach to providing ready access to the many economically, medically, agriculturally, or scientifically important micro-organisms today is to preserve them in offsite collections (19). The following sections assess the techniques required to maintain micro-organisms offsite.

Isolation and Sampling

Isolation of micro-organisms and their incorporation into a collection generally reflects a particular set of goals. Laboratory applications, such as those involved in genetic engineering, require specific strains of micro-organisms that, once obtained, are kept as pure cultures. Micro-organisms have been isolated to study their interrelationships and the way the dynamics between populations influence the entire biological food chain. Some collections represent sampling of specific taxonomic classes of micro-organisms of economic or agricultural importance. The actinomycete collection of the Battelle-Kettering Laboratory (Yellow Springs, OH) and the many collections of varying sizes of *Rhizobium*, the nitrogen-fixing bacteria of legumes, are examples of goal-directed collections. The pathogenic characteristics of a micro-organism, as in the case of a disease-producing virus, can also merit spending funds, time, and expertise to isolate it.

Isolated pure cultures of micro-organisms are necessary for detailed study (7). For some, such as the fungi, a sterile culture of spores on a specially prepared medium may be all that is necessary to obtain a pure culture. Nutritional requirements for various fungi can, however, be specific and difficult to determine. Most bacteria must be cultured on a variety of media that will stimulate growth of possible contaminants, from which pure culture can then be obtained. Viruses are frequently isolated from infected cells or tissues by centrifugation or filtration techniques that separate them from other cellular components. For micro-organisms that consist only of a small piece of genetic material, such as viroids, the newly developed technologies for isolating, multiplying, and characterizing DNA and RNA have been important. The critical determination that an isolated micro-organism is pure can be a lengthy process of repeated culture or separation under varying conditions and can be a research problem in itself (7).

Studies of microbial ecology and microbial diversity are limited by the inability of scientists at the present time to isolate many micro-organisms (19). Identification, for example, generally requires growth in pure culture to allow for nutritional and physiological testing. It is not currently possible to acquire a knowledge of the total microbial diversity in any one environment in a readily definable time period because of this inability to isolate, culture, and characterize every (or even most) micro-organism present. Thus, sampling of diversity is limited to those micro-organisms for which isolation and culture technologies are available.

Microbial Identification

Identification of isolated micro-organisms can be a lengthy and complex task (for details of the principles and procedures, see ref. 15). Preliminary identification involves standard staining procedures and microscopic examination. Analysis of the results of these initial examinations requires extensive knowledge of micro-organisms and the general characteristics of various taxonomic groups. Information regarding the source of the isolate can also play an important role at this stage.

Following preliminary identification, the micro-organism is subjected to more detailed analysis, frequently consisting of examination of growth characteristics on varying substrates and under varying environmental conditions. These tests establish specific physiological characteristics that aid identification. The general protocol is to work with a pure culture and, using selected tests, narrow the range of possibilities. Once identified, the isolate is then compared to a reference sample using selected diagnostic tests (24).

Biochemical analysis of proteins and DNA, as described for plants (see ch. 7), has been used for identification of many strains of micro-organisms (7). Although these technologies routinely identify the micro-organisms used in research laboratories, they are not generally applied in offsite collections. As the field of genetic engineering has developed, however, the capacity to study, compare, and identify the genomes of micro-organisms has improved (10,31). Such techniques could greatly enhance assessment of diversity and facilitate identification of micro-organisms in offsite collections.

Storage of Micro-Organisms

The purpose of preserving micro-organisms is to maintain a strain for an indefinite period in a viable state. The continuous culture of a micro-organism is one way to present it, but it is expensive both in materials and in labor and does not ensure that the genetic stability of the micro-organism will be maintained. Continuously subcultured organisms can adapt to the specialized conditions of the laboratory and take on characteristics different from those for which they were originally isolated. Long-term storage techniques that minimize such effects have been developed.

Storage of micro-organisms involves reducing metabolic rates and, thus, the rate at which micro-organisms multiply and use nutrients (19). All methods that reduce metabolic rates cause loss of a certain percentage of the sample. Methods need to be developed, therefore, that not only reduce metabolic rates but also prevent decline in viability in order to prevent loss of the strain.

An additional time-consuming but crucial task associated with storage technologies is authentication (19). This task involves the maintenance of accurate records about the culture history and diagnostic characteristics of the strains in a collection. It also involves periodically recovering and culturing stored organisms to determine their viability and to confirm purity and genetic stability.

The majority of micro-organisms currently preserved in culture collections are held by freeze-drying or by ultra-freezing (cryogenic storage) (16,19). These two methods permit storage for extremely long periods of time (currently as long as 30 years) (16). Other special methods for storage of micro-organisms are immersion in mineral oil, low-temperature freezing, and desiccation (16).

Freeze-Drying

Freeze-drying, or lyophilization, is now the most commonly used storage technique for culture collections. Healthy microbial cells, grown under optimal conditions, are dispensed in small, sterile vials or ampules at a relatively high concentration (e.g., 10^6 to 10^7 cells per milliliter of solution). The vials are then quickly frozen in a super-cooled liquid solvent bath or in a mechanical ultra-freezer (at −60° C), and these frozen suspensions are placed under vacuum to remove the water in them. The vials are then heat-sealed under vacuum by melting the glass tops with an air-gas torch and stored at temperatures lower than 5° C. Lower storage temperatures (−30° to −70° C) may result in lengthened viability.

Chemical agents (cryoprotectants) that protect cells from damage caused by ice-crystal formation during the initial freezing are commonly added to cells before freeze-drying. The American Type Culture Collection (ATCC) routinely uses 10 percent skim milk or 12 percent sucrose for such purposes. Curators at the Northern Regional Research Laboratory of the USDA's Agricultural Research Service, in contrast, prefer bovine or equine serum as a cryoprotectant for all microbial species (19).

Recovery of the lyophilized cells is simple and straightforward. The sealed vial is opened by scoring, and a small amount of liquid-nutrient medium is added to rehydrate the cells. The contents, once rehydrated, are transferred to a culture vessel containing a medium appropriate for growth.

The initial cost of equipping a laboratory to undertake lyophilization is as much as $25,000 (11). The expense of actually preparing lyophilized cultures, however, is low. The long-term viability of such materials is excellent, and this procedure is thus probably the most cost-effective means of microbial preservation in use today (19). Unfortunately, some microbial species do not fare well under these techniques, and other storage methods must be used.

Ultra-Freezing

Fastidious microbial species (i.e., those with complex nutritional requirements) that do not retain viability under other preservation methods (e.g., plant pathogenic fungi) frequently can be preserved by ultra-freezing (2,19). In this procedure, cells sealed in vials or ampules are frozen at a slow cooling rate (1° C per minute) until they reach −150° C. The vials are then stored at −150° to −196° C using liquid nitrogen freezers.

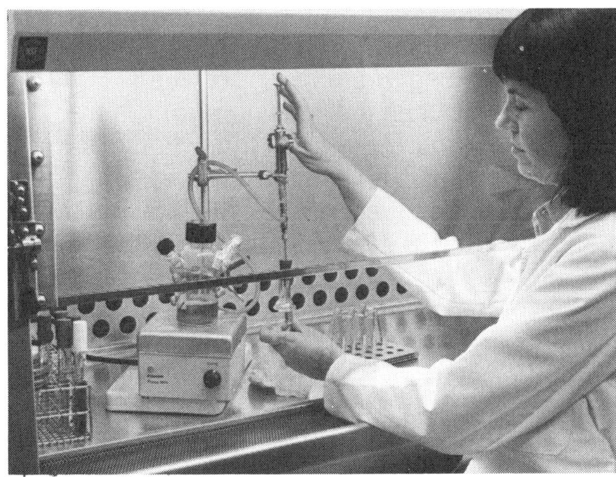

Photo credit: W.H. Siegel

Cells being dispensed into ampules to be frozen and stored in liquid nitrogen (−196° C). The cabinet contains only sterile, filtered air to lessen the chances of contamination of the freeze preparation.

Photo credit: W.H. Siegel

Ampules of freeze-dried or frozen living strains can be stored in mechanical refrigerators at −60° C (−76° F), in walk-in cold rooms at 5° C (40° F), or in vacuum-insulated refrigerators (above) automatically supplied with liquid nitrogen at −196° C (−320° F).

The cryoprotective agents necessary for this procedure differ from the ones used in lyophilization. ATCC routinely uses a mixture of glycerol (10 percent), dimethylsulfoxide (5 percent), and nutrient medium for most bacterial strains. These chemicals are taken into the cells and protect the internal membranes from injury caused by freezing.

Cells stored at cryogenic temperatures must be handled carefully when being recovered. Ice crystals can form as vials are warmed and can kill cells that would otherwise survive the technique. Loss of viability is minimal when the sample is thawed rapidly. Sealed vials are thus put in water at 37° C until all ice melts. Then they are opened and the contents transferred to nutrient medium.

Ultra-frozen cultures must be maintained at very low temperatures at all times during storage. Liquid nitrogen freezers are therefore needed. Proper precautions are important to ensure that such freezers operate properly and have sufficient supplies of liquid nitrogen coolant over long periods of time. Thus, the technique can cost more than freeze-drying, both in labor and in materials necessary to maintain storage temperatures. Ultra-freezing is reserved for microbial species that are not amenable to other, less costly procedures.

Other Methods

Microbial cell cultures can be stored for short periods of time if the culture is overlaid with sterile mineral oil. The oil prevents dehydration and reduces the metabolic rate of the organisms (16). Cells are grown on either nutrient gels or in broth cultures. After incubation and growth, mineral oil is added to the culture to a depth of about 2 centimeters. The cultures are then stored at approximately 4° C. Recovery is by procedures similar to those used for routine subculture. Although cultures preserved in this way have remained viable for as long as 3 years, the method is not considered appropriate for long-term storage of micro-organisms, because cultures have to be recovered, authenticated, and restored every few years (19).

A few micro-organisms, like some of the actinomycetes and some soil-borne spore-forming bacteria, can withstand normal freezing processes and retain both viability and genetic stability. Cultures are stored on a nutrient medium at 0° to −20° C. Cells may remain viable for as long as 2 years, but damage by ice crystal formation is thought to be extensive (19). For certain microbial cells, low-temperature freezing is inexpensive and useful for a short period. Like immersion in mineral oil, however, repetitive recovery, authentication, and restorage make this technique too labor-intensive for long-term maintenance of micro-organisms.

The majority of microbial cells die if they are dried at ambient temperatures (16). But a few can withstand dehydration and remain viable for moderate periods of time. Spore-forming bacteria are particularly suited to this method of storage. Microbial cells are usually transferred to a sterile, solid material, and then dehydrated under vacuum. A soil, paper, or ceramic bead medium is frequently used. Cells also may be suspended in gelatin solution and then drops of the gelatin dried under vacuum. Once dehydrated, the cells must be stored in desiccators but will remain viable longer if refrigerated. Recovery of the cells is by rehydration with nutrient medium and subculture. This method is relatively inexpensive and routinely used for some important bacterial genera (e.g., *Rhizobium*). However, other techniques, if available, are preferred for long-term storage (19).

NEEDS AND OPPORTUNITIES

No major technical constraints limit the storage and use of micro-organisms in the United States and most other industrial nations, although developing countries lag behind in the application, use, and technological knowledge of micro-organisms (28). The most satisfactory long-term storage techniques for micro-organisms are also those that are technologically the most sophisticated. In the case of cryogenic storage, the technique requires a dependable source of liquid nitrogen. Lyophilization of cultures in sealed, evacuated glass-ampules creates culture units with excellent longevity, even when stored at room temperature. The attraction of this method for developing countries is diminished slightly by the relatively high initial cost of equipment and problems keeping such equipment operational (19).

Improvisation in the laboratories of developing-country scientists has resulted in a wide array of variations of standard preservation methods (19). These modified methods are, in many cases, satisfactory to the individual collection curators, though most require micro-organisms to be regularly subcultured. This requirement makes these methods suitable only for relatively small collections, and it increases the likelihood of strains becoming genetically adapted to culture and losing their original characteristics.

Catalog Collections

Maintaining and distributing a current catalog is the goal of virtually every collection curator. Without such a compilation to provide potential users with ready access to its contents, the value of a collection is greatly diminished. That very few collections are adequately cataloged is a reflection of just how onerous this task can be. In a sense, this aspect of collection management has been constrained by lack of an appropriate technology (19). The advent of microcomputers and highly adaptable, user-friendly database management systems software heralds a new era enabling a curator to compile, print, and update a catalog inexpensively and with relative ease (9). Such electronic catalogs would make current collections more accessible and improve their management (4).

One way to catalog the contents of various collections is through creation of a National Microbial Resource Network. Two main obstacles can be anticipated to such a network:

1. the differences in history, traditions, and independence of existing collections; and
2. the difficulty of standardizing technical and informational protocols to assure meaningful interchange of germplasm and data among participating network institutions.

One network of microbiological resource centers (MIRCENs) has been addressing these obstacles for almost a decade (12) (see ch. 10). In practice, however, this endeavor has been frustrated by an inability to come up with standardized data sets that reconcile the different orientations of individual collections worldwide and by financial resources that fall far short of the level required (12).

A more focused attempt to achieve an integrated microbial resource database was initiated in 1984 by UNESCO. This MIRCEN project is still being developed and provides for standardization of a minimum data set for characterized strains of *Rhizobium* (the bacteria involved in nitrogen fixation in soybeans, alfalfa, beans, and other legumes); adoption of compatible database management systems; and periodic publication of an integrated catalog of the collections held at Beltsville, Maryland (USA), Porto Alegre (Brazil), Nairobi (Kenya), and Maui, Hawaii (USA).

An appraisal of the lessons learned in the international MIRCEN effort could greatly benefit establishment of a National Microbial Research Network. Reservations may be expressed about whether the institution-building rationale for the MIRCEN program will mean the collections are of less-than-optimal quality; nevertheless, with limited financial resources, the MIRCEN program has achieved a high degree of network effectiveness, including regular global and regional newsletters, electronic data exchange, and computer conferences.

Develop Methods for Isolation and Culture

An important challenge to maintaining micro-organisms offsite is the development of methods for culture of those organisms that have not yet been isolated in the laboratory (19). Basic research into the isolation and cultivation of these fastidious micro-organisms is essential to further applications of the world's microbial diversity. Research on microbial culture would allow better characterization of diversity in natural environments as well as enable more efficient handling of difficult micro-organisms in existing collections. Efforts to isolate the *Legionella* micro-organism or the retrovirus (human T-cell leukemia-lymphoma virus) associated with acquired immune deficiency syndrome (AIDS) illustrate both the difficulty and importance of such research.

An appreciation of the complexity of the technical barriers faced by microbiologists trying to isolate and culture many micro-organisms is necessary to support research. Basic studies of microbial physiology, through grant programs and in-house research by such agencies as the National Science Foundation, U.S. Department of Agriculture, and the National Institutes of Health (NIH), can improve present abilities to isolate and culture micro-organisms.

Study of Microbial Ecology

Another research priority is that of microbial interactions that permit efficient functioning of the microbial flora of an environment and, ultimately, support higher organisms in that environment. Research into microbial ecology is an integral part of any strategy to preserve micro-organisms. Present understanding of microbial ecology and the extent of microbial diversity in ecosystems is, however, inadequate (19,25). Many plants and animals depend on bacteria and fungi in the environment to survive (25). In some cases, such as digestion in the termite or dairy cattle, microbes are important parts of the organism's basic physiology. Study of the soil micro-organisms that are active in nutrient recycling, such as those associated with nitrogen fixation, are of great potential importance to agriculture.

Grant programs and in-house research at agencies such as NIH, USDA, the Department of Energy, and the Environmental Protection Agency could focus on improved understanding of microbial ecology. Present efforts are spread over several agencies with little coordination. Examination of the overall efforts relating to microbial ecology and diversity could lead to better coordination of research and development of a specific funding program within one agency that would address microbial ecology research.

CHAPTER 8 REFERENCES

1. Allen, E.B., research assistant professor, Utah State University, Logan, UT, personal communication, May 14, 1986.
2. "ATCC Collection Emphasizes Living Plant Pathogenic Fungi," *ATCC Quarterly Newsletter* 3(1):5, 1983.
3. Baird, J.K., Sandford, P.A., and Cottrell, I.W., "Industrial Applications of Some New Microbial Polysaccharides," *Bio/Technology*, November 1983, pp. 778-783.
4. Baker, D., research scientist, Battelle Kettering Laboratory, Yellow Springs, OH, personal communication, Feb. 10, 1986.
5. Brenner, D.J., "Impact of Modern Taxonomy on Clinical Microbiology," *ASM News* 49:58-63, 1983.
6. Brill, W.J., "Agricultural Microbiology," *Scientific American* 245:198-215, 1981.
7. Brock, T.D., *Biology of Micro-Organisms* (Englewood Cliffs, NJ: Prentice-Hall, 1979).
8. Clarke, P.H., "Microbial Physiology and Biotechnology," *Endeavour, New Series* 9:144-148, 1985.
9. Colwell, R.R., "Computer Science and Technology in a Modern Culture Collection," *The Role of Culture Collections in the Era of Molecular Biology*, R.R. Colwell, (ed.) (Washington, DC: American Society for Microbiology, 1976).
10. Colwell, R.R., vice president for academic affairs and professor of microbiology, University of Maryland, Adelphi, personal communication, July 7, 1986.
11. Colwell, R.R., vice president for academic affairs and professor of microbiology, University of Maryland, Adelphi, personal communication, Aug. 26, 1986.
12. Colwell, R.R., "Microbiological Resource Centers," *Science* 233:4762, 1986.
13. Demain, A.L., "Industrial Aspects of Maintaining Germplasm and Genetic Diversity," *The Role of Culture Collections in the Era of Molecular Biology*, R.R. Colwell (ed.) (Washington, DC: American Society for Microbiology, 1976).
14. Dixon, B., *Invisible Allies: Microbes and Man's Future* (London: Temple Smith, 1976).
15. Gerhardt, P., Murray, R.G.E., Costilow, R.N., Nester, E.N., Wood, W.A., Krieg, N.R., and Phillips, G.B. (eds.), *Manual of Methods for General Bacteriology* (Washington, DC: American Society for Microbiology, 1981).
16. Gherna, R.L., "Preservation," *Manual of Methods for General Bacteriology*, P. Gerhardt, et al. (eds.) (Washington, DC: American Society for Microbiology, 1981).
17. Gibbons, N.E., "Reference Collections of Bacteria—The Need and Requirements for Type Strains," *Bergey's Manual of Systematic Bacteriology*, N.R. Krieg and J.G. Holt (eds.) (Baltimore, MD: Williams & Wilkins, 1984).
18. Grimes, D.J., Singleton, F.L., and Colwell, R.R., "Allogenic Succession of Marine Bacterial Communities in Response to Pharmaceutical Waste," *Journal of Applied Bacteriology* 57:247-261, 1984.
19. Halliday, J., and Baker, D., "Technologies To Maintain Microbial Diversity," OTA commissioned paper, 1985.
20. Henahan, J., "Feisty Gram-Negative Aerobes Succumb to New Antibiotic Group," *Journal of the American Medical Association* 248:2085-2086, 1982.
21. Holmes, B., and Owen, R.J., "Proposal That *Flavobacterium breve* Be Substituted as the Type Species of the Genus in Place of *Flavobacterium aquatile* and Amended Description of the Genus *Flavobacterium*: Status of the Named Species of *Flavobacterium*," *International Journal of Systematic Bacteriology* 29:416-426, 1979. *In*: Halliday and Baker, 1985.
22. Klausner, A., "Bacteria Living at 350° May Have Industrial Uses," *Bio/Technology*, 1983, pp. 640-641.
23. Klausner, A., "Microbial Insect Control," *Bio/Technology*, May 1984, pp. 408-419.
24. Krieg, N.R., "Enrichment and Isolation," *Manual of Methods for General Bacteriology*, P. Gerhardt, et al. (eds.) (Washington, DC: American Society for Microbiology, 1981).
25. Margulis, L., Chase, D., and Guerrero, R., "Microbial Communities," *BioScience* 36:160-170, 1986.
26. National Academy of Sciences, *Microbial Processes: Promising Technologies for Developing Countries* (Washington, DC: National Academy Press, 1979).
27. Peele, E.R., Singleton, F.L., Deming, J.W., Cavari, B., and Colwell, R.R., "Effects of Pharmaceutical Wastes on Microbial Populations in Surface Waters at the Puerto Rico Dump Site in the Atlantic Ocean," *Applied and Environmental Microbiology* 41:873-879, 1981.
28. Pramer, D., "Microbiology Needs of Developing Countries: Environmental and Applied Problems," *ASM News* 50(5):207-209, 1984.

29. Sonea, S., and Panisset, M., *A New Bacteriology* (Boston, MA: Jones & Bartlett Publishers, 1983).
30. U.S. Congress, Office of Technology Assessment, *Impacts of Applied Genetics. Micro-Organisms, Plants, and Animals,* OTA-HR-132 (Washington, DC: U.S. Government Printing Office, April 1981).
31. U.S. Congress, Office of Technology Assessment, *Commercial Biotechnology: An International Analysis,* PB #84-173 608 (Springfield, VA: National Technical Information Service, 1984).
32. Walker, J.D., and Colwell, R.R., "Some Effects of Petroleum on Estuarine and Marine Micro-Organisms," *Canadian Journal of Botany* 21:305-313, 1975.
33. White, P., "No Fungus Among Us?" *Sierra*, January/February 1986.

Part III
Institutions

Chapter 9
Maintaining Biological Diversity in the United States

CONTENTS

Tables

Figure

Chapter 9
Maintaining Biological Diversity in the United States

HIGHLIGHTS

- Many U.S. public laws and programs addressing the use of natural resources and the activities of private groups contribute significantly to the conservation of biological diversity. However, diversity is seldom an explicit objective, and where it is mentioned, it is not well-defined. The resulting ad hoc coverage is too disjunct to address the full range of concerns over the loss of diversity.
- Existing laws and programs focus on either onsite or offsite conservation, which impedes establishment of effective linkages between the two general approaches to maintaining diversity. Links help define common interests and areas of potential cooperation between various institutions—important steps in defining areas of redundancy, neglect, and opportunity.
- Personnel of federally mandated programs that deal directly with maintenance of biological diversity, such as the National Plant Germplasm System and the Endangered Species Program, have stretched budgets to meet their mandated responsibilities. It appears, however, that these programs will be unable to continue to meet their mandates without significant increases in funding and staffing.

OVERVIEW

Federal legislation authorizes onsite conservation of species and communities and offsite collection and development of plant and animal species of economic importance. The Federal Government consequently supports programs for agricultural plant and animal conservation and for onsite conservation of selected species, but little consideration is given to a myriad of other diversity maintenance objectives. The numerous Federal onsite programs are not well-coordinated to promote a comprehensive approach. State and private efforts fill some gaps, but in many cases, maintaining diversity is not a specific objective, merely a result.

Many organizations or programs discussed in this chapter focus on one aspect of diversity maintenance: plant seeds, rare animal breeds, or onsite conservation of endangered species. This chapter considers Federal mandates related to diversity conservation, onsite conservation, offsite plant and animal conservation, and microbial conservation. In each case, Federal, State, and private activities are assessed, although these categories are arbitrary and, in fact, biological diversity maintenance programs frequently fall into more than one category.

FEDERAL MANDATES AFFECTING BIOLOGICAL DIVERSITY CONSERVATION

No Federal law specifically mandates the maintenance of biological diversity, either offsite or onsite, as a national goal. The term itself is used only in Title VII of the Foreign Assistance Act of 1983 (discussed in ch. 11). A number of Federal laws require the conservation of resources on Federal lands, however, or require that consideration be given to resources in Federal agency activities. Offsite maintenance of agricultural plant germplasm diversity is mandated indirectly through legislation authorizing the National Plant Germplasm System (discussed later in this chapter). But offsite maintenance of wild plants, wild animals, and microbial resources is not explicitly mandated by Federal legislation.

The lack of a comprehensive Federal onsite policy leads to uncoordinated programs, frequently leaving important gaps in conservation. Generally, Federal agencies coordinate conservation activities onsite for species that are specifically mentioned in Federal protection laws, but this coordination frequently does not extend to nonlegislated species. For example, onsite conservation can be coordinated among Federal agencies for threatened and endangered species under the Endangered Species Act of 1973 (Public Law 93-205). But no formal institutional mechanism exists to coordinate conservation of thousands of plant, animal, and microbial species not recognized as threatened or endangered.

Offsite germplasm conservation mandates are equally vague. For example, the Agricultural Marketing Act of 1946 is intended to "promote the efficient production and utilization of products of the soil" (7 U.S.C.A. 427), but it is interpreted narrowly by the Agricultural Research Service to mean domesticated plant species and varieties. Little consideration has been given to conservation of wild plant species.

Federal mandates give even less attention to offsite conservation of domesticated and wild animals. Legislative authority is vague and provides little direction to the Agricultural Research Service.

Table 9-1 lists the major Federal mandates pertinent to diversity maintenance. Species protection laws authorize Federal agencies to manage specific animal populations and their habitats onsite. Legislation on the protection of natural areas authorizes the acquisition or designation of habitats and communities that help maintain a diversity of natural areas under Federal stewardship. Federal laws for offsite maintenance of plants authorize conservation and development (or enhancement) primarily of plant species that demonstrate potential economic value. Offsite maintenance of domestic animal germplasm is authorized indirectly by the Agricultural Marketing Act of 1946. The Endangered Species Act of 1973 is in both categories of table 9-1 because it authorizes wild plant and animal species protection, habitat protection, and offsite conservation for those species considered threatened or endangered in the United States.

Although the National Forest Management Act of 1976 (Public Law 94-588) is the only Federal legislation that includes in its mandate the onsite conservation of a "diversity of plant and animal communities," it offers no explicit congressional direction on the meaning and scope of onsite maintenance of biological diversity. Interpretation of this provision has been a difficult process and has involved lengthy consultation with scientists and managers around the country (50). The U.S. Forest Service ultimately decided the law gave them a mandate to maintain terrestrial vertebrate species diversity and the structural timber stands on all Forest Service lands in conjunction with planning and management processes (44 F.R. 53967-53779). Whether this interpretation fulfills the congressional intent on conserving diversity has not been challenged.

Such terms as *biological resources*, *wildlife*, *animals*, and *natural resources* can and have been interpreted differently by Federal agen-

Table 9-1.—Federal Laws Relating to Biological Diversity Maintenance

Common name	Resource affected	U.S. Code
Onsite diversity mandates:		
Lacey Act of 1900	wild animals	16 U.S.C. 667, 701
Migratory Bird Treaty Act of 1918	wild birds	16 U.S.C. 703 et seq.
Migratory Bird Conservation Act of 1929	wild birds	16 U.S.C. 715 et seq.
Wildlife Restoration Act of 1937 (Pittman-Robertson Act)	wild animals	16 U.S.C. 669 et seq.
Bald Eagle Protection Act of 1940	wild birds	16 U.S.C. 668 et seq.
Whaling Convention Act of 1949	wild animals	16 U.S.C. 916 et seq.
Fish Restoration and Management Act of 1950 (Dingell-Johnson Act)	fisheries	16 U.S.C. 777 et seq.
Anadromous Fish Conservation Act of 1965 (Public Law 89-304)	fisheries	16 U.S.C. 757a-f
Fur Seal Act of 1966 (Public Law 89-702)	wild animals	16 U.S.C. 1151 et seq.
Marine Mammal Protection Act of 1972	wild animals	16 U.S.C. 1361 et seq.
Endangered Species Act of 1973 (Public Law 93-205)	wild plants and animals	7 U.S.C. 136 16 U.S.C. 460, 668, 715, 1362, 1371, 1372, 1402, 1531 et seq.
Magnuson Fishery Conservation and Management Act of 1977 (Public Law 94-532)	fisheries	16 U.S.C. 971, 1362, 1801 et seq.
Whale Conservation and Protection Study Act of 1976 (Public Law 94-532)	wild animals	16 U.S.C. 915 et seq.
Fish and Wildlife Conservation Act of 1980 (Public Law 96-366)	wild animals	16 U.S.C. 2901 et seq.
Salmon and Steelhead Conservation and Enhancement Act of 1980 (Public Law 96-561)	fisheries	16 U.S.C. 1823 et seq.
Fish and Wildlife Coordination Act of 1934	terrestrial/aquatic habitats	16 U.S.C. 694
Fish and Game Sanctuary Act of 1934	sanctuaries	16 U.S.C. 694
Historic Sites, Buildings, and Antiquities Act of 1935	natural landmarks	16 U.S.C. 461-467
Fish and Wildlife Act of 1956	wildlife sanctuaries	15 U.S.C. 713 et seq. 16 U.S.C. 742 et seq.
Wilderness Act of 1964 (Public Law 88-577)	wilderness areas	16 U.S.C. 1131 et seq.
National Wildlife Refuge System Administration Act of 1966 (Public Law 91-135)	refuges	16 U.S.C. 668dd et seq.
Wild and Scenic Rivers Act of 1968 (Public Law 90-542)	river segments	16 U.S.C. 1271-1287
Marine Protection, Research and Sanctuaries Act of 1972 (Public Law 92-532)	coastal areas	16 U.S.C. 1431-1434 33 U.S.C. 1401, 1402, 1411-1421, 1441-1444
Federal Land Policy and Management Act of 1976 (Public Law 94-579)	public domain lands	7 U.S.C. 1010-1012 16 U.S.C. 5, 79, 420, 460, 478, 522, 523, 551, 1339 30 U.S.C. 50, 51, 191 40 U.S.C. 319 43 U.S.C. 315, 661, 664, 665, 687, 869, 931, 934-939, 942-944, 946-959, 961-970, 1701, 1702, 1711- 1722, 1731-1748, 1753, 1761-1771, 1781, 1782
National Forest Management Act of 1976 (Public Law 94-588)	national forest lands	16 U.S.C. 472, 500, 513, 515, 516, 518, 521, 576, 581, 1600, 1601-1614
Public Rangelands Improvement Act of 1978 (Public Law 95-514)	public domain lands	16 U.S.C. 1332, 1333 43 U.S.C. 1739, 1751- 1753, 1901-1908
Offsite diversity mandates:		
Agricultural Marketing Act of 1946 (Research and Marketing Act)	agricultural plants and animals	5 U.S.C. 5315 7 U.S.C. 1006, 1010, 1011, 1924-1927, 1929, 1939-1933, 1941-1943, 1947, 1981, 1983, 1985, 1991, 1992, 2201, 2204, 2212, 2651-2654, 2661-2668 16 U.S.C. 590, 1001-1005 42 U.S.C. 3122
Endangered Species Act of 1973 (Public Law 93-205)	wild plants and animals	7 U.S.C. 136 16 U.S.C. 460, 668, 715, 1362, 1371, 1372, 1402, 1531 et seq.
Forest and Rangeland Renewable Resources Research Act of 1978 (Public Law 95-307)	tree germplasm	16 U.S.C. 1641-1647

NOTE: Laws enacted prior to 1957 are cited by Chapter and not Public Law number.

SOURCE: Office of Technology Assessment, 1986.

cies. *Wildlife*, for example, has been defined in a number of ways, including the following:

- mammals that are hunted or trapped (game);
- all mammals—the word *animal* is sometimes used interchangeably with mammal;
- all animals, both vertebrates and invertebrates, excluding fish; and
- all animals, both vertebrates and invertebrates, including fish (65).

These definitional differences are further evidence of the lack of a comprehensive Federal approach to these issues.

ONSITE BIOLOGICAL DIVERSITY MAINTENANCE PROGRAMS

U.S. onsite programs seem to have one of three main objectives: 1) maintenance of diverse habitats or ecosystems, 2) preservation of individual species through habitats' protection, and 3) restoration of habitats to their natural condition. These objectives are not necessarily exclusive. Safeguarding communities and ecosystems could help protect rare species. Protecting the habitat of a species may conserve an ecosystem or community. And restoring habitats could enhance the diversity of species within an ecosystem.

Ecosystem Diversity Maintenance

Maintaining ecosystems is the only way to ensure the continued viability and evolutionary processes of the organisms within these areas (see ch. 5). Numerous mechanisms exist at the Federal, State, and local level to manage land and water areas for their maintenance. The net result is the continued existence of a diversity of ecosystems in the United States.

Ecosystem diversity maintenance within Federal, State, and private holdings depends on the degree of protection given to the area, its size, and the impact of external influences. Protection of ecosystem diversity within land and water designations ranges from scant to strict. The use of land and waters in the National Wilderness Preservation System is greatly restricted—generally, motorized vehicles and long-term human activities are prohibited. Some wilderness areas are regularly patrolled and violators cited. Others receive little regulatory attention. At the other extreme, estuarine sanctuaries are not required to have any Federal protection; jurisdiction over any use is determined exclusively by the States. One preliminary assessment concluded that privately owned, legally secured, single-purpose nature reserves offer the greatest protection to biological diversity (10).

The size of a designated area and proximity to other land designations also influence its contribution to onsite diversity (10). Some Research Natural Areas (RNAs), for example, are well-protected but may be very small (the smallest is only 2 acres). Numerous vertebrates and larger plants would not be able to survive and reproduce successfully in a small "island" habitat; therefore, small RNAs contribute little to community diversity maintenance.

Natural areas are influenced by human activities on surrounding land that reduce the area's ability to sustain natural biological communities. The National Park Service has reported that 55 percent of the threats to park natural resources come from influences outside park boundaries (64). The National Wildlife Refuge System also noted that influences from adjacent areas were harming the fish and wildlife within refuges (63). Concern over such threats has prompted introduction of legislation to minimize negative effects of activities conducted in adjacent areas.

Table 9-2 provides a summary of the Federal ecosystem conservation programs in which designated areas are maintained in a relatively natural condition. The land designations included are only some of more than 100 categories used by Federal agencies. Some programs involve more than one agency, such as the Re-

Table 9-2.—Examples of Federal Ecosystem Conservation Programs

Program title and responsible Federal agency or agencies	Number of units	Acres (millions)
National Natural Landmarks		
National Park Service	10	0.95
U.S. Forest Service	48	0.69
Bureau of Land Management	45	10.56
Fish and Wildlife Service	3.15	
Federal Aviation Administration	1	0.003
Department of Energy	1	0.13
Department of Defense	16	0.19
Department of Transportation	3	0.014
Bureau of Reclamation	1	0.032
Research Natural Areas		
National Park Service	66	2.3
U.S. Forest Service	151	0.184
Department of Energy	2	0.75
Fish and Wildlife Service	194	1.94
Bureau of Land Management	18	0.048
Department of Defense	4	0.006
Tennessee Valley Authority	4	0.0001
Bureau of Indian Affairs	1	0.0009
Wild and Scenic Rivers		
National Park Service	23	1,927 miles
Bureau of Land Management	15	1,367 miles
U.S. Forest Service	23	2,098 miles
Fish and Wildlife Service	7	1,043 miles
Biosphere Reserves[a] *(Man and the Biosphere)*		
National Park Service	25	25.09
U.S. Forest Service	18	1.63
Fish and Wildlife Service	4	2.81
Bureau of Land Management	1	0.034
Agriculture Research Service	2	0.209
National Oceanic and Atmospheric Administration	3	0.633
Wilderness Areas		
U.S. Forest Service	332	31.84
Bureau of Land Management	22	0.37
National Park Service	38	36.78
Fish and Wildlife Service	65	19.33
National Parks		
National Park Service	337	79.44
National Monuments		
National Park Service	77	4.72
National Preserves		
National Park Service	12	21.10
National Rivers		
National Park Service	4	0.359
National Forests		
U.S. Forest Service	152	190.4
Experimental Forests, Ranges, and Watersheds		
U.S. Forest Service	88	0.240
Experimental Ecological Reserves		
U.S. Forest Service	27	0.219
Fish and Wildlife Service	2	0.057
Bureau of Land Management	2	0.022
National Oceanic and Atmospheric Administration	1	0.006
Agriculture Research Service	4	0.100
National Park Service	1	0.093
Tennessee Valley Authority	1	0.069
Smithsonian Institution	1	0.001
National Wildlife Refuges		
Fish and Wildlife Service	424	89.9
Outstanding Natural Areas Management		
Bureau of Land Management	37	0.377
Areas of Critical Environmental Concern		
Bureau of Land Management	236	1.94
Marine Sanctuaries		
National Oceanic and Atmospheric Administration	7	2,322 (sq. nautical miles)
Estuarine Sanctuaries		
National Oceanic and Atmospheric Administration	17	268,762 (sq. nautical miles)
National Environmental Research Parks		
Department of Energy	5	1.15

NOTE: When more than one agency has responsibility for an area, acreage has been divided equally and each agency receives credit for an area.
[a]Because biosphere reserves are managed by several agencies simultaneously, the total number (53) in the table exceeds the actual number of reserves (43).

SOURCE: Adapted from W.D. Crumpacker, "Status and Trends of U.S. Natural Ecosystems," OTA commissioned paper, 1985; M. Bean, "Federal Laws and Policies Pertaining to the Maintenance of Biological Diversity on Federal and Private Lands," OTA commissioned paper, 1985.

search Natural Area Program. Other programs are under the jurisdiction of just one agency, such as the National Forest System.

Few programs are designed specifically to maintain biological diversity, even though some programs may indirectly have this as one of their objectives. One exception is the Man and the Biosphere Program, coordinated through the United Nations Educational, Scientific, and Cultural Organization (UNESCO), which considers the onsite maintenance of biological diversity a major goal (17). The U.S. network of 43 biosphere reserves provides a framework for linking complementary protected areas in particular biogeographical regions and for conducting research on strategies for managing ecosystems to conserve diversity (22). The U.S.

Photo credit: National Park Service

Great Smoky Mountains National Park, site of 1 of 43 biosphere reserves in the United States.

program, unlike programs in other countries, is strictly voluntary; designation is used mainly to encourage cooperation and increase use for scientific and educational purposes.

Research Natural Areas and Experimental Ecological Areas are designated by appropriate Federal agencies and the National Science Foundation, respectively, to conserve natural ecological communities for research in natural community manipulation. A Federal Committee on Ecological Reserves was established in 1974 to coordinate designation of these sites, in part to ensure that each community type was included in the system (4). The coordinating committee still exists nominally, but it no longer provides an advisory function. Designations of Research Natural Areas are currently determined independently by each Federal agency.

A variety of management options exist within programs that consider diversity an objective. For example, national forests are directed by law (National Forest Management Act) to be managed in a way that sustains plant and animal diversity. At the individual forest level, supervisors have flexibility in determining how and to what extent vertebrate species diversity will be considered in forest operations.

Similarly, National Wildlife Refuges and National Parks consider maintaining diversity an objective, although this attitude is not supported by specific mandate. National Wildlife Refuge managers may try to maintain a diversity of species with the existing habitat or may manipulate areas to create a diversity of habitats. In some cases, refuges are managed exclusively for a single species. National Parks have to bal-

Photo credit: National Park Service

Hearts Content Scenic Area National Natural Landmark, part of the U.S. Forest Service's Allegheny National Forest, protects a rare example of virgin, old-growth forest dominated by white pines of up to 400 years old.

ance maintenance of diversity with recreation opportunities. In some cases, these goals conflict, as they have with managing the grizzly bears at Glacier National Park.

Contributions of these Federal programs to maintaining diversity depends on the degree of protection offered for each designation (10). For example, National Natural Landmarks are designated to identify and conserve unique, rare, or representative communities in the United States. Designation of these sites does not, however, include protecting the site from human alteration. Approximately half the National Natural Landmarks exist on private lands, where conservation depends on the good will of the individual landowner. National Natural Landmarks on Federal- or State-controlled lands require the cooperation of the authorized agency to ensure that protection is considered in the area's management.

An attempt has been made to identify the amount of potential ecosystem diversity that is protected in Federal landholdings. Potential ecosystem diversity is that which would be expected to develop on a site under natural conditions. According to an assessment that considered areas of approximately 23,000 acres or larger, lands of four agencies failed to include 22 percent of the recognized ecosystem types (i.e., 69 out of 315). These four agencies were the National Park Service, Forest Service, Fish and Wildlife Service, and Bureau of Land Management. Another 29 percent of these ecosystem types were only minimally included (9). Since this analysis assessed only potential diversity, it probably underestimates existing ecosystem diversity in the landholdings (9). The largest number of unrepresented types were in Texas and Oklahoma, which have relatively large amounts of ecosystem diversity but relatively few Federal lands.

Another analysis of the same Federal holdings, using a different classification scheme for potential ecosystem diversity, obtained similar results (12). These two studies indicate that any attempt to include all ecosystem types within Federal programs would require considerable expansion of existing holdings. For the national wilderness preservation system, however, almost half the unrepresented types in that system could be added from existing Federal agency holdings (12).

Natural area management programs also occur at the State level. State parks, forests, and protected sites may be managed for one or a few resources, but they help preserve some remnants of diversity, particularly when they are managed in conjunction with private or Federal reserves. State designations could also be wildlife areas, fishing areas, university research stations, botanic sites, school and other public lands, or special districts (e.g., a water management district) (10).

Private holdings also contribute to maintaining diversity, especially through the protection of remnant areas. Many of the remaining tall grass prairies in the Midwest, for example, are privately owned by individuals or as railroad right-of-ways. Private land trusts lease parcels of land for biological or historical significance, which may contribute to onsite diversity maintenance. (For further details, see ref. 55.) Many of the land parcels are small and isolated, with little attention given expressly to diversity maintenance, but they do contribute to the patchwork of natural areas in the United States. An assessment of the protection associated with all these Federal, State, local, and private land designations is under way (11).

One private institution with an explicit goal of natural area preservation is The Nature Conservancy (TNC). TNC is a nonprofit organization with chapters in most States. Its objectives are to identify species and community diversity onsite, purchase areas or work with landholders to protect the species or community, and manage areas to ensure the continued existence of the species or community.

TNC, through State Natural Heritage Programs (discussed in ch. 5 and in the next section), conducts field investigations of rare, threatened, or endangered organisms and communities across the Nation. The information generated from these surveys helps identify organisms that should be given Federal or State protected status, as well as habitats and communities where special attention is necessary.

TNC, one of the largest private landholders in the United States, owns 895 preserves (39). In addition, it works with Federal, State, and local governments to designate protected areas. Thus, the organization, through a grassroots approach, is effectively identifying and maintaining a diversity of rare species or community types in the United States.

Species Habitat Protection

The most comprehensive national program for the protection of species diversity and their habitats is the Endangered Species Program, authorized under the Endangered Species Act of 1973. The program authorizes the Secretary of the Interior, through the U.S. Fish and Wildlife Service (FWS), and the Secretary of Commerce, through the National Marine Fisheries Service (NMFS), to protect endangered and threatened species of plants and animals in the United States and elsewhere.

The program, administered by the Office of Endangered Species, has several phases: listing species, developing recovery plans, and managing species' habitats. Species are listed as threatened or endangered when sufficient information on the status and distribution of the species suggests significant declines in population or range or both and when an extensive public review has been completed. In general, a species is considered a candidate between the time a petition to propose a species is received and the listing process is completed. In addition, many candidate lists are put together through expert review by the regional and Washington offices of the FWS.

Lists of candidates are published periodically in the *Federal Register*. (The most recent lists were published in September 1985 for vertebrates and plants and in May 1984 for invertebrates.) To date, approximately 3,900 species and subspecies of plants, vertebrates, and invertebrates are candidates compared with approximately 385 species already listed (18).

When a species is listed, the next step is development of a formal recovery plan outlining the responsibilities of all parties with jurisdiction over the species' habitat and their management roles. Recovery plans are advisory documents to the Secretary of the Interior, not binding agreements. Recovery plans are approved or awaiting approval for approximately two-thirds of the species listed (see table 9-3). Implementation of recovery activities, however, has been slow (13).

The thrust of the Endangered Species Program is protection through proper management of a species' habitat. Most management activities are carried out by Federal and State agencies with jurisdiction over the habitats, not by FWS (unless the species occur on National

Table 9-3.—Number of U.S. Species at Various Stages of Listing and Recovery as of 1985

Species identified as candidates for listing	3,908
Candidates with completed status research	964
Species listed as threatened or endangered	383
Species with approved recovery plans	223
Species recovering	22

SOURCE: J. Fitzgerald and G.M. Meese, *Saving Endangered Species* (Washington, DC: Defenders of Wildlife, 1986).

Wildlife Refuges). These habitats, often called critical habitats—because the species depend on these areas for survival or reproduction—are designated either in conjunction with, or subsequent to, the listing of a species. To date, critical-habitat designations have been made for only about 70 listed species (13).

A federally listed threatened or endangered species is protected from any federally authorized activity that may jeopardize its continued existence, even when the activity occurs entirely on private land (4). Any Federal agency undertaking or authorizing a project in the range of an endangered species must consult with FWS or NMFS to ensure that the impacts on a listed species will be minimal. The consultation requirement is one of the most effective parts of the program in protecting threatened or endangered species (4). It is one of the least well-funded areas of the Endangered Species Program, however.

To a limited degree, efforts to manage a threatened or endangered species involve off-site techniques, such as artificial propagation of plants and captive breeding programs for animals. Efforts to recover several species of large birds (e.g., the peregrine falcon and whooping crane) demonstrate the success of such techniques. In some cases, captive breeding programs provided the opportunity for species to be reintroduced into their historic range.

In addition to Federal activities, State agencies may receive Federal funding to implement species-specific recovery and management efforts. To date, 41 States have approved programs for animals, and 17 have programs for plants (4).

Overall, the Endangered Species Program effectively maintains species already listed and protected under the law, but it provides insufficient protection for those that are candidates. The program is criticized for the slow pace of candidate review in the listing process. Some animals and plants may have become extinct between the time they were proposed as candidates and their review by FWS (4,18). The Texas Henslow's sparrow and the Schweinitz's waterweed are two such examples (31). This delay underscores the need to list species or take other action in time to prevent their loss.

By publishing lists of candidates in the *Federal Register*, the Endangered Species Office has succeeded in bringing public attention to these candidates. Now the office is working with other Federal agencies to promote consideration of candidate species in agency planning. However, no legislative authority currently protects candidates from adverse impacts of Federal agency actions.

Underfunding and understaffing of the Office of Endangered Species hampers its ability to implement listing, recovery, and consultation objectives (18). With an increased budget, resources would be applied initially to develop recovery plans for all listed species. Consultation among agencies is also severely underfunded. Any funding increase to the Endangered Species Program could be gradual, over 5 years perhaps, so the office could expand existing program efforts. Program growth might involve annual increases of $2 million for State grant programs, $500,000 for species listing, $1.5 million for consultation, and $4 million for recovery plans (54).

Other programs identify and protect selected species, sometimes known as public trust resources, designated in Federal mandates. Examples include migratory bird management and anadromous fish hatchery programs. The programs provide little protection for overall species diversity. The Fish and Wildlife Service, the major Federal agency with authority to manage biological resources, is currently focusing its limited personnel and budget allocations primarily on public trust resources.

One Federal program focusing on public trust resources is the National Wildlife Refuge System, administered by FWS (3). Many of the refuges have been created by revenues from annual waterfowl hunting permits. Consequently, most refuges are purchased to protect habitats for migratory birds. Refuges may also protect habitats of threatened or endangered species (e.g., Atwater Prairie Chicken National Wildlife Refuge in Texas) or large mammals (National Bison Range in Montana), with funding from the Land and Water Conservation Fund, a land trust funded by the sales of grazing leases, offshore oil, mineral rights, and other sources on Federal lands. The Land and Water Conservation Fund is the principal source of money for land purchases by Federal agencies.

Refuges may provide habitats for a diversity of species, but the designation of the refuge is to benefit one or a few species of special interest. Woodland habitats along some east coast refuges, for example, have been converted to grassland-wetland habitats to enhance waterfowl at the expense of overall diversity.

State programs also tend to focus on selected species of fish, wildlife, and plants, although the emphasis differs somewhat from Federal programs. States generally receive revenues from hunters, fishermen, and Federal grants, for management and conservation of harvested species. Interest in nongame species is increasing, however. State agencies, through referendums, are expanding their fish and wildlife programs to a wider array of species' conservation efforts. Public pressure to conserve and manage nongame populations and increased budgets to implement programs (62) are increasing State efforts. However, State nongame programs are funded by add-on monies from tax checkoffs, which hampers the ability of most States to adequately fund or maintain personnel for their nongame programs. In addition, this type of funding severely hampers long-range planning and implementation of nongame projects. An alternative to this type of funding is to provide monies from the State's general fund, as is being done by the Florida Fresh Water Game and Fish Commission (31).

Another State activity is the Natural Heritage Program, a set of public and private programs to protect diversity in each State (46). Each program develops an inventory of the State's rare species and ecosystems and identifies priority actions. The Nature Conservancy establishes and initially supports the programs. In some instances, States will take over the program devised by TNC and incorporate activities into the State government. In other instances, States and TNC share responsibilities.

State Natural Heritage Programs and databases are designed to be compatible so national information on species diversity can be collated. As of February 1986, 44 States had contracted for the program and 26 of these had assumed administration of the program from TNC (30). The Conservancy also maintains four noncontracted programs and has separate contracts with the Tennessee Valley Authority, the Navaho Nation, and Puerto Rico.

Programs' abilities to protect diversity are limited by their resources and the degree of influence they have in the State governments. The Rhode Island program, for instance, although part of the State government, receives its funding from the Federal Fish and Wildlife Service. Thus, its inventory is primarily limited to species identified by the Endangered Species Act. A lack of resources and influence hamper this program's ability to comment on State and Federal developments and State land-acquisitions. The South Carolina program, also part of the State government, is supported by a $400,000 State grant fund and income-tax checkoff (21). With these resources, the program maintains a larger inventory, buys and manages land, and comments on all relevant State and Federal developments.

Programs that are not part of a State government have fewer resources and opportunities to affect Federal and State decisions. Two further constraints are the limited information that programs are able to collect and the lack of a national classification system for natural ecosystems.

Nevertheless, State Natural Heritage Programs perform a function unfulfilled by existing institutions. The continuing inventory of rare species and ecosystems enables the protection of the most important biologically diverse lands and the early identification and modification of potentially destructive development plans.

A variety of private conservation organizations work to protect species of particular interest. The National Audubon Society, for example, maintains some 60 refuges to protect the habitat of endangered species. Many of the first refuges were designated to protect marine and coastal waterbird colonies (15). More recently, sanctuaries are being acquired to protect inland habitats and to restrict development. These areas provide refuge for an array of species, in addition to the key species for which the sanctuary was purchased.

Conservation organizations such as Izaak Walton League of America help maintain diversity through an advocacy role. These groups work with the U.S. Congress and Federal and State agencies to develop laws and programs that reflect the importance of maintaining species. Like Federal programs, diversity conservation is not a stated objective of most nonprofit organizations (except TNC), but their efforts aid in maintaining species and habitat diversity onsite.

Additional groups working for species preservation include single-species organizations or foundations, such as the Carolina Bird Club, Desert Fishes Council, or Trout Unlimited (47). These offices work to promote habitat protection for these organisms, manage habitats for particular species, and advocate survival of these species through Federal and State agencies. The net result is species maintenance and conservation of particular components of biological diversity.

A multitude of nonprofit organizations also function at the local and State level. These groups tend to be small, poorly financed, and focused on a particular area or species of concern. (For further discussion, see ref. 59.) Such organizations generally do not have biological diversity as an exclusive objective, but they contribute to the maintenance of biological diversity through their achievements of preserving a specific species of concern or its habitat.

Onsite Restoration

Another facet of maintaining biological diversity is the restoration of degraded sites. The field is relatively new, few institutions have well-developed programs, and complete restoration has been difficult to achieve. (See ch. 5 for discussion of restoration technologies.)

A key Federal legislation that directly provides for revegetation after a disturbance is the Surface Mining Control and Reclamation Act of 1977, also known as SMCRA (Public Law 95-87). Section 515(b)(19) states that mining operations shall:

> . . . establish . . . a diverse, effective, permanent vegetation cover of the same seasonal variety native to the area of land to be affected.

The number and composition of species is often suggested by past management practices. The Bureau of Land Management (BLM), Forest Service, and Soil Conservation Service have each developed vegetative mixtures for various types of disturbances that can be economically managed and are likely to succeed. There is a problem, however, with the definition of "native." BLM, for instance, interprets native to include introduced exotics that have been established within the area before the project was assessed.

Section 515(b)(2) states that the mine operation shall:

> . . . restore the land affected to a condition capable of supporting the use which it was capable of supporting prior to any mining, or higher or better uses.

Thus, SMCRA provides an incentive to develop techniques for establishing native plant species. The natural diversity aspect of SMCRA could be strengthened at the State level by requiring the use of native species in revegetation mixtures.

A few Federal agencies are initiating restoration efforts. The Forest Service is mandated by the National Forest Management Act of 1976 to replant all lands in the National Forest System that do not regenerate naturally after timber harvesting. Tree monocultures are most likely to be planted, thus reducing diversity instead of restoring the area's original diversity. The National Park Service (NPS) has instituted small-scale restoration projects, mainly for areas affected by past tourist use or other disturbances (32). An exception to the typical small-scale NPS restoration project is the legislatively mandated (Public Law 95-250) Redwood Creek rehabilitation project in Redwood National Park. The project is developing rehabilitation techniques for 36,000 acres of previously logged and seriously eroded slopes in the redwood-mixed conifer ecosystem.

The Fish and Wildlife Service and Environmental Protection Agency identify water bodies polluted by chemicals or acid rain that are suitable for restoration. In lakes damaged by acid rain in the Northeast, for example, FWS has spent $5 million in a liming effort to reduce lake acidity and restore aquatic life (32). The U.S. Army Corps of Engineers is researching wetland restoration techniques to mitigate development projects in wetlands.

One future opportunity to restore diversity is by the U.S. Department of Agriculture's (USDA) implementation of the conservation reserve provision of the Food Security Act of 1985 (Public Law 99-198). The conservation reserve:

> . . . authorizes USDA to contract with farmers to remove 40 million acres of erodible land from row crop production. . . . The retired acres would be planted to grasses, legumes, and trees to reduce erosion and enhance wildlife (66).

This provision could be strengthened if restoration of vegetation in riparian areas were included in the legislation. The reconstruction of debt portion of this bill may be more beneficial to diversity. It allows the farmer to offer up land for not less than 50 years to be used to lower the debt.

Private efforts may be the leading contributors to restoring biological diversity. Although much reclamation is being carried out by industries and consulting firms in compliance with regulations, work is also being done by small organizations and individuals motivated by esthetic and environmental interests. Universities also are conducting research to develop techniques for restoring different ecosystems. Recently, restoration has been identified as a focus of research at the Cary Arboretum in New York and at the Center for Restoration Ecology at the University of Wisconsin (32).

OFFSITE DIVERSITY MAINTENANCE PROGRAMS

Federal programs to maintain diversity offsite generally involve germplasm of agriculturally or economically important plants and animals. Less attention is given to wild plants and animals at the Federal level than at the State or private level. State efforts to conserve a diversity of plants, animals, or micro-organisms offsite are poorly documented and tend to be widely dispersed. Private institutions conduct numerous activities directly related to the maintenance of biological diversity offsite. Consequently, offsite conservation of many biological resources occurs only as a result of private efforts. For ease in discussion, offsite maintenance of biological diversity is divided into plant, animal, and micro-organism programs, although programs overlap considerably.

Plant Programs

Historically, responsibilities for maintaining plant resources at the Federal level included only domesticated plants under the jurisdiction of USDA. Although recent legislation has included some wild plant species (e.g., Forest and Rangeland Renewable Resources Research Act, Rural Development Act), the focus of USDA is still reflected in programs to maintain crop-related germplasm.

Agricultural Plants

The most significant program is the National Plant Germplasm System (NPGS)—a diffuse network of USDA, State, and private institutions, private industry, and individuals. NPGS activities include acquiring, maintaining, and improving plant germplasm. Various components of the system also conduct research that supports preservation of genetic diversity, acquisition of new materials, and use of stored germplasm (see figure 9-1).

Work done by NPGS is in response to specific national needs. Agricultural plant exploration and development of new crop species led to a formal Federal program (Section of Seed and Plant Introduction) in 1898 within USDA (28). Recognizing that germplasm resources were being lost due to inadequate maintenance facilities, Congress enacted the Agricultural Marketing Act in 1946, authorizing regional centers to maintain and develop plant germplasm (27).

Federal contributions to NPGS currently are administered through the Agricultural Research Service (ARS) and the Cooperative State Research Service (CSRS). The ARS National Program Staff in Beltsville, MD, coordinates these various activities and facilities:

- *Advisory Committees:* The National Plant Germplasm Committee and individual crop advisory committees provide both policy and technical advice to administrators and curators of NPGS. The National Plant Genetic Resources Board advises the Secretary of Agriculture on resource issues and serves as liaison between NPGS and the International Board for Plant Genetic Resources.
- *Plant Genetics and Germplasm Institute:* This USDA/ARS facility includes the following:
  —the Plant Introduction Office that coordinates the acquisition of new materials, assignment of introduction numbers, and distribution to appropriate facilities;
  —the Plant Molecular Genetics Laboratory, devoted to developing methods for using germplasm to improve crops;
  —the Germplasm Resources Information Network (GRIN) Database Management Unit, responsible for developing and maintaining the computer-based system that is intended to contain passport, evaluation, and inventory information on NPGS germplasm; and
  —the National Small Grains Collection.
- *National Seed Storage Laboratory:* The National Seed Storage Laboratory (NSSL) in Ft. Collins, CO, is designed to be the principal storage facility for agricultural crop seeds in the NPGS. Ideally, all plant varieties are stored at NSSL as base collections. NSSL is responsible for monitoring the viability of seeds within its collections as well as seeds stored in active collections. The laboratory does not evaluate its samples, however, and depends on other facilities in the network to regenerate samples when germination declines.
- *Germplasm Collections:* National responsibility for maintaining major crops is divided among four Regional Plant Introduction Stations (RPISs). Many important collections are not associated with an RPIS, such as those for soybeans, cotton, sugar crops, and small grains. Germplasm that must be clonally maintained is the responsibility of the five newly established and four developing national clonal repositories. Several collections of genetic or mutant stocks that possess specific traits exist. Although not generally used in breeding, such stocks have been important resources for research on cytogenetics, physiology, biochemistry, and molecular genetics of crops.

The mission of NPGS is to acquire, maintain, evaluate, and make accessible as wide a range of genetic diversity as possible in the form of seed and clonal materials to crop breeders and plant scientists (60). The scientific expertise on germplasm maintenance is among the best available.

Assessments of NPGS during the past 5 years have highlighted shortcomings in coordination, communication, storage facilities, maintenance of seed viability, and staffing levels (7,56,57,60). Facilities such as NSSL, for example, have been criticized for inadequately maintaining seed stocks and for storage limitations. A 1981 study by the General Accounting Office (GAO) found that NPGS curators sent only half the seeds

Figure 9-1.—Principal Components of the National Plant Germplasm System

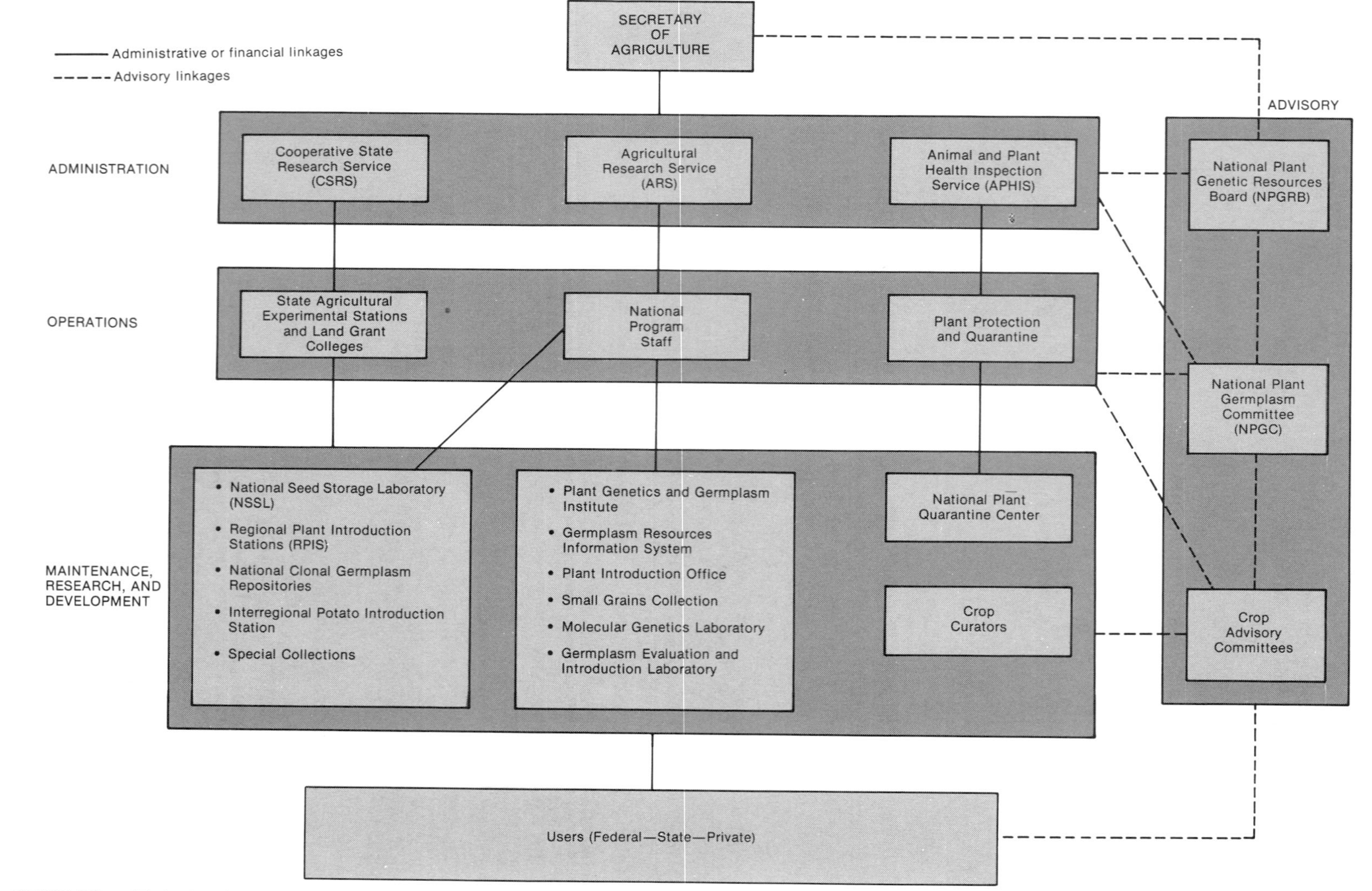
SECRETARY OF AGRICULTURE
Administrative or financial linkages
Advisory linkages
ADMINISTRATION
Cooperative State Research Service (CSRS)
Agricultural Research Service (ARS)
Animal and Plant Health Inspection Service (APHIS)
ADVISORY
National Plant Genetic Resources Board (NPGRB)
OPERATIONS
State Agricultural Experimental Stations and Land Grant Colleges
National Program Staff
Plant Protection and Quarantine
National Plant Germplasm Committee (NPGC)
MAINTENANCE, RESEARCH, AND DEVELOPMENT
• National Seed Storage Laboratory (NSSL)
• Regional Plant Introduction Stations (RPIS)
• National Clonal Germplasm Repositories
• Interregional Potato Introduction Station
• Special Collections
• Plant Genetics and Germplasm Institute
• Germplasm Resources Information System
• Plant Introduction Office
• Small Grains Collection
• Molecular Genetics Laboratory
• Germplasm Evaluation and Introduction Laboratory
National Plant Quarantine Center
Crop Curators
Crop Advisory Committees
Users (Federal—State—Private)

SOURCE: Office of Technology Assessment.

from active collections to the NSSL (56). And the study determined that approximately 63 percent of the active germplasm collections were stored in inadequate containers or in undesirable climates. The result, GAO concluded, may be the loss of at least one-fourth of the germplasm resources held by NPGS.

Efforts have been made to address some of these deficiencies through reallocation of resources, construction of new facilities, and centralization of responsibilities, but the need to improve germplasm maintenance remains. Recommendations to improve NPGS have been hampered by the diffuse nature of the network and by inadequate resources.

The system has been cited as needing a clearer division of responsibility for maintaining and evaluating germplasm collections (7). Because it is a cooperative network, lines of authority are frequently unclear, and there may be too many levels of authority to adequately administer a national program on germplasm (60). The result is a general lack of understanding of how decisions concerning NPGS are made by ARS. Such decisions can be further complicated by the competing interests and concerns of other cooperative Federal (i.e., CSRS) or State agencies that may provide program support.

The ARS staff has recently increased its input into budget allocations for Federal facilities and has attempted to centralize program responsibilities into one office (42). Further centralization could provide increased coordination of the system's collections, improved communication on available germplasm diversity (especially through the GRIN database), and more effective identification of funding priorities.

One area that has received insufficient funds is regeneration of seeds with reduced viability. Although NSSL monitors seed viability, it sends seeds that germinate poorly to another facility for growing-out. If viability is found to be low, it may be difficult to obtain a regenerated sample. If NPGS does not have specific responsibility for a sample, NSSL must locate a willing donor, but it does not have funds to pay for grow-outs. A comprehensive system to support regeneration of stored seed has been hampered by competing interests for available resources.

The crop advisory committees (CACs) in NPGS were developed to improve communication about crop-specific needs (see table 9-4). CACs are comprised of scientists from NPGS, private industry, and the academic community. They provide technical expertise to the National Plant Genetic Resources Board, the National Plant Germplasm Committee, ARS staff, and NPGS curators. In some cases, such as the pear collection at the national clonal repository in Corvallis, OR, CACs advise the facility in charge of a particular crop (7).

CACs are growing in importance and influence within NPGS (45). Committees have been

Table 9-4.—Existing and Proposed Crop Advisory Committees of the National Plant Germplasm System

Existing committees	Proposed committees
Alfalfa	Asparagus
Barley	Florist crops
Carya	Leafy vegetables
Citrus	Tropical fruit and nuts
Clover	Woody ornaments
Cotton	
Crucifer	
Grass	
Juglans	
Maize	
Malus	
Oats	
Pea	
Peanut	
Phaseolus	
Potato	
Prunus	
Pyrus	
Rice	
Root and bulb	
Small fruits	
Sorghum	
Soybean	
Sugar beet	
Sugarcane	
Sunflower	
Sweet potato	
Tomato	
Vigna	
Vine crops	
Vitis	
Wheat	

SOURCE: U.S. Department of Agriculture, Agricultural Research Service, Plant Genetics and Germplasm Institute, Germplasm Resources Information Network, Progress Update, February 1986.

asked to identify gaps in the diversity of crop species, coordinate collection and maintenance needs, develop priorities for crops, and assess the data available on accessions. However, no provision exists within NPGS to ensure that the necessary meetings of a CAC will be held or reports developed. To date, NPGS has relied on the dedication and commitment of the scientists involved to accomplish these tasks. Although some CACs have achieved a great deal, others have been slow to organize and develop their activities. ARS has argued that funding or other support for CACs is unnecessary, but OTA has found the committees feel they would be more effective if funds for frequent and regular meetings were available.

The diverse nature of NPGS can also be seen in its different roles of providing service functions of maintaining germplasm and undertaking research programs. Functions such as growing out seeds, evaluating accessions, assessing viability, and managing information are service-oriented. Many Federal and State scientists within NPGS, however, are evaluated on a system that can provide disincentives for such activities. The problem can become acute when decreased funding means that research staff must handle service functions.

The need for more personnel and funding has increased with the amount of germplasm held by NPGS facilities. Concern about characterization and evaluation of accessions has created additional burdens for many facilities. Therefore, proposed changes should consider improved support of the basic operations along with plans for new construction.

The National Seed Storage Laboratory continues to need improvement (7,56,57,58,60). Within 2 years, the existing facility will exceed its storage capacity. Collections at the RPISs and other facilities are witholding some accessions from NSSL. But keeping them creates an additional burden for facilities not equipped for long-term storage. Without expanded space, NSSL cannot provide the necessary backup storage for NPGS germplasm collections.

In addition, NSSL storage rooms were built before the use of subfreezing and cryogenic storage. OTA found that the NSSL collections require upgraded facilities with access to modern storage technologies and backup refrigeration systems. One proposed NSSL facility would quadruple present storage capacity and enable use of modern technologies. Funds for construction, however, are not available in the Administration's current budget (45).

Although many long-standing deficiencies have been addressed by administrative changes such as creation of the crop advisory committees, future improvements of NPGS will require additional funds for facilities, as well as personnel, equipment, and supplies to support basic operations.

Most States do not formally fund offsite germplasm maintenance activities independent of NPGS. California, an exception, began a program in 1980 to conserve the genetic diversity of important plant and animal species within the State (33). The California Gene Resources Conservation Program, which is currently inactive, raised awareness of the need to conserve germplasm resources. A program at the University of California at Davis will conduct research on germplasm resources in the State and provide funds for orphan collections, those that may be vulnerable due to the death or retirement of principal curators (49).

Private individuals and grassroots organizations are preserving a significant amount of agricultural crop diversity not found in governmental collections (20,44,59). The Seed Savers Exchange, for example, helps preserve heirloom vegetable varieties and other vegetable seeds not available from the Federal Government or commercial producers. The exchange of seeds among its 450 members helps ensure the survival of some 3,500 plant varieties, most of which can be found only within the organization. (For further discussion of Seed Savers Exchange and other grassroots efforts to preserve agricultural plant germplasm, see ref. 59.)

Private industry also maintains plant germplasm in conjunction with developmental programs for new crop varieties or as marketed seed varieties. United Brands, for instance, maintains the most extensive collection of

Photo credit: Seed Savers Exchange

Several of the rare fruits of moon and stars watermelon are displayed by Kent Whealy, Director of the Seed Savers Exchange. This private organization works toward saving heirloom and endangered garden seeds from extinction.

banana germplasm (23). Although the objective in most cases is not the maintenance of genetic diversity, industries could maintain germplasm resources that contribute to the overall plant diversity in the United States.

Support can be provided by private industry by granting funds, equipment, facilities, or land. The Rhododendron Species Foundation, for example, maintains an extensive collection of wild rhododendrons at a facility donated by the Weyerhaeuser Co. (59). A grant to NPGS by Pioneer Hi-Bred International, Inc., of $1.5 million over 5 years will support the evaluation of Latin American corn varieties (2)—work not possible under present NPGS budgets.

Wild Plants

No Federal equivalent to NPGS exists for wild plant species. Although NPGS maintains some wild plant germplasm, this is clearly a secondary function and generally involves relatives of cultivated crops or species economically valuable, such as ornamentals or florist crops. Most wild plant diversity is stored in living collections such as botanic gardens and arboretums.

Federal programs that make some contribution to maintaining wild plant diversity do not cover the majority of plant diversity. USDA's Soil Conservation Service (SCS) maintains some wild species (those with known or suspected value to soil or water conservation) in its Plant Materials Centers. Species not being used in plant development programs are sent to NSSL (52).

The Forest Service maintains germplasm of tree species with known or potential commercial value (5). The Smithsonian Institution maintains an extensive collection of North American wild plant species. The Office of Endangered Species provides some funding for offsite maintenance and propagation of threatened or endangered plant species.

The contributions of current State efforts are unclear. Generally, State programs are coordinated through the State Department of Agriculture and focus on species with some economic importance to the State—e.g., timber varieties, shrubs, and grasses useful in land reclamation, along with important wildlife foods.

The most significant offsite programs for germplasm are financed and managed in the private sector (59). One such effort, the Center for Plant Conservation (CPC), is beginning a network of botanic gardens and arboretums to conserve all threatened and endangered wild plant species. CPC, located at the Arnold Arboretum in Massachusetts, has solicited the participation of 14 major botanic institutions across the country to act as regional centers for wild plant diversity. By establishing a data network, CPC hopes to identify plant species

Photo credit: L. McMahan

Greenhouses of the Berry Botanic Garden, Portland, OR. Botanic gardens are becoming increasingly important to the effort to conserve diversity.

for inclusion into the national program (16). CPC also has agreements with NSSL for long-term storage of selected wild species (40).

Arboretums and botanic gardens historically have not considered the maintenance of wild plant diversity a goal (37). They have generally provided display gardens—areas where showy flowers or unique plants are presented—with a secondary objective of preserving wild plant species. Interest in maintaining diversity is increasing, however (37,59). In some cases, reproductive individuals of rare plant species may be found only in arboretums or botanic gardens. Yet, aquatic plants are underrepresented in botanic institutions, and few aquatic gardens exist to conserve such species.

Regardless of their objectives, these botanic institutions and the individuals who run them contribute to the maintenance of plant diversity. However, no coordination exists for information exchange or evaluation of contributions. Although it is too early to assess results, the Center for Plant Conservation may provide significant coordination of such efforts.

One possible way to improve offsite wild plant maintenance is to expand NPGS to include nonagricultural varieties. The objective is to take advantage of existing Federal, State, and private cooperation. Crop advisory committees could be established for species that are important but have little market value. NSSL could be expanded to serve as a repository for wild species' seeds that may have future economic or ecological significance. Existing plant centers and scientists could play a larger role in propagation and reintroduction programs for wild plant species, particularly threatened or endangered species.

The underlying responsibilities of NPGS would need to be changed to accommodate nonagricultural species. Biological differences between agricultural and wild species, such as dormancy and seed production barriers, would increase the need for research to prepare plans for storage, germination, and regeneration.

Expanding the role of the system to include wild plant species could reduce already insufficient funding for existing programs, however. The Agricultural Research Service budgeted nearly $16 million (gross) for germplasm work in 1986, but one report has estimated that by the 1990s, annual allocations of almost $40 million (1981 dollars) will be needed to support programs (43,60). Adding approximately 20,000 new plant species (perhaps millions of accessions to represent the diversity of each species) would severely strain an already underfunded program.

Animal Programs

The United States has no organized program for maintaining diversity in agricultural animals (6). Federal activities to conserve genetic resources are minimal, and private efforts, though more substantial, are so disperse that it is difficult to assess gaps or overlaps.

Neither the Federal Government nor State governments have programs designed to maintain wild animal diversity offsite. It is minimally supported by Federal contributions to private sector programs, but no overall Federal plan exists and funding is erratic. Thus, the private sector is currently making the most significant contributions to maintaining domestic and wild animal diversity.

Domestic Animals

USDA was authorized to collect, maintain, and develop animal genetic resources under the same legislation that provides authority for the National Plant Germplasm System's components (Agricultural Marketing Act of 1946). However, USDA contributions to domestic animals did not evolve along with its agricultural plant activities.

The department has concentrated on identifying foreign germplasm of potential importance in U.S. livestock production. Beginning in the mid-1960s, a substantial number of foreign breeds were introduced into the United States (6). The importation of cattle was emphasized, but several breeds of sheep and swine were also introduced. Breeds were chosen for their likely contribution to U.S. agriculture and without particular attention to the degree of endangerment in their country of origin. Several of these stocks have since become firmly established within the United States.

USDA evaluated the breeds and in some cases (especially for sheep and swine) initiated their importation. A key group in this effort was the Germplasm Evaluation Program of the Roman L. Hruska U.S. Meat Animal Research Center (MARC) in Nebraska, which compared more than 20 foreign and domestic cattle breeds (61). Current efforts at MARC deal with development of composite gene-pool stocks for new and more productive breeds of sheep and swine.

Within the private sector, breed associations —loose unions of individuals who produce a particular livestock breed—have been formed for common species (e.g., cattle, pigs, sheep, goats, and horses) to record pedigrees and production of individuals within livestock breeds available in the United States (see table 9-5). These groups do not consider maintenance of biological diversity as a goal, although they may contribute to maintenance of animal genetic resources (25,59). A diversity of livestock breeds will be maintained only if an association exists for each breed.

Most programs that deal with germplasm conservation as such (i.e., separate from efforts to use that diversity within the livestock industry) are undertaken and funded by the private sector. Many minor livestock breeds in the United States are maintained by one person or a few individuals, working relatively independently (25,59).

The American Minor Breeds Conservancy (AMBC), a nonprofit organization, is currently seeking to identify these people and open lines of communication among them. (For further discussion of AMBC and the contributions of individuals and breed associations to domestic animal genetic diversity, see ref. 59.) AMBC recently completed a census of North American livestock that identifies some 80 breeds, including cattle, pigs, sheep, donkeys and

Table 9-5.—Active Breed Associations in the United States

Species	Number of associations	Number of registrations[a]		
		Minimum	Maximum	Average
Beef cattle	18	297	195,267	43,976
Dairy cattle	6	4,085	425,385	89,382
Sheep	8	4,568	58,994	18,675
Swine	10	382	245,423	61,050
Horses	15	631	68,346	22,260

[a]For fiscal year 1983.

SOURCE: National Society of Livestock Record Associations, 1983.

mules, and goats, needing special attention to ensure their survival (26).

Private companies also make significant contributions to animal germplasm maintenance. For example, the majority of poultry germplasm is maintained by firms that operate both domestically and internationally (6). Several maintain unselected, random-bred control lines that serve as reservoirs of genetic diversity. These lines, however, are vulnerable to changes in economic conditions, and their maintenance does not currently represent public or industry policy.

Artificial insemination (A.I.) firms control and distribute the majority of U.S. dairy cattle germplasm. These companies have formed pools of individual breeders involved in planned matings, testing progeny of specific germplasm strains, and development of improved breeding lines (6). Companies focus almost entirely on Holstein cattle because the market is so large. Increased emphases on planned matings among superior individuals have been required to maximize genetic improvement within the dairy industry because of intense competition among A.I. organizations.

As a result, new bulls for use in artificial insemination often represent the offspring of a small sample of bulls from the previous generation. For example, of the 6 to 7 million dairy cows bred each year in the United States, about 65 percent are impregnated by only 400 to 500 A.I. sires. In addition, of the approximately 1,000 performance-tested dairy bulls in a given year, nearly half are sons of the 10 best bulls of the previous generation (67). This process tends to maximize rates of genetic improvement and almost certainly will result in an excessive narrowing of the genetic base.

Researchers affiliated with universities and Agricultural Experiment Stations help identify genetic resources or help maintain and develop germplasm resources, although not as much as breed associations or private industries do. For example, one researcher at the University of Connecticut has produced an international registry of poultry genetic stocks that is annually updated and acts as an important catalog of existing poultry resources (53). University animal or veterinary science departments may maintain small breeding populations of livestock for experimental and educational purposes (26).

U.S. universities with programs for domestic animal research and utilization also play a role at the international level. The International Sheep and Goat Institute associated with Utah State University, for example, works with researchers and livestock operators in other countries to identify and propagate genotypes of sheep and goats. Although the focus of the institute is to assist countries in the production of sheep and goats best-suited to local environments, its members are also involved in training international institutions in the storage and management of sheep and goat genetic resources (29).

Even with these various efforts, the overall diversity within many domestic animal breeds is declining (6). In summary:

- Storage facilities do not exist for *in vitro* maintenance of sheep, swine, or poultry genetic stocks.
- Breed associations report that although a few breeds of sheep in the United States have declined to very small numbers, global diversity of sheep germplasm remains adequate.
- Genetic diversity in dairy and meat goats does not appear to be changing significantly.
- Because relatively few competitive strains of highly specialized egg and meat chickens, turkeys, and waterfowl account for much of the world poultry populations, there is concern about maintaining adequate genetic diversity for future needs.
- The increasing emphasis on whole-milk production dairy cattle favors the adoption of Holstein breeds among milk producers, causing the decline of other minor dairy breeds.
- Genetic diversity appears to be stabilized or increasing slightly in beef cattle in the United States.

Public awareness of the potential problems associated with loss of genetic diversity and institutional concern about the issue are not as evident for domestic animal species as they are for agricultural crop species. Concern about loss of agricultural animal diversity is increasing, however, at the international level, where a perception exists that a significant amount of genetic diversity is disappearing (see ch. 10). Insufficient information exists on the status and trends of domestic animal breeds at the global level to substantiate this belief (19). But it is the unregistered and unrecognized breeds that are in the greatest danger of becoming extinct.

Wild Animals

Federal efforts to maintain wild animals offsite occur only through the captive breeding programs of the U.S. Fish and Wildlife Service. Individual specimens of critically endangered species may be selected for captive breeding programs at the Patuxent Wildlife Research Center in Patuxent, MD. The center has been responsible for the captive breeding and reintroduction of more than 60 species of birds, mammals, and reptiles native to the United States (37).

Endangered fish species have been propagated at the Fish and Wildlife Service's National Fish Hatchery in Dexter, NH. Additionally, FWS provides nominal funding for captive breeding and reintroduction programs for endangered animal species to the private sector, and in one case, to the State of Wyoming to recover the black-footed ferret. Overall, however, programs for the offsite maintenance of diversity in wild animals are controlled and financed primarily by universities and institutions in the private sector (37).

Zoos are well-known storehouses for wild animal species, although historically they made few contributions to maintaining biological diversity. But their role in this area is becoming significant, especially in terms of public education. More institutions are identifying the need for expanded activity in research and technology development to maintain genetic diversity of zoo animals.

In one case, zoos are working together to maintain viable populations of wild animals bred in captivity. The American Association of Zoological Parks and Aquariums coordinates breeding programs for selected endangered wild species. These programs, known as Species Survival Plans (SSPs), are being implemented for some 30 species that are critically endangered in the wild, that have sufficient numbers at various zoos to ensure genetic viability within a captive breeding program, and that have a sufficient nucleus of professionals at the cooperating institutions to carry out the plan (1). (For a discussion of captive breeding techniques, see ch. 6.)

Breeding programs are designed by experts with knowledge of the species and carried out by scientists within the zoological community (1). Since more animal species meet the SSP criteria than zoos realistically have resources to implement, further criteria exist for determining which species to include:

1. a high probability of successful implementation of the plan,
2. a high relative degree of endangerment, and
3. a high relative degree of uniqueness within the animal kingdom.

Species Survival Plans are designed to overcome the space and population limitations of most zoos. For many institutions, adequate facilities to maintain a viable breeding population of at least 250 animals simply do not exist. The SSP outlines agreements between participants in the program for the translocation of breeding adults or their reproductive products (e.g., eggs, sperm, or embryos) among zoos to simulate a much larger breeding population than could exist at any one facility. Information on the breeding programs must be carefully recorded and entered into a master database, the International Species Inventory System (ISIS). These programs are too new to assess their effectiveness in maintaining genetic diversity.

ISIS was developed at the Minnesota Zoo to catalog information about the genetic makeup of individual animals from more than 200 zoological institutions worldwide. One goal in the database development was to address the problem of inbreeding among species within zoos. ISIS acts as a computerized matching service, helping zoos around the world identify other institutions that have distinct bloodlines in breeding populations of a particular wild animal (14). Other goals include identifying captive management problems, monitoring the captive status of some 2,500 species, and providing information to managers. It appears that ISIS is widely used by zoological institutions and therefore makes important contributions to maintaining genetic diversity.

A large number of zoos are not involved with the SSP or ISIS, yet still provide offsite maintenance of selected wild animal species. These institutions may support populations of locally endemic wild animals or include individuals of internationally rare species. Maintenance of a diversity of species or of individuals within a species is generally not an objective at these institutions, however.

Much of the work undertaken by zoos to preserve species involves internationally endangered ones, with less attention given to threatened and endangered species found in the United States. The focus on species from elsewhere in the world or exotic animals is due, in part, to the degree of endangerment of these animals. Those that are critically endangered in the United States, such as the California condor or the black-footed ferret, are also the focus of active captive breeding programs at U.S. zoos (8,36). Compared with zoos, most aquariums accord the maintenance of aquatic species diversity a low priority. Almost no work has been done at U.S. aquariums to maintain the diversity of species found in U.S. waters (38). When they need specimens, they generally collect them from the wild (37).

Fairly large collections of breeding wild animals are maintained by individuals. In many cases, these people establish societies around a particular species or group of species to exchange information and breeding stock among society members. Their efforts range from the small-scale activities of individuals that breed exotic birds or reptiles to the management of large herds of Asian and African antelope species by Texas game ranchers. (For further discussion, see ref. 59.)

Microbial Resource Programs

No U.S. institution or institutional mechanism addresses the preservation of microbial diversity. Numerous collections of micro-organisms exist in the United States in both the public and private sectors. Most were established to study a particular taxonomic group of micro-organisms, and they represent detailed sampling within that group. Several hundred specialized working collections of microbial germplasm are part of the basic and applied research programs of scientists working in both the public and private sectors (7).

Federal Institutions

The largest public microbial culture collection in the United States is the Northern Regional Research Laboratory (NRRL) collection held by USDA's Agricultural Research Service. It is an archival collection with a taxonomically broad range of micro-organisms stored for long-term preservation. NRRL does not publish a catalog of its holdings and does not encourage general distribution of the germplasm it holds, in part because of the high cost of distribution. No moderately sized collections (3,000 to 10,000 accessions) of micro-organisms function as national repositories or resource collections for a range of microbial classes (24).

Several collections supported by the U.S. Government are devoted to assembling microbial strains within a particular taxonomic group. The largest of these is held by the Neisseria Reference Laboratory of the U.S. Public Health Service. Similar taxonomically specific collections supported by USDA, such as the cereal rust collections at the Universities of Minnesota and Kansas, distribute microbial germplasm on request, but they generally do not catalog their holdings (24).

Like many scientific institutions, organizations holding culture collections are currently suffering from financial cutbacks (24). Funding from USDA has been reduced or redirected to other areas at the expense of the network of archival collections (table 9-6). The result is a diminished capacity to maintain the recordkeeping, authentication, and taxonomic characterization necessary for a collection. Expansion of existing collections is restricted by such financial constraints.

State Collections

No organized State efforts to collect and maintain microbial diversity, apart from specialized collections, seem to exist. The many specialized collections that exist at State universities and colleges are typically the responsibility of individual scientists. Some have gained institutional support and achieved national significance. Pennsylvania State University supports the major U.S. collection of *Fusarium* species, a plant fungus of major interest to breeders (7). Such efforts commonly depend on the continued interests and abilities of individuals who initiated them.

Unless sources of support and personnel are available, institutional commitments to microbial collections, where they exist, may not continue after key individuals leave, retire, or die—a problem noted earlier with regard to agricultural plant germplasm collections. When the curator of an extensive collection of *Rhizobium* germplasm died in 1975, his university was unable to provide future management for the extensive collection—at that time considered the richest collection in the world of soil bacteria in this group (24). It would have been lost had it not been acquired by the University of Hawaii as part of a U.S. Agency for International Development (USAID) research project in 1976.

It was not until 1981 that the university agreed to accept responsibility to maintain the collection in perpetuity as part of an international

Table 9-6.—Microbial Culture Collections in the United States With More Than 1,000 Accessions

Collection	Sponsor	Number of cultures
Living resource:		
American type culture collection (Rockville, MD)	Private	27,630
Reference and archival:		
Northern Regional Research Center (Peoria, IL)	Government	63,000
USDA Rhizobium Culture Center (Beltsville, MD)	Government	1,200
Neisseria Repository, School of Public Health (Berkeley, CA)	Government	1,700
Neisseria Reference Laboratory (Seattle, WA)	Government	30,000
NiFTAL Rhizobium Germplasm Resource (Paia, HI)	Government and university	2,000
Plasmid Reference Center (Stanford, CA)	Government and university	2,000
Education and research:		
Fungal Genetics Stock Center (Arcata, CA)	Government	7,755
L.L. Collection Waksman Institute of Microbiology (Piscataway, NJ)	University	3,070
Industry:		
Microbial and Fermentation Products Research (Indianapolis, IN)	Industry	66,060
Upjohn Culture Collection (Kalamazoo, MI)	Industry	7,755

SOURCE: V.F. McGowen and V.B.D. Skerman, *World Directory of Collections of Cultures of Microorganisms* (Brisbane, Australia: World Data Center, 1982).

agreement designating them as an international Microbiological Resource Center, under the auspices of UNESCO and the United Nations Environment Programme / International Cell Research Organization Panel on Microbiology (see ch. 10). Such an agreement would not have been possible if the University of Hawaii had not obtained USAID funding.

Private Collections

The best microbial resource reference collection in the United States is maintained by the American Type Culture Collection (ATCC). ATCC is a national nonprofit repository for medical, industrial, and agricultural microbial germplasm, as well as a national and international repository for patented microbes. Its collection of approximately 36,000 strains includes bacteria, fungi, clamydiae, rickettsiae, protozoans, algae, cell lines, and viruses. Although its holdings are smaller and it charges for each culture sent, actual distributions from ATCC far exceed those of the government-sponsored NRRL (24). Of the U.S. collections of more than 30,000 accessions, only ATCC distributes a catalog.

Many collections are also held for specific purposes by U.S. corporations or universities. These are usually personal collections of individual scientists and may receive little or no direct financial support. Private specialized microbial collections, like their counterparts in universities, usually begin as personal collections accumulated and maintained over a career. Although typically holding a limited number of microbial genera, they are unequaled for taxonomic detail and are an important facet of the total microbial diversity conservation effort.

The *Frankia* culture collection, for example, held at the Battelle-Kettering Laboratory in Yellow Springs, OH, began as a personal commitment by one scientist to isolate and culture frankiae, the symbiotic actinomycetes of some nitrogen-fixing plants (24). This internationally respected collection is not supported by an institutional commitment or extramural funding and thus depends on the dedication and resourcefulness of its curator. When a curator leaves a company, or when business considerations force redirection of that person's efforts, a specialized collection like this can be lost.

The costs of maintaining a collection pose significant constraints for commercial collections. In fact, recordkeeping and distribution expenses are important factors in many corporate decisions not to make materials from their collections generally available (24). Specialized collections are vulnerable to deterioration if funding cannot be obtained for their upkeep. For commercial collections such as ATCC, cultures must be maintained on a no-loss basis. Accessions that are not requested or used frequently and have no current intrinsic value may be discarded.

Quarantine

Maintaining biological diversity frequently involves the importation of foreign materials to increase the available genetic base for crop and livestock species or for offsite maintenance in zoos, botanic gardens, or arboretums. Quarantine regulations are designed to prevent accidental introduction by imports of exotic pests and diseases that could be harmful to U.S. agriculture. For both plants and animals, quarantine procedures are a combination of regulatory requirements controlling importation and distribution of germplasm and inspection or testing procedures designed to detect pests and diseases (for specific testing methods, see chs. 6 and 7). Regulations, administered by the Animal and Plant Health Inspection Service (APHIS) of USDA, have been viewed by some as restrictive with regard to importation of new genetic diversity (34,45).

Quarantine regulations classify both plant and animal germplasm ranging from materials considered to be of low risk of carrying disease organisms to those prohibited entry due to the extreme hazard they pose of introducing disease. Rice is prohibited from all countries and sorghum from many countries—except for germplasm that may enter under a USDA permit specifying the safeguard conditions, which often result in extensive delays.

Likewise, limited numbers of cloven-footed animals can be imported with heavy restrictions from areas known to harbor foot-and-mouth disease (41). For most materials, some degree of restriction is required and the plants or animals must enter through a designated port of entry for inspection. Most plants, for example, must enter the United States through one of 14 APHIS Plant Inspection Stations, where they are inspected, treated if necessary, and released within 1 to 3 days (35).

Some materials enter under conditions of post-entry quarantine, whereby they are inspected at an appropriate facility and released to the importer under an agreement that regulates their maintenance and release. Such agreements exist for some zoo animals, including most ungulates. Animals subject to post-entry quarantine are permanently consigned to the designated facilities, and only their offspring can be distributed or moved to other institutions. Plant materials are generally exempt from restriction if no pests or diseases are detected during the normal detention period of 2 years (35).

Importation of wild animal species posing threats to agricultural livestock can be lengthy, expensive, and difficult. For example, most bird species are very difficult to import due to APHIS concerns over the introduction of Newcastle's disease, a serious threat to domestic poultry stocks.

State programs regulating movement of plants and animals vary and are most stringent for States with economies based heavily on agricultural crops (e.g., California and Florida). Efforts to prevent the spread of pests and diseases can include restrictions on the carrying of fruits and vegetables into a State or requirements for treatment of potentially infected materials before entry. For importation of germplasm from outside the country, States cannot place restrictions on materials that are greater than those imposed by the Federal Government (35).

Biological diversity maintenance has not been a concern in the establishment of quarantine regulations. Although such regulations and procedures can be important to protecting agricultural diversity, they also, paradoxically, inhibit the development of new varieties by restricting or slowing the flow of new materials into the country.

NEEDS AND OPPORTUNITIES

A variety of activities in the United States address the maintenance of some aspect of biological diversity. These efforts are carried out by both government and the private sector. Benefits from maintaining diversity, such as improvements in agriculture and the ecological processes that support life, accrue to all individuals though they seldom pay for them. The public nature of these benefits makes it the major responsibility of the public sector to maintain.

Private sector activities, nonetheless, complement government efforts in important ways. Activities of some groups and individuals may back up national programs. In other cases, private activities maintain diversity in ways that the public sector does not, cannot, or will not. A number of private groups supplement the National Plant Germplasm System by maintaining heirloom and endangered commercial varieties of vegetables, for example, including many that are not contained in existing national collections. Private crop breeders have been influential in elevating the issue of genetic diversity loss to a national concern and have been providing increasing input in public germplasm maintenance activities. To date, these private activities have received little recognition, and minimal effort has been made to encourage and support private initiatives. Increased cooperation between public and private efforts could not only strengthen the latter but also improve maintenance of diversity.

The various laws and programs of Federal, State, and private organizations provide an elaborate framework on which a concerted biological diversity effort could be built. But because few of these activities cite the maintenance of biological diversity as an explicit objective, the goal is not considered in a comprehensive or coherent manner. Duplication of effort, conflicts in goals, and gaps in geographic and taxonomic coverage consequently exist.

One means of addressing biological diversity maintenance in a comprehensive way is to develop a national strategy. The process of developing a plan would help pinpoint areas where activities overlap or are lacking. At the least, such a process would initiate coordination of Federal agencies' activities. Those administering programs related to biological diversity would have to provide detailed reports on how programs are being implemented to conserve diversity. In particular, they would need to identify measures being undertaken to reduce program overlap, minimize jurisdictional problems, and identify areas for new initiatives. The latter is most evident in the lack of a national animal genetic resources program and of a system of protected representative ecosystems in the United States.

Any strategy, no matter how good it appears on paper, cannot be effectively implemented without adequate resources. Sustained long-term funding, in turn, requires consistent commitment to the process. The inconsistent funding and staffing of many existing programs illustrate the complexity and accompanying uncertainty of the political process.

An examination of trends in Federal budget allocations for natural resource conservation—including pollution control, water resources, public lands, recreation, and soil conservation—reveals a considerable decline over the last decade. This decline stands in contrast with real spending increases between 1978 and 1986 for defense (50 percent), payments to individuals—e.g., social security, Medicare, veterans' benefits, food stamps, and so on—and a tripling of interest payments on the national debt (51). The proportion of U.S. Government research and development (R&D) expenditures devoted to environmental R&D has also declined in recent years and assumes a smaller proportion of total government R&D funding compared with other industrial countries (48).

Programs of particular relevance to biological diversity maintenance, namely the Endangered Species Program and the National Plant Germplasm System, have been stretched to the point of being unable to adequately meet their objectives. These programs are able to prevail, in light of the constraints, mainly because of the dedication and ingenuity of the individuals working in them. The National Plant Germplasm System, for instance, has been underfunded for years. Within 2 years, the National Seed Storage Laboratory's storage facilities will be full, and aging equipment and buildings at NSSL and other facilities require major repair, upgrading, or replacement. Similarly, the effectiveness of the Endangered Species Program in preventing extinctions has been hindered by a shortage of resources.

CHAPTER 9 REFERENCES

1. American Association of Zoological Parks and Aquariums, *Species Survival Plan* (Wheeling, WV: undated).
2. Anonymous, "Block Announces Pioneer Grant," *Diversity* 7:9-10, fall 1985.
3. Bean, M. J., *The Evolution of National Wildlife Law* (New York: Praeger Publishers, 1983).
4. Bean, M., "Federal Laws and Policies Pertaining to the Maintenance of Biological Diversity on Federal and Private Lands," OTA commissioned paper, 1985.
5. Bonner, F., "Technologies To Maintain Tree Germplasm Diversity," OTA commissioned paper, 1985.
6. Council for Agricultural Science and Technology, *Animal Germplasm Preservation and Uti-*

lization in Agriculture, Report No. 101 (Ames, IA: 1984).

7. Council for Agricultural Science and Technology, *Plant Germplasm Preservation and Utilization in U.S. Agriculture,* Report No. 106 (Ames, IA: 1985).
8. Crawford, M., "Condor Recovery Effect Hurt by Strategy Debate," *Science* 231:213-214, Jan. 17, 1986.
9. Crumpacker, W.D., *Potential Diversity and Current Protection Status of Major Natural Ecosystems in the United States: A Preliminary Report to the Heritage Conservation and Recreation Service* (Washington, DC: U.S. Department of the Interior, Heritage Conservation and Recreation Service, 1979).
10. Crumpacker, W.D., "Status and Trends of U.S. Natural Ecosystems," OTA commissioned paper, 1985.
11. Crumpacker, W.D., University of Colorado, personal communication, June 1986.
12. Davis, G.D., "Natural Diversity for Future Generations: The Role of Wilderness," *Proceedings Natural Diversity in Forest Ecosystems Workshop,* J.L. Cooley and J.H. Cooley (eds.) (Athens, GA: University of Georgia, 1984). *In*: Crumpacker, 1985.
13. Drabelle, D., "The Endangered Species Program," *Audubon Wildlife Report, 1985* (New York: National Audubon Society, 1985).
14. Dresser, B., and Leibo, S., "Technologies To Maintain Animal Germplasm," OTA commissioned paper, 1986.
15. Dunstan, F., assistant director, Sanctuary Program, National Audubon Society, personal communication, May 18, 1984.
16. Falk, D., and Walker, K., "Networking To Save Imperiled Plants," *Garden,* January-February 1986, pp. 2-6.
17. Fernald, E.A., et al., *Guidelines for Identification, Evaluation, and Selection of Biosphere Reserves in the United States,* OES/ENR, U.S. Man and the Biosphere Report No. 1 (Washington, DC: U.S. Department of State, 1983). *In*: Crumpacker, 1985.
18. Fitzgerald, J., and Meese, G.M., *Saving Endangered Species* (Washington, DC: Defenders of Wildlife, 1986).
19. Fitzhugh, H., Getz, W., and Baker, F., "Status and Trends of Domesticated Animals," OTA commissioned paper, 1985.
20. Fowler, C., "Report on Grassroots Genetic Conservation Efforts," OTA commissioned paper, 1985.
21. Greeter, S., South Carolina State Heritage Trust, personal communication, Apr. 7, 1985.
22. Gregg, W.P., and Fretz, P.R., "Biosphere Reserves: A Background Paper for the Task Force on Conserving Gene Pools," unpublished U.S. National Park Service report, 1986.
23. Gulick, P., and van Sloten, D. H., *Directory of Germplasm Collections: Tropical and Subtropical Fruits and Tree Nuts* (Rome: International Board for Plant Genetic Resources, 1984).
24. Halliday, J., and Baker, D., "Technologies To Maintain Microbial Diversity," OTA commissioned paper, 1985.
25. Henson, E., "An Assessment of the Conservation of Animal Genetic Diversity at the Grassroots Level," OTA commissioned paper, 1985.
26. Henson, E., *1985 North American Livestock Census* (Pittsboro, NC: American Minor Breeds Conservancy, 1985).
27. Hougas, R.W., "Plant Germplasm Policy," *Plant Genetic Resources: A Conservation Imperative,* C.W. Yeatman, D. Kafton, and G. Wilkes (eds.), AAAS Selected Symposium 87:15-30 (Washington, DC: American Association for the Advancement of Science, 1984).
28. Hyland, H.L., "History of Plant Introduction in the United States," C.W. Yeatman, D. Kafton, and G. Wilkes (eds.), AAAS Selected Symposium 87: 5-14 (Washington, DC: American Association for the Advancement of Science, 1984).
29. International Sheep and Goat Institute, untitled pamphlet (Logan: Utah State University, undated).
30. Jenkins, R., science director, The Nature Conservancy's State Natural Conservancy Program, personal communication, Feb. 6, 1987.
31. Jones, B., Office of Endangered Species, Fish and Wildlife Service, U.S. Department of the Interior, Washington, DC, personal communications, August 1986.
32. Jordan, W., Peters, R., and Allen, E., "Ecological Restoration as a Strategy for Conserving Biodiversity," OTA commissioned paper, 1986.
33. Kafton, D., "The California Gene Resource Conservation Program," C.W. Yeatman, D. Kafton, and G. Wilkes (eds.) AAAS Selected Symposium 87:55-62 (Washington, DC: American Association for the Advancement of Science, 1984).
34. Kahn, R.P., "Plant Quarantine: Principles, Methodology, and Suggested Approaches," *Plant Health and Quarantine in International Transfer of Plant Genetic Resources,* W.B. Hewitt and L. Chiarappa (eds.) (Cleveland, OH: CRC Press, 1977).

35. Kahn, R.P., "Technologies To Maintain Biological Diversity: Assessment of Plant Quarantine Practices," OTA commissioned paper, 1985.
36. Kline, L., Office of Endangered Species, Fish and Wildlife Service, U.S. Department of the Interior, personal communication, 1986.
37. Lucas, G., and Oldfield, S., "Living Collections (Plants and Animals)," OTA commissioned paper, 1985.
38. Lynch, M.P., and Ray, C., "Preserving the Diversity of Marine and Coastal Ecosystems," OTA commissioned paper, 1985.
39. Matia, W.T., director of stewardship, "Public Information Flyer on Stewardship Program," The Nature Conservancy, 1986.
40. McMahan, L., "Participating Gardens Vital to Center's Work," *The Center for Plant Conservation* 1:1, summer 1986.
41. Moulton, W. M., "Constraints to International Exchange of Animal Germplasm," OTA commissioned paper, 1985.
42. Murphy, C., National Germplasm Program Coordinator (acting), Agricultural Research Service, U.S. Department of Agriculture, personal communication, Dec. 16, 1985.
43. Murphy, C., National Program Staff, Agricultural Research Service, U.S. Department of Agriculture, personal communication, Aug. 11, 1986.
44. Nabhan, G., and Dahl, K., "Role of Grassroots Activities in the Maintenance of Biological Diversity: Living Plants Collection of North American Genetic Resources," OTA commissioned paper, 1985.
45. National Plant Genetic Resources Board, Minutes of the Meeting of the National Plant Genetic Resources Board, Washington, DC, May 14-15, 1986.
46. The Nature Conservancy, *Preserving Our Natural Heritage* (Washington, DC: U.S. Government Printing Office, 1977).
47. Norse, E.A., "Grassroots Groups Concerned With *In-Situ* Preservation of Biological Diversity in the United States," OTA commissioned paper, 1985.
48. Organization for Economic Cooperation and Development, *The State of the Environment 1985* (Paris: 1985).
49. Qualset, C., Agriculture Department, University of California at Davis, personal communication, 1986.
50. Salwasser, H., Thomas, J.W., and Samson, F., "Applying the Diversity Concept to National Forest Management." *In*: J.L. Cooley, J.H. Cooley (eds), *Natural Diversity in Forest Ecosystems: Proceedings of the Workshop*, Nov. 29-Dec. 1, 1982, Institute of Ecology, Athens, GA, 1984.
51. Sampson, N., "Natural Resources Get the Axe," *American Forests* 92:10-11, 58-59, 1986.
52. Sharp, C., National Plant Materials Specialist, Soil Conservation Service, U.S. Department of Agriculture, personal communications, Jan. 7, 1986.
53. Somes, R.G., Jr., *International Registry of Poultry Genetic Stocks,* Bulletin 469, Storrs Agricultural Experiment Station (Storrs, CT: University of Connecticut, 1984). *In:* Henson (OTA commissioned paper), 1985.
54. Spinks, J., Office of Endangered Species, U.S. Fish and Wildlife Service, U.S. Department of the Interior, personal communication, January 1986.
55. Stone, P., *1985-1986 National Directory of Local and Regional Land Conservation Organizations* (Bar Harbor, ME: Land Trust Exchange, 1985).
56. U.S. Congress, General Accounting Office, *Better Collection and Maintenance Procedures Needed To Help Protect Agriculture's Germplasm Resources* (Washington, DC: U.S. Government Printing Office, 1981).
57. U.S. Congress, General Accounting Office, Comptroller General, *The Department of Agriculture Can Minimize the Risk of Potential Crop Failures* (Washington, DC: U.S. Government Printing Office, 1981).
58. U.S. Congress, National Seed Protection Act of 1985, H.R. 3973, Dec. 17, 1985.
59. U.S. Congress, Office of Technology Assessment, *Grassroots Conservation of Biological Diversity in the United States—Background Paper #1,* OTA-BP-F-38 (Washington, DC: U.S. Government Printing Office, March 1986).
60. U.S. Department of Agriculture, "The National Plant Germplasm System—Current Status, Strengths and Weaknesses, Long-Range Plans" (Washington, DC: 1981).
61. U.S. Department of Agriculture, "Germplasm Evaluation Program," Progress Report No. 11, USDA/ARS-1, 1984.
62. U.S. Department of the Interior, Fish and Wildlife Service, *1980 National Survey of Fishing, Hunting and Wildlife Associated Recreation* (Washington, DC: U.S. Government Printing Office, 1982).
63. U.S. Department of the Interior, Fish and Wildlife Service, Division of Refuge Management,

"Preliminary Survey of Contaminant Issues of Concern on National Wildlife Refuges," (Washington, DC: 1986).

64. U.S. Department of the Interior, National Park Service, Office of Science and Technology, *State of the Parks 1980: A Report to the Congress* (Washington, DC: U.S. Government Printing Office, 1980).

65. U.S. Library of Congress, Congressional Research Service, "Mandates To Federal Agencies To Conduct Biological Inventories," OTA commissioned paper, 1985.

66. Williamson, L.L. (ed.), "Senate Approves Farm Bill," *Outdoor News Bulletin* 39(29):1, Nov. 29, 1985.

67. Young, C.W., "Inbreeding and the Gene Pool," *Journal of Dairy Science* 67:472-477, 1984.

Chapter 10
Maintaining Biological Diversity Internationally

CONTENTS

Tables

Figure

Box

Chapter 10

Maintaining Biological Diversity Internationally

HIGHLIGHTS

- Existing international laws and programs to maintain biological diversity are too disconnected to address the full range of concerns over the loss of biological diversity. As a result, redundancies and gaps exist.
- Concerns over free flow of genetic resources have led to heated political controversy in international fora. However, debates have been largely counterproductive and could benefit from a more informed and less impassioned analysis of the issues.
- Intergovernmental and nongovernmental organizations are making significant contributions to maintaining biological diversity worldwide. These organizations, however, have different strengths and weaknesses; those of nongovernmental groups are largely the converse of intergovernmental groups.

OVERVIEW

International laws and programs relevant to maintaining biological diversity have evolved on an ad hoc basis. Efforts tend to be focused on particular species or habitat types and undertaken in relative isolation from other conservation and development activities. Consequently, overall efforts fail to deal comprehensively with diversity maintenance concerns. Redundancy and gaps in coverage result, and benefits of interactions between different activities go unrealized.

As a relatively new platform, biological diversity maintenance has yet to achieve prominence on international agendas. Increasingly, however, international conservation and development organizations, both public and private, are redefining their activities around the concept of diversity maintenance. What remains to be accomplished is an overall accounting of the scope and effectiveness of this increased activity to determine gaps in the current system and methods to fill them.

Onsite laws and programs have their roots in early 19th century Europe and a narrow constituency concerned with the protection of certain bird species (10). Since World War II, however, the number of organizations, legal instruments, and scope of activities in the international arena has increased dramatically. There has also been a shift in focus from protecting particular species to recognizing the importance of habitat in species maintenance. Programs for maintaining genetic resources offsite are barely a decade old, and increased attention has led to efforts to define national obligations to maintain and provide access to genetic resources. Growing realization of the threats to diversity has also focused attention on the importance of cooperation between onsite and offsite programs. Efforts to control trade in endangered species on an international scale and initiatives to link conservation activities of zoos and botanic gardens with onsite conservation programs highlight these potentials.

The diversity of international institutions and programs dealing with conservation defies complete enumeration or simple categorization. In general, however, their principal functions encompass one or more of the following: problem identification, monitoring and evaluation, data gathering, risk estimation and impact assessment, information exchange and dissemination, national and international program coordination, standard setting and rulemaking, standards and rules supervision, and operational activities (62).

This chapter outlines the major international laws and programs with particular bearing on maintaining biological diversity. International laws are described by the breadth of diversity they cover, ranging from global conventions on ecosystems to treaties concerned with particular species. International conservation programs and institutional networks are also highlighted. Onsite and offsite program activities are addressed separately because of the distinctness of their operations.

INTERNATIONAL LAW

Public international law governs relations between countries, compared with private international law, which governs relations between individuals. Public international law provides a variety of direct and indirect tools for maintaining onsite biological diversity. Most are part of broader conservation objectives, commonly focused on protection of single species, groups of species, or habitats.

The instruments of international law dealing with conservation have varying levels of binding obligation. The terms "hard" and "soft" law are used to distinguish levels of legal significance (52). "Hard" law refers to binding obligations reflected either in treaties or customary international law. "Soft" law refers to instruments that have little legally binding force but may carry persuasive influence and policy guidance for state conduct (e.g., international declarations and resolutions from international conferences or intergovernmental organizations).

The effectiveness of international law depends on the support, implementation, and enforcement at the national level. The uneven distribution of diversity creates major complexities in promulgating binding international law in this area. The difficulty is compounded because regions with the greatest diversity are often those with the most limited financial and technical capacities to devote to these efforts.

In international law, a state has authority over all natural resources within its territory. When a state ratifies a treaty, however, it voluntarily restricts some of its rights and assumes certain obligations. The early development of international conservation law was inspired by interests in protecting large game mammals and birds. Less attention has been paid to onsite conservation of wild plants, unless they are indirectly protected by international traffic controls to protect commercial and agricultural plants from pests or pathogens. An exception is the convention to control trade in endangered wild species of fauna and flora (discussed later in this chapter).

The extensive array of international laws that deal with various aspects of biological diversity maintenance should not be interpreted as evidence that concerns for diversity loss have been adequately addressed. As noted previously, many laws deal with specific species or habitat types. Comprehensive coverage is lacking. Further, it is important to consider the degree of obligation (e.g., hard v. soft law) and effectiveness of legal instruments (e.g., existence or adequacy of a secretariat or other operational support).

The following discussion of international law examines onsite and offsite maintenance, the former being the focus of the majority of legal instruments. The onsite discussion examines various hard-law treaties and several soft-law documents. Although international laws related to offsite maintenance are scant, several soft-law agreements exist. Relevant hard-law agreements deal with tangential issues of international patenting and quarantine.

International Laws Relating to Onsite Maintenance

The existence of an internationally recognized and established obligation to conservation can be of substantial importance to maintaining biological diversity onsite at national and international levels. Increasingly, international obligations are providing national conservation authorities with the extra justification needed to strengthen their own conservation programs. Particularly because of this growing role, international conservation conventions and soft-law documents are important legal and policy tools to be used with other technical, administrative, and financial measures.

Global and regional treaties are also important tools for long-term conservation, although many are not effectively implemented. For some treaties, lack of institutional machinery, such as a secretariat and a budget, is a major drawback. Many are difficult to enforce because incentives are weak and early signs of success are hard to identify, making retaliation difficult if a party chooses to ignore the treaty or fails in its obligations. Some global conventions have too few non-European parties. Finally, in many developing countries in particular, technical and financial resources for implementation are scarce (47).

Global Conventions

Of the five conventions discussed here, the first four are commonly referred to as the "big four" wildlife conventions and are the most important for protection of flora, fauna, and their habitats (47). The Law of the Sea Convention is also included because of its global scope. (For texts of these international and regional treaties, see ref. 11; for summaries of major environmental treaties, see ref. 73.)

The *Convention on International Trade in Endangered Species of Wild Fauna and Flora (CITES)*, established in 1973, controls international trade in wild species of plants and animals listed in the convention appendices as endangered or threatened. With 91 countries now party to it (48), CITES has been called the most successful international treaty concerned with wildlife conservation (52).

The convention has been reinforced by U.S. legislation. U.S. importation of wildlife taken or exported in violation of another country's laws was prohibited by amendments in 1981 (Public Law 97-79) to the Lacey Act of 1900. This legislation supports other nations' efforts to conserve their wildlife resources and the international controls under CITES. It provides a powerful tool for wildlife conservation throughout the world because of the significant amount of wildlife imported by the United States.

The *Convention on Wetlands of International Importance Especially as Waterfowl Habitat* (commonly known as Ramsar, after the town in Iran where the convention was signed), passed in 1971, established a wetlands network and promotes the wise use of all wetlands with special protection for those on the List of Wetlands of International Importance. As of mid-1985, there were 40 contracting parties to the convention and about 300 wetland sites, covering some 20 million hectares, on the List of Wetlands of International Importance (47). Once a site is on the list, the party concerned has a legal obligation to conserve the site (article 3(1)).

The *Convention Concerning the Protection of the World Cultural and Natural Heritage*, signed in 1972, established a network of protected areas and provides a permanent legal, administrative, and financial framework for identification and conservation of areas of outstanding cultural and natural importance. It organized a World Heritage Committee, a World Heritage List, a List of World Heritage in Danger, and a World Heritage Fund to help achieve convention goals. (The World Heritage program is discussed later in this chapter.)

The *Convention on the Conservation of Migratory Species of Wild Animals* (commonly cited as the Bonn Convention), passed in 1979, provides strict protection for migratory species in danger of extinction throughout all or a significant part of their range, and encourages range states to conclude agreements for management of species that would benefit from international cooperation. Fifteen states were party to the convention as of 1984, and the first meeting of the parties in October 1985 established machinery for implementing the convention.

Marine conservation also has received increased attention, particularly in the past two decades. The *Convention on the Law of the Sea*, adopted in 1982 at Montego Bay and yet to come into force, identifies a number of general obligations relevant to conservation. Article 192 imposes an obligation on states to protect and preserve the marine environment. Coastal states are obliged to ensure through proper conservation and management measures that living resources in their exclusive economic zones are not endangered by exploitation (article 61(2)). Activities outside national jurisdiction are to be undertaken "in accordance with sound principles of conservation" (article 15(b)).

Regional Conventions

Other regional treaties have emphasized conservation of habitat through creation of protected areas and other programs. The major conventions in force are the *Convention on Nature Protection and Wildlife Preservation in the Western Hemisphere* from 1940; the *African Convention on the Conservation of Nature and Natural Resources* from 1968; the *Convention on the Conservation of European Wildlife and Natural Habitats* from 1979; and the *ASEAN Agreement on the Conservation of Nature and Natural Resources* from 1985. With habitat destruction being a principal threat to biological diversity, treaties that call for protection of flora and fauna through habitat protection are particularly important and need long-term support. The Western Hemisphere and African conventions, however, have had difficulties with implementation and enforcement at the national level, largely due to financial and technical limitations. The more recently developed Association of Southeast Asian Nations (ASEAN) Convention involved regional consultations to incorporate management and conservation techniques and therefore elicits greater hopes for success.

The regional seas programs developed by the United Nations Environment Programme (UNEP) in cooperation with other agencies, particularly the Food and Agriculture Organisation of the United Nations (FAO) and the International Meteorological Organization, involve 10 regions encompassing about 120 of the 130 or so coastal states in the world. (The 10 regions are the Caribbean, Mediterranean, Persian Gulf, West and Central Africa, East Africa, East Asia, Red Sea and Gulf of Aden, South Pacific, South-East Pacific, and South-West Atlantic.) The objective is to reduce pollution and conserve biological resources through cooperative management efforts. The legal mechanisms include action plans and regional conventions. The regional seas conventions include articles on pollution from ships, aircraft, and land-based sources; pollution monitoring; and scientific and technological cooperation. Protocols are authorized in each convention text and address specific approaches to certain problems. Technical annexes provide standards for regulatory or cooperative activity.

Protocols are also being explored for protection of easily disrupted marine ecosystems and for habitats of depleted or endangered marine life through the creation of protected areas. The *Convention for the Protection and Development of the Marine Environment of the Wider Caribbean Region*, signed in Cartagena, Colombia in 1983, is generating government discussion on protected areas and wildlife in this region. Resolutions adopted call for preparation of draft protocols (19). U.S. technical support could be a key factor in the ratification and implementation of such protocols (20).

The *Convention on the Conservation of Antarctic Marine Living Resources*, passed in 1980, contains important innovations on the conservation of biotic resources. It obliges parties to adopt an ecosystem approach to exploitation of Antarctic resources, thus requiring consideration of impacts on interdependent species and the marine system as a whole when setting harvest limits. Article I(2) of the convention defines marine living resources to include all species of living organisms, including birds, found south of the Antarctic convergence (where the warm and cold waters of the Antarctic Ocean meet).

Species-Oriented Treaties

A group of species-oriented treaties focus on controlling exploitation of specific wildlife, such as polar bears, vicúna, northern fur seals, whales, and Antarctic seals (52). Although these treaties are concerned primarily with controlling harvesting, attention to specific species commonly extends to concerns for their habitat, thus potentially serving biological diversity more broadly. The major species-oriented treaties are listed in table 10-1.

Declarations and Resolutions

The United Nations Conference on the Human Environment, held in Stockholm, Sweden in 1972, adopted a *Declaration on the Human Environment* that remains a key soft-law document on international environmental issues. The Stockholm Declaration contained 26 principles to guide the international effort to protect the environment. Principle 2 addresses conservation of the Earth's biological resources:

> The natural resources of the earth including the air, water, land, flora and fauna, and especially representative samples of natural ecosystems must be safeguarded for the benefit of present and future generations through careful planning or management as appropriate.

Another important soft-law is the *World Conservation Strategy* (WCS), a comprehensive document prepared by the International Union for Conservation of Nature and Natural Resources (IUCN). Advice, cooperation, and financial assistance for the preparation of WCS were provided by UNEP and the World Wildlife Fund, with collaboration from FAO and the United Nations Educational, Scientific, and Cultural Organization (UNESCO).

The strategy was launched worldwide in 1980 in some 30 countries. It provides broad policy guidelines for determining development priorities to secure sustainable use of renewable resources, and it links conservation and development. The World Conservation Strategy has three principal objectives: 1) maintenance of essential ecological processes and life-support systems, 2) preservation of genetic diversity, and 3) sustainable use of species and ecosystems.

Introductory sections of the WCS define conservation as:

> . . . the management of human use of the biosphere so that it may yield the greatest sustainable benefit to present generations while maintaining its potential to meet the needs and aspirations of future generations (43).

Development is defined as:

> . . . the modification of the biosphere and the application of human, financial, living, and non-living resources to satisfy human needs and improve the quality of human life.

As defined and used in the WCS, conservation and sustainable development are mutually dependent processes.

A key WCS priority is the promotion of national conservation strategies. These conservation planning tools are now completed or in preparation in 29 countries (see table 10-2). Their long-term purpose is to integrate conservation and development planning and provide an important tool for all stages of development.

The *World Charter for Nature* offers a third example of soft law that is becoming increasingly influential in development. This document, the result of 7 years of effort by international organizations and the United Nations, proclaims 24 principles of conservation by which all human conduct affecting nature is to be guided and judged. In 1982, the United Nations General Assembly, by a vote of 111 to 1, adopted the charter sponsored by the Government of Zaire and 35 other nations.

The United States, the only dissenting vote, objected to the mandatory language contained in the supposedly nonbinding document (14).

Table 10-1.—International Treaties and Conventions for Onsite Maintenance

Title	Established	U.S. signed
Global conventions		
Convention on Wetlands of International Importance Especially as Waterfowl Habitat	1971	pending
Convention Concerning the Protection of the World Cultural and Natural Heritage	1972	1973
Convention on International Trade in Endangered Species of Wild Fauna and Flora	1973	1975
Convention on the Conservation of Migratory Species of Wild Animals	1979	
Convention on the Law of the Sea	1982	not signed
Regional conventions		
Convention on Nature Protection and Wildlife Preservation in the Western Hemisphere	1940	1942
African Convention on the Conservation of Nature and Natural Resources	1968	NA
Convention on the Conservation of European Wildlife and Natural Habitats	1979	NA
Convention on the Conservation of Antarctic Marine Living Resources	1980	1982
ASEAN Agreement on the Conservation of Nature and Natural Resources	1985	NA
Convention for the Protection of Mediterranean Seas Against Pollution	1976	NA
Kuwait Regional Convention for Cooperation on Protection of Marine Environment and Pollution	1978	NA
Convention for Cooperation in the Protection and Development of the Marine and Coastal Environment of the West and Central African Regions	1981	NA
Convention for the Protection of Marine Environment and Coastal Areas of Southeast Pacific	1981	NA
Convention for the Conservation of Red Sea and Gulf of Aden Environment	1982	
Convention for Protection and Development of Marine Resources of the Wider Caribbean Region	1983	1983
Convention for Protection and Development of the Natural Resources and Environment of the South Pacific Region	1985	pending
Species-oriented treaties		
Birds:		
Convention for the Protection of Birds Useful to Agriculture (Europe)	1905	NA
Convention for the Protection of Migratory Birds (Canada/U.S.A.)	1916	1916
Convention for the Protection of Migratory Birds and Game Animals (Mexico/U.S.A.)	1936	1936
International Convention for the Protection of Birds (Europe)	1950	NA
Benelux Convention on the Hunting and Protection of Birds (Europe)	1972	NA
Convention for the Protection of Migratory Birds and Birds in Danger of Extinction and Their Environment (Japan/U.S.A.)	1972	1972
Convention for the Protection of Migratory Birds and Birds Under Threat of Extinction and on the Means of Protecting Them (U.S.S.R./Japan)	1973	NA
Agreement for the Protection of Migratory Birds and Birds in Danger of Extinction and Their Environment (Japan/Australia)	1974	NA
Convention Concerning the Conservation of Migratory Birds and Their Environment (U.S.S.R./U.S.A.)	1976	1976
Directive of the Council of the European Economic Community on the Conservation of Wild Birds (EEC)	1979	NA
Polar bears		
Agreement on the Conservation of Polar Bears	1973	1976
Seals		
Interim Convention on the Conservation of North Pacific Fur Seals	1957	1957
Agreement on Measures To Regulate Sealing and To Protect Seal Stocks in the Northeastern Part of the Atlantic Ocean	1957	
Agreement on Sealing and the Conservation of Seal Stock in the Northwest Atlantic	1971	
Convention for the Conservation of Antarctic Seals	1972	1978
Vicuña:		
Convention for the Conservation of Vicuña	1969	NA
Convention on the Conservation and Management of Vicuña	1979	NA
Agreement Between the Bolivian and Argentinian Governments for the Protection and Conservation of Vicuña	1981	NA
Whale:		
Convention for the Regulation of Whaling	1931	1935
International Convention for the Regulation of Whaling	1946	1948

SOURCES: Simon Lyster, *International Wildlife Law* (Cambridge, England: Grotius Publications Ltd., 1985); Barbara Lausche, "International Laws and Associated Programs for *In-Situ* Conservation of Wild Species," OTA commissioned paper, 1985; Federal Interagency Global Issues Work Group, *U.S. Government Participation in International Treaties, Agreements, Organizations and Programs, In the Fields of Environment, Natural Resources and Population,* 1984; United Nations Environment Programme, *Regional Seas Achievement and Planned Development of UNEP's Regional Seas Programmes and Comparable Programmes Sponsored by Other Bodies,* UNEP Regional Seas Reports and Studies, No. 1, 1982; and M. Wecker, Council on Ocean Law, personal communication, 1986.

Table 10-2.—Countries Where National Conservation Strategies Are Being Developed

Australia	Madagascar	Sierra Leone
Bangladesh	Malawi	Spain
Belize	Malaysia	Sri Lanka
Botswana	Mauritania	St. Lucia
Great Britain	Nepal	Switzerland
Canada	Netherlands	Togo
Costa Rica	New Zealand	Uganda
Fiji	Norway	Vanuatu
Guinea Bissau	Oman	Venezuela
Honduras	Pakistan	Zambia
Indonesia	Panama	Zimbabwe
Italy	Philippines	
Jordon	Senegal	

SOURCE: Mark Halle, deputy director, Conservation for Development Center, International Union for Conservation of Nature and Natural Resources, Gland, Switzerland, personal communication, Oct. 17, 1986.

That is, the document used "shall" rather than "should," despite a general recognition that "by its very nature, the charter could not have any binding force, nor have a regime of sanctions attached to it" (83).

The charter includes several principles relevant to biological diversity:

- The genetic viability on the Earth shall not be compromised; the population levels of all life forms, wild and domesticated, must be at least sufficient for their survival, and to this end, necessary habitats shall be safeguarded.
- The allocation of areas of the Earth to various uses shall be planned, and account shall be taken of the physical constraints, the biological productivity and diversity, and the natural beauty of the areas concerned.
- The principles set forth in the present charter shall be reflected in the law and practice of each State, as well as at the international level.
- All planning shall include among its essential elements the formulation of strategies for the conservation of nature, the establishment of inventories of ecosystems, and assessments of the effects on nature of proposed policies and activities; all of these elements shall be disclosed to the public by appropriate means in time to permit effective consultation and participation.

Other documents include action plans and recommendations from international organizations, such as the UNESCO Action Plan for Biosphere Reserves (discussed later in this chapter), the IUCN Bali Action Plan and Recommendations (resulting from the 1982 World National Parks Congress), and IUCN General Assembly Resolutions. A recently developed tropical forests action plan (84) has also been receiving increased recognition by various countries and intergovernmental and international nongovernmental agencies.

International Laws Relating to Offsite Maintenance

The scope of international law addressing offsite maintenance of diversity is far more limited than that for onsite maintenance. Growing international concern over loss of genetic resources and recognition of the increased importance of offsite maintenance in supporting national and international conservation initiatives have focused attention on this gap in international law (21,47).

To date, attention has been largely focused on defining national responsibilities with regard to crop germplasm maintenance and exchange between countries. Tangentially related international legal instruments deal with international patent protection of biological material and processes, as well as international quarantine as it relates to the flow of plants, animals, and microbes between countries.

Germplasm Maintenance and Exchange

Issues of offsite germplasm maintenance, control, and exchange have assumed a prominent, if controversial, position in international debates in recent years. Declarations of the importance of genetic diversity can be traced to the 1972 Stockholm Conference on the Human Environment. In addition to the Stockholm Declaration mentioned earlier, the conference produced 106 recommendations as tasks and guidelines that should be adopted by governments and international organizations (76). Recommendation 39 called on governments to agree to an international program to preserve genetic resources. This recommendation has been implemented most actively with offsite conservation of cultivated and domesticated materials, particularly crop germplasm. In fact, the creation of the international plant germplasm system that now exists has been credited, in large part, to the Stockholm conference (63).

The only other international agreement dealing specifically with offsite maintenance of germplasm is the *FAO International Undertaking on Plant Genetic Resources*. This initiative to establish, among other things, an international convention dealing with the maintenance and free flow of plant germplasm has been controversial since its inception in 1981. Although initiated as a binding convention, for political expediency it emerged as a nonbinding agreement in 1983, although efforts to make it binding continue (3). As outlined in article 1 of the resolution (26):

> The objective of this undertaking is to ensure that plant genetic resources of economic and/or social interest, particularly for agriculture, will be explored, preserved, evaluated, and made available for plant breeding and scientific purposes. This undertaking is based on the universally accepted principle that plant genetic resources are a heritage of mankind and consequently should be available without restriction.

Subscription to the FAO undertaking has been polarized along industrial and developing-country lines, with some exceptions on both sides (68,82). Developing-country charges that industrialized countries have been capitalizing on Third World genetic resources without remuneration is central to the debate (53). The most hotly contested aspect, however, is free access to private breeders' germlines. Industrial countries with plant breeders' rights legislation (discussed later in this chapter), which include the United States, are unable, if not unwilling, to subscribe to the undertaking without major reservations.

The issues of control and free flow of genetic resources are likely to be debated further in international fora. A closer examination of the U.S. position and options is needed. (Further consideration of this issue is provided in the following discussion of international offsite programs.)

International Patent Law

International patent law is tangentially relevant to genetic resource maintenance because the proprietary status that patenting living organisms provides is central to the debate on international access to germplasm. Current debate focuses on plant patenting, although it could well extend to microbial patenting, for example, in the future. Advances in biotechnology have brought increased attention to patenting living organisms because of the lucrative possibilities the technology offers and because of the likelihood that these advances will accelerate trends toward patenting (e.g., through the ability to establish genetic "signatures" on human-altered organisms). The ability of legislation to keep pace with rapidly evolving biotechnologies is uncertain and raises serious questions for policymakers (67). Effects of patenting on genetic diversity in agricultural crops raise further concerns (see box 10-A).

The expansion of plant patenting into international law occurred with the establishment in 1961 of the *International Union for the Protection of New Varieties of Plants* (IUPOV), consisting of countries party to the international convention on this issue. The convention itself does not provide global patent protection. However, the parties to the convention—almost exclusively industrial countries—agreed to enact plant breeders' rights (PBR) legislation and to guarantee citizens the right to obtain protection under their respective national patent systems (6).

The system does not affect the free flow of germplasm as such. It typically permits protected material to be used for research and breeding by nations that have obtained the rights. Further, there is currently no legal obstacle in using the same material in other countries (6).

Critics charge that since the inception of IUPOV, there has been a concerted effort to

Box 10-A.—Patent Law And Biological Diversity

Patent law essentially entitles inventors to profit from their inventions for a specified period in return for disclosing the secrets of the invention in the public domain, presumably to allow others to build on it. Although the patent system has engendered much controversy since it was formalized, legislation enabling the patenting of living organisms has become one of its most controversial aspects (8,81).

The U.S. Congress passed the Plant Patent Act of 1930 covering asexually propagated plant species. Coverage was extended to sexually propagated species with the 1970 Plant Variety Protection Act (PVPA). With the Supreme Court decision in *Diamond* v. *Chakrabarty* in 1980, microbes became patentable products under the basic patent act (Section 101). A recent decision by the Board of Patent Appeals of the U.S. Patent and Trademark Office has now extended patentability under Section 101 of the Patent Act to included plant material, a reversal of an earlier decision (4).

European countries have a similar system of plant varietal protection, commonly referred to as plant breeders' rights (PBR). In addition to providing patent protection, however, the European system establishes a system of seed control using common catalog requirements to establish legitimate cultivars that can be grown legally (5,61). The European control system is cited as having greater detrimental implications for biological diversity, by increasing uniformity and reducing crop genetic variability, than the basic legislative protection that exists in the United States (9).

Since the emergence of PBR, concerns have been expressed that the proprietary controls it provides may create undesirable trends in the agricultural economy, including several of consequence to biological diversity. Specifically, concerns exist that such legislation is contributing to a consolidation in the seed industry, a reduction of sharing of germplasm and information among researchers, and the loss of genetic diversity.

A review of studies on these linkages (12,49,50,56) reveals different interpretations of their magnitudes, with most analysts agreeing that a strong link is not apparent or at least is difficult to determine. Most pronounced is the degree of consolidation in the seed industry, but separating the specific impact of PBR from other factors is difficult. Studies do, however, reveal that plant patents tend to be concentrated among larger companies and for certain types of crops. Some of these companies have petrochemical interests, which has raised concerns that their research will be directed by efforts to promote agrochemical sales (e.g., emphasizing development of pesticide-tolerant or fertilizer-dependent plant varieties). With regard to the other concerns (reduced exchange of germplasm and research information, or loss of genetic diversity), evaluation is hampered by a lack of objective measures. The conclusion is that careful monitoring in each of these areas seems warranted (8,9).

Perhaps more important is the finding that plant breeding by public agencies plays a critical role in countering the potential negative consequences of the patent system by contributing to competition in the seed industry, to the flow of information and germplasm, and to crop diversity (12). This finding suggests that continued support of national and international (e.g., International Agricultural Research Centers) plant breeding programs is important for maintaining and enhancing genetic diversity. Their contributions should be considered in the context of concerns that interest in biotechnology has detracted from emphasis on traditional breeding and cultivar development (9).

encourage developing countries to adopt plant breeders' rights laws and become members of the union (56). The trade-offs for a developing country enacting a plant patent system, however, are different from those for industrial countries (5). The arguments for adoption are that it would encourage private breeders to develop varieties suited to conditions in each country and that private firms would be less reluctant to export seeds to countries having such legislation (2,5). Without adopting PBR, however, a country would still be able to take advantage of publicly developed varieties, which constitute the most important source of improved seeds for most crops (81). In addition, developing countries are not restricted from using seeds protected under IUPOV.

The extent to which private investment would be encouraged by instituting PBR, given that markets and infrastructures in many developing countries are weak and thus unattractive to many private seed companies, is not clear (5). Concerns also are expressed over the impact that PBR would have on research activities at international agricultural research centers. In the final analysis, whether a country decides to adopt PBR will depend on how governments perceive their own best interests given these and other considerations.

Microbes are patentable (at least for specific process applications) in most industrial nations. The *Treaty on the International Recognition of Deposit of Micro-organisms for the Purposes of Patent Protection* (known as the Budapest Treaty), however, supports a degree of internationalization of the microbial patenting system. This treaty, established in 1977, was instituted in part as a means to provide "enabling disclosure" (as required under patent law) that permits third parties to understand an invention and presumably build on it. It establishes an agreement among participants to recognize deposit of a micro-organism in another country as adequate for patenting purposes. The Budapest Treaty also sets standards and procedures for such depositories (6). This system has engendered much less controversy than the IUPOV system, which may reflect the current limited concern among developing countries over microbe patenting, although this may change in the future (5).

International Quarantine Restrictions

Plant and animal quarantine rules, actions, or procedures are established by governments to prevent entry of pests or pathogens in or on articles imported along pathways created by humans. Regulated articles include plants, animals, propagative material (e.g., seeds, cuttings, cultures, sperm, and embryos), commodities, soil, packing materials, nonagricultural cargo, and used vehicles and farm equipment, as well as their containers and means of conveyance.

The legal umbrella under which international plant quarantine activities are covered is the *International Plant Protection Convention* (IPPC) of 1951 (known as the Rome Convention). The IPPC provided the international model for the phytosanitary certificate that accompanies certain articles in transit (45) and proposed creation of inspection services (6). However, the program seems to have suffered from lack of funds and attention (13,35). Since the mid-1970s, FAO has explored the possibility of establishing a special phytosanitary certificate for the international transfer of germplasm (6).

Though no equivalent to IPPC exists for animals, many countries have signed bilateral agreements on import health requirements of animals, including the establishment of protocols. In general, these international treaties or commissions between governments deal with the movement of live animals or specific animal products such as meat or semen. Restrictive requirements for commerce are generally under the jurisdiction of the respective veterinary services because of hazards related to disease prevention and control (57).

Policies on international commerce in live animals have generally been established and accepted. Research has been considerable and will likely continue, and relaxation of current health-related constraints is anticipated. Policies on international shipment of animal semen are still largely based on the health status of the donors. The technology of embryo transfer is now at the point where research could facilitate international transfer of animal germplasm (57).

INTERNATIONAL PROGRAMS AND NETWORKS

There are so many different organizations involved in international programs to maintain biological diversity, it is difficult to generalize about their effectiveness. Nonetheless, the strengths and weaknesses of two basic categories of organizations—intergovernmental and nongovernmental—are evident. Three intergovernmental organizations, all part of the United Nations, are most prominent:

1. FAO, by virtue of its interest in crops, livestock, forestry, and wildlife (the latter primarily in terms of exploitable resources);
2. UNESCO, whose involvement in biological diversity emphasizes a more scientific and cultural approach (reflected in the Man in the Biosphere concept, outlined below); and
3. UNEP, which extends intergovernmental involvement into more traditional conservation activities (10).

Perhaps the greatest asset of intergovernmental organizations is their ability to elevate issues to international prominence, based largely on the organizations' access to top-level authorities. They may also be able to influence national agendas in various ways. Funding to support activities is central to the influence of these organizations. In recent years, however, the functions and effectiveness of certain offices have been questioned, particularly in the case of UNESCO. Of concern have been the costs of programs in relation to their accomplishments and the politicization of activities and rhetoric, reflecting the dominance of a number of developing countries with an anti-Western bias (62). In general, however, there has been less political volatility and controversy where scientific activity and personnel are central elements of particular intergovernmental initiatives. In fact, UNESCO's most important program dealing with onsite maintenance of biological diversity, Man in the Biosphere, has been singled out for its integrity.

Nongovernmental organizations (NGOs) are most effective as catalysts of international conservation activities. The early work of institutions such as the International Council for Bird Preservation (ICBP) and the International Council of Scientific Unions influenced the evolution of international environmental organizations (10). Considerable activity on maintaining diversity has also resulted from extending national programs to the global arena, a trend that continues and that supports the maxim "environmentalism breeds globalism" (10).

The strengths and weaknesses of NGOs are largely the obverse of those of intergovernmental organizations (62). Their major advantage is the ability to adopt a problem-oriented approach outside a governmental framework, thus minimizing problems associated with political interests and conflicts. This is not to imply that such activities should ignore the political nature of conservation activities. As one analyst has noted:

> For the conservationist to argue that nature is apolitical can be a useful strategy. For him actually to believe this is a recipe for ineffectiveness (10).

Lack of financial resources is the major limiting factor of international NGOs. Yet, limited funds are likely to be applied in a more flexible and responsive way than in intergovernmental institutions, and NGOs often benefit from the voluntarism and enthusiasm characteristic of such groups. However, what is lacking is the ability to influence national governments directly. International NGOs must be cautious to avoid the impression that they are meddling in the affairs of state or impinging on national sovereignty.

The emergence of the International Union for the Conservation of Nature and Natural Resources marked a departure from the traditional dichotomy. IUCN is unique not only because of its emphasis on biological diversity but because of a membership arrangement that combines a number of state and government agencies with an array of national and international conservation groups and scientific organizations. In a sense, it reflects a hybrid institution. The linkages IUCN has cultivated with FAO, UNESCO, and UNEP reinforce its dual nature. Certain advantages are evident in such an arrangement:

> The combination of the two types of organizations provides two approaches to the resolution of problems: the individual scientists working in the non-governmental organization are able to provide a problem-oriented approach with an analysis of the studies being undertaken that is independent and has a minimum of political bias, while the intergovernmental organization can provide political and finan-

cial support for programmes and can make available the time of scientists working in the national research councils and national institutes (23).

Cooperation between intergovernmental organizations and NGOs has not been without conflict, however, especially over how to treat conservation concerns within the context of development. The rapid increase in U.N. membership that occurred in the 1960s, as many developing countries became independent, led to an increasing emphasis on development issues. The landmark 1972 U.N.-sponsored Conference on the Human Environment emphasized the need to incorporate economic development concerns in conservation activities. IUCN responded gradually at first but today the integration of conservation and development has emerged as a central theme of IUCN activity as reflected in its development of such documents as the World Conservation Strategy and the emergence of its Conservation for Development Center.

Although IUCN and the affiliated World Wildlife Fund represent the central international NGOs, a large number of actors are present in the international conservation arena. These organizations vary greatly in size, function, constituency, approach, and focus. Although the contributions of these many groups is acknowleged, the following discussion is necessarily restricted to the largest and most prominent international organizations.

Onsite Programs

Ecosystem and Species Maintenance

Among the array of international programs dealing with onsite diversity maintenance, several stand out for their breadth of coverage. Under the umbrella of UNESCO are two independent programs involved in protection of specific sites, partially chosen for and indirectly concerned with protection of biological diversity. The Man in the Biosphere Program (MAB) supports conservation of sites representing the Earth's different ecosystems, based on the Udvardy system described in chapter 5. The World Heritage Convention mentioned earlier promotes preservation of sites that have outstanding examples of nature.

The *Man and the Biosphere Program* is an international scientific cooperative program supporting research, training, and field investigation. Research focuses on understanding the structure and function of ecosystems and the environmental impacts of different types of human intervention. The program involves disciplines from the social, biological, and physical sciences; it is supervised by an International Coordination Council and is tied to the field through national-level scientific MAB committees.

Launched in 1971, MAB took as one of its themes the "conservation of natural areas and the genetic material they contain." The concept of biosphere reserve was introduced as a series of protected areas linked through a global network that could demonstrate the relationship between conservation and development. Building this network has formed a focus for implementing the program through national-level scientific committees. The first biosphere reserves were designated in 1976. At present, the network consists of 252 reserves in 66 countries (see figure 10-1) (30).

In view of their joint interests, UNESCO, FAO, UNEP, and IUCN convened the First International Biosphere Reserve Congress in 1983 to review experiences and lessons and to develop general guidance for future action. One result of the congress was the preparation of an Action Plan for Biosphere Reserves, which has three main thrusts:

1. improving and expanding the biosphere reserves network;
2. developing basic knowledge for conserving ecosystems and biological diversity; and
3. making biosphere reserves more effective in linking conservation and development, as envisioned by the World Conservation Strategy (71).

The biosphere reserve concept is being applied in a number of cases, but evaluation of

Figure 10-1.—Distribution of Biosphere Reserves Worldwide

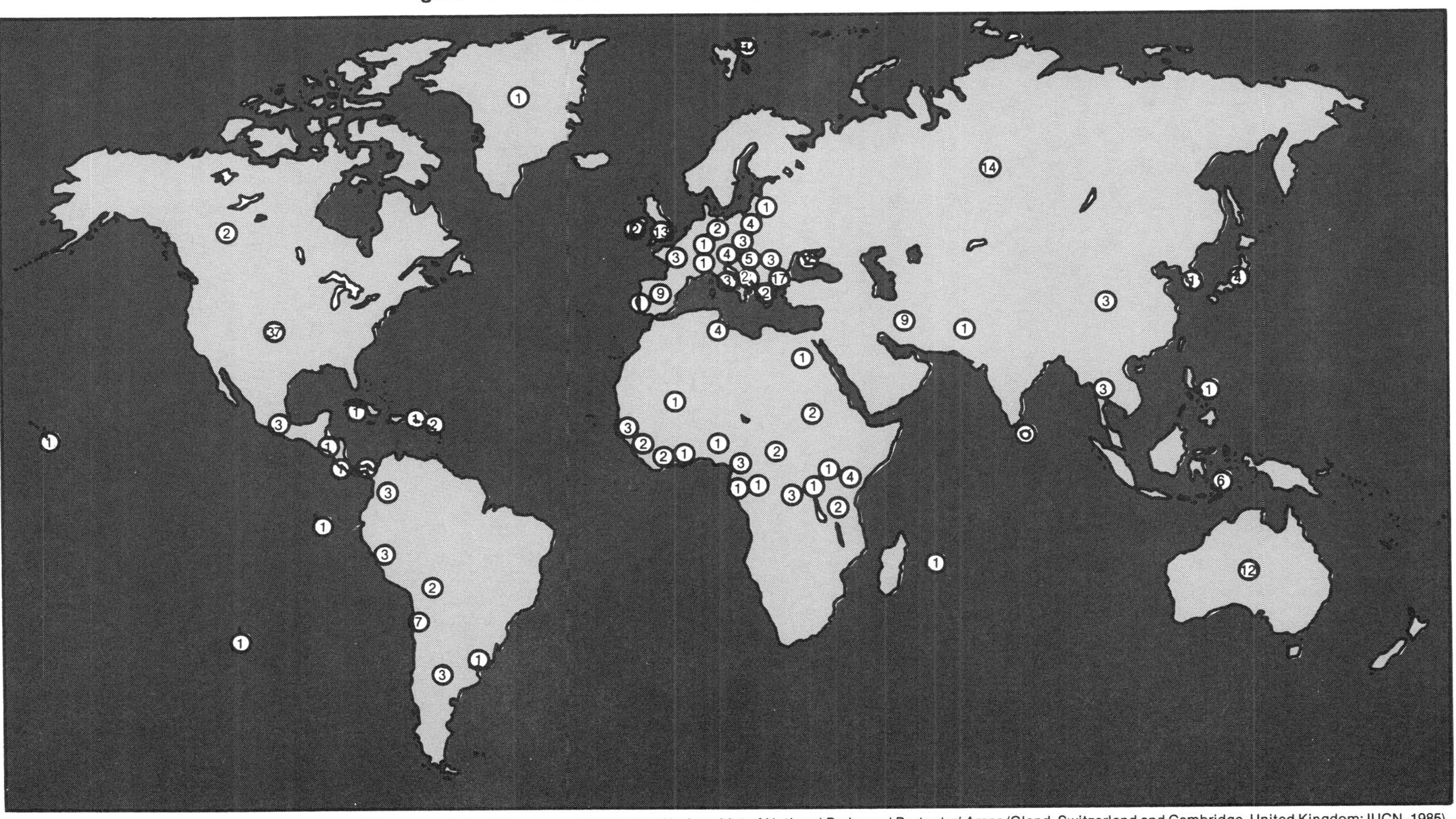

SOURCE: International Union for the Conservation of Nature and Natural Resources, *1985 United Nations List of National Parks and Protected Areas* (Gland, Switzerland and Cambridge, United Kingdom: IUCN, 1985).

success is premature. Full application of the concept, essentially as a conservation and development tool, presents complex problems both legally and administratively. The program has not required special legislation, which leaves each country to adapt existing laws, which are often too weak and too segmented for the kind of integrated multiple-use planning and conservation required (ranging from core areas receiving strict protection to buffer zones in agricultural or other compatible uses).

Moreover, because large areas are involved, generally with some human settlement, application of such a concept necessarily involves many levels of government as well as several technical agencies. Most government administrations tend to be sector-oriented and inexperienced in coordinating jurisdiction and program reponsibilities in such areas as public health, agriculture, forestry, wildlife conservation, and public works—all of which may be required for an effective long-term biosphere reserve program. Special councils or committees of governmental and nongovernmental representatives may need to be formed to play this coordinating role.

Notwithstanding the program's practical problems, the planning and management principles in the biosphere reserves concept reflect what an international conservation program needs to endorse—"conservation as an open system," where areas of undisturbed natural ecosystems can be surrounded by areas of "synthetic and compatible use," and where people are considered part of the system (71).

A number of more recent developments suggest that the MAB program will become an increasingly important investment opportunity for biological diversity maintenance. First, the concept and purpose of biosphere reserves has been sharpened and clarified to reflect pragmatic lessons learned over the 10 years since the first biosphere reserve was established (7). The establishment of the Scientific Advisory Panel for Biosphere Reserves in 1985 promises a more informed, consistent, and structured approach to the MAB system. Current directions also suggest that MAB will continue to stress the important work in research on human needs and impacts within its conservation approach as reflected in its four recently approved research areas (72):

1. ecosystem functioning under different intensities of human impact,
2. management and restoration of resources affected by humans,
3. human investment and resource use, and
4. human response to environmental stress.

Critical review of the existing system has prompted greater attention to ensuring that all three basic elements of biosphere reserves are incorporated into existing and future reserves. These basic elements are the following:

1. **Conservation Role:** conservation of genetic material and ecosystems.
2. **Development Role:** association of environment with development.
3. **Logistic Role:** international network for research and monitoring (7).

Placing greater emphasis on the last two roles, as opposed to the first role which has been predominant to date, will likely contribute increased opportunities and benefits to the biosphere reserve system. Finally, the MAB program may be able to provide important contributions and cooperations within the most recently launched international environmental program, the International Geosphere/Biosphere Program, being formulated by the International Council of Scientific Unions (54).

The United States withdrawal from UNESCO has had a number of implications for U.S. participation in MAB (59). An evaluation of the impacts of this withdrawal suggests that, because MAB activities are largely undertaken as national projects or bilateral arrangements, the short-term impacts on MAB are not very significant. Long-term impacts, however, could seriously compromise the effectiveness and potential of international MAB unless alternatives can be found to provide U.S. scientific and financial participation (59).

The *Convention Concerning the Protection of the World Cultural and Natural Heritage* began in 1975 and at present has 85 member states and 52 natural sites—8 in the United States. Sites are selected by domestic committees, technically reviewed by IUCN, and then evaluated and described by the Bureau of the World Heritage Committee. Approved sites are placed on the World Heritage List.

Site-selection criteria do not specifically mention biological diversity but include areas of ongoing biological evolution, areas of superlative natural phenomena, and habitats of endangered species important to science and conservation. Only exceptional sites are chosen, and the focus is on well-known animals, especially mammals. Sites must have domestic protection in place before being listed. Most nations select already protected sites, such as national parks, rather than new ones. Managers of such sites may have different priorities than those of the convention. The impact of becoming a World Heritage Site on management practices is not fully known.

In signing the convention, members agree to protect their properties and those of other nations. Although the language in agreement is strong, its legal strength has not been established and member governments often ignore provisions. Members are assigned a fee or voluntarily contribute to a World Heritage Fund. Resources are used for training, equipment purchases for members with few resources, and assistance in identifying candidate sites. This support, though small, can be crucial to identifying and protecting sites especially in less well-off nations.

The convention's annual budget averages $1 million. The United States, one of the forces behind the convention's founding, normally contributes at least one-fourth of the budget. In fiscal years 1977 and from 1979 through 1982, U.S. voluntary contributions averaged $300,000. No contributions were made the following two years. The United States contribution in fiscal year 1985 was $238,903. In fiscal year 1986, $250,000 was appropriated (cut to $239,000 under budget-reduction legislation), but no money has yet been contributed. Unless Congress agrees to an Office of Management and Budget request for recision of the entire amount, the contribution will be made, which means the United States, having contributed for two consecutive years, can run for a seat on the World Heritage Committee.

IUCN, is the central nongovernmental organization dealing with onsite diversity maintenance on a global scale. As noted earlier, IUCN is actually a network of governments, nongovernmental organizations, scientists, and other conservationists, organized to promote the protection and sustainable use of living resources. Founded in 1948, IUCN's membership now includes 57 governments, 123 government agencies, 292 national NGOs, 23 international NGOs, and 6 affiliates in at least 100 countries. Developing-country representation has become a more visible component of the network in recent years, although limited active participation by African, Asian, and Latin American countries remains a problem.

Establishment of IUCN resulted from a desire to open up channels of communication between different countries and to serve as an umbrella for various organizations and individuals active in international conservation. Early initiatives focused on research and education activities, in part reflecting the initial funding provided through UNESCO. With the establishment of the World Wildlife Fund (WWF) in 1961 (largely to serve as a fund-raising initiative for IUCN and ICBP) and of UNEP in 1972 (which provided contract work for IUCN), the emphasis shifted back to species and habitat conservation. Today, IUCN and WWF have emerged as central actors in international environmental policy, with influence in both intergovernmental and national conservation work (10). IUCN support for national programs includes the following:

- provision of aid and technical assistance to countries and organizations;
- development of a series of policy aids, particularly in relation to the creation and management of national parks and other protected areas, the framing of legislative instruments, and the making of development policy; and
- preparation, on request from governments,

of specific policy recommendations pertaining to conservation and development plans (10).

Several components of IUCN are particularly relevant to international conservation efforts. These include the three centers that form part of the IUCN network—the Conservation for Development Center (Gland, Switzerland), the Conservation Monitoring Center (Cambridge, England), and the Environmental Law Center (Bonn, West Germany). Central to IUCN prominence and legitimacy in international conservation are its six commissions of experts on threatened species, protected areas, ecology, environmental planning, environmental policy, law and administration, and environmental education.

The Conservation for Development Center has emerged as one of IUCN's most successful components. In particular, its role in assisting countries in the development of national conservation strategies has received growing support. The growth in the program reflects not only the importance of integrating conservation and development interests but IUCN's growing commitment to following this approach.

The Environmental Law Center has been indexing national and international environmental legislation since the early 1960s. Some 20,000 titles are now part of the center's Environmental Law Information System. The center has recently developed a species law index that codes protected species of wild fauna to the corresponding national legislation. This index is computerized, allowing manipulation by species, region, or country, and it is becoming a valuable databank for program and policy planning by governments and NGOs when used in conjunction with scientific information about endangered species, ranges, and protection needs.

Ecosystem and Species Monitoring

Information on the status and trends in loss of the world's fauna and flora is a critical element in defining strategies and priorities. For this reason, a number of international organizations are involved in the inventory and monitoring of biological diversity. Most prominent are the efforts of UNEP, FAO, UNESCO, IUCN, WWF, and ICBP.

UNEP has an assessment arm, Earthwatch, whose function has been to acquire, monitor, and assess global environmental data. At the heart of Earthwatch is the Global Environment Monitoring System (GEMS), an international effort to collect data needed for environmental management. GEMS current activities are divided into monitoring renewable natural resources, climates, health, oceans, and long-range transport of pollutants. These activities are coordinated from the GEMS Programme Activity Center in Nairobi which, like UNEP, works mainly through the intermediary of the specialized agencies of the United Nations—notably FAO, the International Labour Organization, UNESCO, the World Health Organization, and the World Meteorological Organization—together with appropriate intergovernmental organizations such as IUCN (15).

To provide access to the databanks, UNEP-GEMS has begun a 2-year pilot project to set up a computerized Global Resource Information Database (GRID) (74). If successful, GRID may prove to be a powerful tool for international inventory and monitoring, not only of biological diversity but of other areas too (15). GRID will provide a centralized data-management service within the U.N. system, designed to convert environmental data into information usable by decisionmakers. The main data-processing facility will be in Geneva, Switzerland, but it will be controlled from UNEP headquarters in Nairobi.

The pilot phase of GRID is to result in an operational system with preliminary results and the training of some personnel. An initial evaluation could be expected by the end of UNEP's 1986/87 biennium. A full assessment of the system is unlikely before several more years of operations (74).

Inventory and monitoring activities at the species level are also undertaken by the Conservation Monitoring Center (CMC), one of several centers operating under the auspices of the

IUCN. Its mandate is to analyze and disseminate information on conservation worldwide and provide services to governments and the conservation and development communities. CMC supplies information in the form of books, specialist publications, and reports. Major output includes Red Data Books on endangered species, protected-area directories, conservation site directories and reports, threatened plant and animal lists, U.N. Lists of National Parks and Equivalent Reserves, preliminary environmental profiles of individual areas (by request), comparative tabulations of transactions under CITES, and analyses of wildlife trade data for individual countries and taxonomic groups (15).

The International Council for Bird Preservation (ICBP) takes responsibility for ornithological aspects of IUCN's activities and shares the IUCN database at CMC. ICBP is also in the process of establishing an oceanic-islands database to identify areas where action is required for numerous threatened endemic bird species. The initial target is to collect details about some 160 islands that support endemic species of birds, especially islands smaller than 20,000 square kilometers (15).

An important supplement to these initiatives is the growing number of national organizations taking an international perspective in their data collection efforts. Of particular importance is The Nature Conservancy International (TNCI), based in the United States. TNCI has developed a regional database on distribution of fauna and flora in the neotropics that is the most comprehensive of its kind and is promoting establishment of country-level conservation data centers (see ch. 11).

International Network

The emergence of IUCN as a recognized network of conservation specialists has both galvanized international conservation activities and established conservation programs as scientific initiatives. It also established two major functions of the organization:

1. promoting contacts among institutes and individuals, primarily by acting as a device for the exchange of information; and
2. setting up some kind of procedure whereby common platforms and goals could be articulated and, ultimately, a measure of influence exerted on public policy (10).

The Ecosystem Conservation Group (ECG), consisting of FAO, UNEP, UNESCO, and IUCN, was established in 1975 to advise on planning and execution of international conservation activities by the four organizations (75). ECG has recently begun to take a more active role in conservation. ECG agreed at its 11th General Meeting, held in Rome in February 1984, to institute an ad hoc Working Group on Onsite Conservation of Plant Genetic Resources (40). The working group consists of FAO (lead agency), UNESCO, UNEP, IUCN, and the International Board for Plant Genetic Resources (IBPGR). The first meeting of the working group was held at IUCN headquarters in April 1985, during the 12th General Meeting of ECG. The charge to the working group was twofold:

1. review ongoing and planned activities in onsite conservation in light of recommendations of the First Session of the FAO Commission of Plant Genetic Resources, UNESCO's Action Plan for Biosphere Reserves (see earlier discussion), and the IUCN Bali Action Plan; and
2. identify ways to strengthen action and cooperation in response to these recommendations, with particular attention to improving information flow and promoting pilot demonstration activities (40).

Six major goals for coordination and action were recognized at the first meeting of the ECG working group and activities were identified within the framework of these goals. This development signifies an important step among involved organizations to focus their programs on plant genetic resources within a common framework, and, as reflected by the addition of IBPGR, begin to build a mechanism to integrate onsite and offsite efforts.

Offsite Programs

International institutions dealing with offsite maintenance are most easily considered under the separate headings of plant, animal, and microbial genetic resources. The level of existing international activity between and within these categories of organisms varies considerably. Major factors determining the level of attention devoted to offsite maintenance include economic importance, threat of loss, and ability to maintain viable collections offsite.

Plant Diversity

International programs and networks are differentiated by the types of plants they deal with. By far, the most developed institutions are those concerned with major agricultural plants. For the most part, these offsite collections are maintained in association with agricultural research institutions. Concern over loss of wild species of nonagricultural plants in their natural habitats has prompted the establishment of an international network of botanic institutions for preserving rare and endangered species in living collections.

The focus, extent, and effectiveness of international genebank efforts in recent years have been largely shaped by International Agricultural Research Centers (IARCs) supported by the Consultative Group on International Agricultural Research (CGIAR). This organization was founded in 1971 and consists of donors that fund a network of centers doing research on increasing agricultural productivity, primarily in developing countries (see table 10-3). Impetus to form the group stemmed from early successes of two institutes (later to become the first members of the CGIAR system), the International Maize and Wheat Improvement Center (better known by its Spanish acronym CIMMYT), and the International Rice Research Institute (IRRI). Both programs were the outgrowth of research centers supported by the Rockefeller and Ford Foundations.

Financial obligations soon became too great for the two U.S. foundations as budget costs grew with the establishment of two more centers. The desire on the part of several international development institutions, including FAO, UNEP, and the World Bank, to expand the system into a network of international centers led to the formation of the CGIAR, supported by a group of government and international donor agencies.

Table 10-3.—International Agricultural Research Centers Supported by the Consultative Group on International Agricultural Research

CIAT	—Centro Internacional de Agricultura Tropical Cali, Columbia
CIMMYT	—Centro Internacional de Mejoramiento de Maiz y Trigo Mexico City, Mexico
CIP	—Centro Internacional de la Papa Lima, Peru
IBPGR	—International Board for Plant Genetic Resources Rome, Italy
ICARDA	—International Center for Agricultural Research in the Dry Areas Aleppo, Syria
ICRISAT	—International Crops Research Institute for the Semi-Arid Tropics Hyderabad, India
IFPRI	—International Food Policy Research Institute Washington, DC, U.S.A.
IITA	—International Institute of Tropical Agriculture Ibadan, Nigeria
ILCA	—International Livestock Centre for Africa Addis Ababa, Ethiopia
ILRAD	—International Laboratory for Research on Animal Diseases Nairobi, Kenya
IRRI	—International Rice Research Institute Manila, Philippines
ISNAR	—International Service for National Agricultural Research The Hague, Netherlands
WARDA	—West Africa Rice Development Association Monrovia, Liberia

SOURCE: Consultative Group on International Agricultural Research, *Summary of International Agricultural Research Centers: A Study of Achievements and Potential* (Washington, DC: 1985).

Today, most CGIAR centers have specific responsibilities in crop varietal development and germplasm conservation, and in certain cases serve as international base and active collections for specific crops (see table 10-4). A number of IARCs also operate outside the CGIAR system, including several with responsibilities for germplasm maintenance. This group includes the International Soybean Program in Urbana, Illinois and the Asian Vegetable Research and Development Center in Shanhua, Taiwan (18).

The most prominent international institution dealing with offsite conservation of plant genetic diversity is IBPGR. Established in 1974 by CGIAR, it serves as a focal point for governments, foundations, international organizations, and individual researchers with interests in maintaining genetic diversity of crop spe-

Table 10-4.—International Agricultural Research Centers Designated as Base Seed Conservation Centers for Particular Crops

Center	Crop	Nature of collection
AVRDC	mungbean (*Vigna radiata*)	global
	sweet potato (seed)	Asia
CIAT	beans (*Phaseolus*): cultivated species	global
	cassava (seed)	global
CIP	potato (seed)	global
ICARDA	barley	global
	chickpea	global
	faba bean (*Vicia faba*)	global
ICRISAT	sorghum	global
	pearl millet	global
	minor millets (*Eleusine, Setaria, Panicum*)	global
	pigeon pea	global
	groundnut	global
	chickpea	global
IITA	rice	Africa
	cowpea (*Vigna unguiculata*)	global
	cassava (*Manihot esculenta;* seed)	Africa
IRRI	tropical rice (wild species and cultivated varieties)	global

SOURCE: Consultative Group on International Agricultural Research, *Summary of International Agricultural Research Centers: A Study of Achievements and Potential* (Washington, DC: 1985).

cies. IBPGR is a small group; part of the secretariat is provided by FAO. Its mission has been a coordinating one, of setting priorities and creating a network of national programs and regional centers for the conservation of plant germplasm. It has provided training facilities, supported research in techniques of plant germplasm conservation, sponsored numerous collection missions, and provided limited financial assistance for conservation facilities (see ch. 11). It does not operate any germplasm storage facilities itself, however.

As envisioned by IBPGR, collection efforts were to focus on crop plants, based on priorities set by the board and reflecting the economic importance of the crop, the quality of existing collections, and the threat that diversity would disappear. The collected materials were to be kept in national programs and duplicated outside the nation in which they were collected. A global base collection was to be established for major crops, and there were hopes of creating regional programs.

The achievements of IBPGR are impressive, measured in its own terms and against the list of objectives. Ten years after IBPGR was established, the network for base collection storage included 35 institutions in 28 countries. Regional maize collections exist in Japan, Portugal, Thailand, and the United States, for example, and one in the Soviet Union is under negotiation. For rice, a global collection has been established in Japan and the Philippines, and regional collections are found in Nigeria and the United States (27). National programs were created during IBPGR's first 10 years in about 50 countries, and by 1986 some 50 base collection centers had been designated for about 40 crops of major importance (38,79). The program has limited itself to a particular group of plants and has been successful in coordination, in encouragement of national programs, and in scientific and educational assistance (33). In all, IBPGR has links with more than 500 institutes in 106 countries (79).

In part, due to the success of IBPGR in focusing attention on the need to conserve genetic diversity, the issue has become embroiled in political controversy. IBPGR regards itself as a technical and scientific organization. But a number of critics regard the issue of plant genetic resources as much more political. They maintain that IBPGR is implicitly working for the corporate and agribusiness interests of the industrial world, particularly the United States (36,56). Critics also argue that the current genetic material exchange system is inadequate to ensure that material will continue to be available, particularly to developing countries. The debate has become quite acrimonious, with proponents of IBPGR emphasizing their scientific and pragmatic approach to the issue, and critics emphasizing their fear that multinational corporations will gain control over plant germplasm. Plant patenting and access to plant genetic resources are also important elements in the current controversy (see earlier section) (6).

This entire controversy helped catalyze a move toward deeper FAO involvement in the germplasm area and toward a new international approach (70). FAO argued that it should be taking the lead in plant genetic conserva-

IBPGR Network

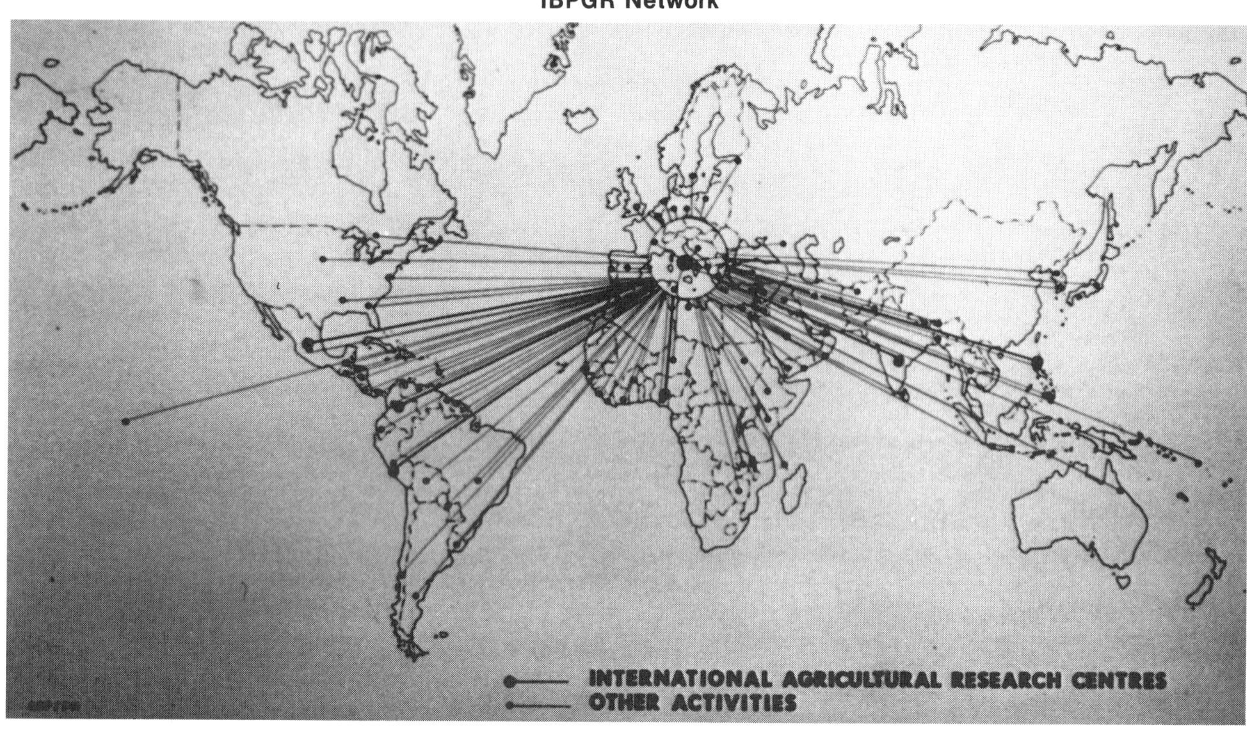

Photo credit: IBPGR

The International Board for Plant Genetic Resources has had a catalytic effect on efforts to conserve dwindling plant genetic resources.

tion; it could provide the framework for developing nations to obtain a greater political voice in the international conservation structure. It was further argued that IBPGR was not a formal organization, and it would therefore have only limited legal ability to enforce any commitment to make germplasm available (26). This legal status argument is questionable, however (6). IBPGR proponents responded that the board's technical emphasis works effectively, and it is in fact an asset in surmounting political problems and dealing with nations outside FAO.

The alternative approach that evolved consisted of an undertaking and a new commission. The *International Undertaking on Plant Genetic Resources* was negotiated within the framework of the FAO. (The United States and a number of other developed countries reserved their positions.) The undertaking was nonbinding, probably to increase participation in such a controversial area. It called for an international germplasm conservation network under the auspices of the FAO, stated a duty of each nation to make all plant genetic material—including advanced breeding material—freely available, and called for development of a procedure under which a germplasm conservation center could be placed under the auspices of FAO. IBPGR was to continue its current work, but it would be monitored by FAO (6).

The other part of the new FAO system is the *Commission on Plant Genetic Resources*, a group established to meet biannually to review progress in germplasm conservation. The commission held its first meeting in March 1985, with the United States present as an observer. Much of the discussion focused on concerns expressed in the FAO undertaking and on issues that had regularly been dealt with by IBPGR, such as base collections, training, and information systems. In addition, discussions and resolutions paid significant attention to onsite conservation and emphasized the importance of this area, which has received little attention from IBPGR.

It is not yet clear whether a practical and cooperative division of responsibilities between the two entities can be developed. One approach that has been suggested is to have each entity assume different responsibilities, such as giving IBPGR responsibility for offsite maintenance and letting FAO focus on onsite genebanks. An alternative would be to have IBPGR assume responsibility for technical aspects of germplasm collection and maintenance and give FAO responsibility for legal and political factors (28).

Botanic gardens and arboretums are increasingly viewed as important for conservation of wild plant species. Efforts to establish an international network of botanic gardens for the purpose of conserving threatened plant species were formalized in an international conference at the Royal Botanical Garden at Kew (United Kingdom) in 1978. IUCN's Species Survival Commission set up a Botanical Gardens Conservation Coordinating Body (BGCCB). This body, established in 1979, is coordinated by the Threatened Plants Unit of IUCN's Conservation Monitoring Center and now has 136 members. In addition, the Moscow Botanic Garden coordinates for BGCCB the response of 116 gardens in the Soviet Union (51). The function of such a network was reviewed at an IUCN conference in Las Palmas in 1985. Representatives of the botanic gardens meetings recommended a new conservation secretariat with IUCN support to coordinate their conservation activities and the establishment of a Botanic Garden Conservation Strategy (39).

Representation of developing nations is poor in BGCCB. The Montevideo Botanic Garden is the only South American member, for example. Efforts are being made to involve more developing-country institutions and to encourage twinning arrangements between institutions, whereby expertise in seed maintenance, curation, and fund raising could be promoted. Mechanisms to fund such activities, however, are not well established (39).

The planning of conservation collections by collaborating botanic gardens is encouraged by BGCCB by drawing attention to rare and threatened species that are poorly represented or not in cultivation at all. This is done through the provision of reports and an annual computer printout for each member garden, detailing the conservation plans IUCN has been provided by the garden. The printouts allow an analysis of the garden's holdings in relation to other gardens. Members are encouraged to propagate and distribute species that are represented, especially if they are endangered or extinct in the wild. BGCCB also has circulated lists of threatened plants to its members and stores the information on holdings in the CMC database (51).

IUCN has located 3,948 threatened plant species in cultivation by members of BGCCB, which is at least one-quarter of the known threatened plants in the CMC computerized database (78). However, these collections constitute only a tiny proportion of the genetic range of threatened species. They also represent only a small proportion of the biological diversity maintained by botanic gardens, which implies that greater emphasis on cultivation of rare and threatened species could be undertaken (51). Although it may be theoretically possible for the botanic gardens of the world to grow the estimated 25,000 to 40,000 threatened species of flowering plants, cultivating sufficient populations to maintain diversity is unrealistic. Consequently, protecting a diversity of wild species will rest on maintaining them in the wild.

Animal Diversity

Just as institutions split offsite maintenance of plants into agricultural and nonagricultural species, offsite maintenance of animals is broken down into categories of domesticated and wild species. The former category has fallen under international agricultural institutions, such as FAO or regional institutions, such as the International Livestock Centre for Africa. Responsibility for offsite maintenance of wild species has been almost exclusively assumed by an international network of zoos.

Concern over loss of genetic diversity in agricultural animals has been much less pronounced than that for agricultural plants. Consequently, no analog to IBPGR currently exists. Growing concern over the loss of potentially valuable genetic diversity for livestock, however, has prompted limited efforts in this area.

FAO and UNEP launched a pilot project in 1973 to conserve animal genetic resources. Initial efforts focused on developing a preliminary list of endangered breeds and of those with economic potential, especially for developing countries. A 1980 FAO/UNEP Technical Consultation extended this work by defining requirements for creating "supranational infrastructure resources for animal breeding and genetics" (37). These covered a range of efforts to develop animal genetic resources. Of particular significance were guidelines in the following areas (37):

- databanks for animal genetic resources, which would also identify endangered breeds;
- genebanks to store semen and embryos of endangered breeds; and
- training of scientists and administrators in genetic resources conservation and use.

Endangered livestock breeds can be maintained either in living collections or through cryogenic storage of semen or embryos (see ch. 6). Although the former option has proved viable in certain European countries (34), widespread success is unlikely. Thus, cryogenic storage will become increasingly important as threats to livestock increase. Concern over loss of livestock diversity is greatest for developing countries, but creating cryogenic genebanks in many countries would be very difficult. Thus, the value of establishing supranational storage facilities becomes apparent.

International networking for conservation of living collections of wild animals is largely restricted to the zoological community, although IUCN's Species Survival Commission has been involved in formulating conservation plans that include captive breeding (51). Zoos have traditionally been established for public education and entertainment. But in recent years, a number of larger zoos have concentrated on breeding rare or endangered species, usually birds and mammals. These efforts have also extended to the creation of regional and international networks to enhance the effectiveness and collective conservation potential of the zoological community.

An International Species Inventory System (ISIS) was created in 1974 in response to major problems of inbreeding in zoo populations and in recognition of the fact that, for an increasing number of wild animals, captive populations held the best hope for survival of the species. Coverage has grown from 55 facilities to 211 as of 1985. About 65,000 living specimens of 2,300 species are included. Information currently comes from facilities in 14 countries, but coverage is best for U.S. and Canadian institutions (see ch. 9). The system is not restricted to endangered species (25).

ISIS publishes biannual survey reports. These include information on the sex ratio and age distribution; the proportion of captive-bred; and the birth, death, and import trends for all mammals and birds held in captivity by the members. The system has also recently begun to incorporate information on holdings of reptiles and amphibians (25).

The American Association of Zoological Parks and Aquariums (AAZPA) set up the Species Survival Plan (29) in September 1980. AAZPA has identified certain species in need of immediate attention and has established a committee for each, consisting of a species coordinator and propagation group. A major committee function is to provide direction for maintenance of studbooks.

A studbook is an international register that lists and records all captive individuals of species that are rare or endangered in the wild. The concept, initially developed for the selective breeding of domesticated animals, was first used on a wild species (the European bison) in 1932. Studbooks are now kept for about 40 endangered species and are a valuable tool in international cooperation in captive breeding, permitting intelligent recommendations to zoos around the world concerning such things as optimal pairings, trades, and management (24). Official studbooks are those recognized and endorsed by IUCN's Survival Commission and the International Union of Directors of Zoological Gardens, and they are coordinated by the editor of the International Zoo Yearbook.

Microbial Diversity

A directory of institutions maintaining microbial culture collections was published in 1972, under the sponsorship of UNESCO, the World Health Organization, and the Commonwealth Scientific and Industrial Research Organization. The directory was revised and updated in 1982 (55) and remains the primary comprehensive source of information on international culture collections. In addition, the American Phytopathological Society convened a panel of scientists to discuss the importance and future of microbial culture collections (1). Information from these sources indicates that probably 1,200 to 1,550 collections exist throughout the world. A brief history of several of the more important collections is available (64).

In 1985, UNEP, the International Cell Research Organization, and UNESCO recognized the need for moderately sized culture collections. Each collection as envisioned would have a special purpose and together they would form a network of collections around the world (6). The establishment of these microbiological resource centers (MIRCENs) began at that time and the specialized collections are now located in 15 locations (17): including Brisbane, Australia; Stockholm, Sweden; Bangkok, Thailand; Nairobi, Kenya; Porto Alegre, Brazil; Guatemala City, Guatemala; Cairo, Egypt; Paia, Hawaii, United States; and Dakar, Senegal (32).

MIRCENs were established to develop and enhance an infrastructure for a world network of regional and interregional laboratories. This network provides a base of knowledge in microbiology and biotechnology to support the biotechnology industry in developed and developing countries. Activities of MIRCENs include collection, maintenance, testing and distribution of *Rhizobium*, and training of personnel (46). Training has perhaps been the most important activity towards developing research capabilities and diffusion of technology, especially in developing countries (16). Though each MIRCEN works according to its own set of priorities, they share a common goal of working together to strengthen the network and advance the knowledge in microbiology and biotechnology. In doing so, MIRCENs provide incentives to develop and maintain offsite microbial collections in support of national programs. They also offer a framework that could provide a secure custodial system for national and international microbial resources.

NEEDS AND OPPORTUNITIES

The United States has historically played an important leadership role in international conservation initiatives. The establishment of Yellowstone Park in 1872 heralded the international movement to create national parks worldwide. The United States also was a central actor in the 1972 Stockholm Conference on the Human Environment, in the creation of the United Nations Environment Programme, the World Heritage Convention, and numerous other initiatives (see previous sections). In recent years, U.S. leadership in international conservation has waned, which is reflected in funding and personnel support for international programs. A number of opportunities exist whereby the United States could reestablish itself as a leading actor in international efforts to promote the maintenance of biological diversity.

Onsite Activities

A major problem in developing a coherent strategy to address concerns over loss of biological diversity is the uncertainty that surrounds the issue. Estimates of the scope of species diversity vary by orders of magnitude, which illustrates obvious impediments to defining and addressing the problem. Further, limited and unreliable data on the rates and impacts of habitat conversion exacerbate the problem of refining a strategy and determining the level of resources that should be directed to address concerns. Clearly, biological diversity in certain regions is acutely threatened and deserves priority attention. However, attention is also needed on gaining a better grasp on defining the scope of diversity and the degree to which it is threatened.

Many questions remain even as understanding of the magnitude of threats to diversity continues to improve. Critics suggest that a better grasp of the situation is needed before large amounts of resources are devoted to the problem (66). It should be noted, however, that funds currently spent on diversity maintenance are relatively small and are not likely to increase dramatically. More important perhaps is the realization that funding, both public and private, continues to be directed to well-defined threats. That is, the situation as it currently exists is essentially reactionary—responding to acute threats that have already materialized. Recognition of the importance of biological diversity has yet to assume the prominence that would make most national governments take systematic and preemptive approaches to threatened diversity, which in the long run might prove less costly. Increased attention and recognition of national and regional conservation strategies as important elements of integrated development planning may represent movement to adopt this approach.

Considerable discussion among international conservation organizations has been directed toward the need to develop an international network of protected areas that would include representative and unique ecosystems. To date, however, organizing, implementing, and supporting such a system remains difficult. Efforts to establish such a system have not suffered from lack of creativity, as reflected in two large-scale proposals: one to create a major international program to finance the preservation of 10 percent of the remaining tropical forests (65) and another to establish a world conservation bank (69).

It may be possible to establish an international network of protected areas within the framework of existing programs, specifically UNESCO's Man in the Biosphere and World Heritage programs. To do so, however, would require adopting a more organized and strategic policy, further invigorating both programs, and providing increased resources. This would require a more concerted effort on the part of national governments, intergovernmental agencies, and the participation of specific international non-governmental groups (especially IUCN).

Two other issues are prominent with respect to the effectiveness of international laws and programs (47). First, there is debate over the value of a global treaty to fill what some perceive as a serious gap in hard law. Second, alternatives to conventional protected areas need to be considered to provide protection beyond such areas or at sites where the conventional approach is not feasible.

The notion of a world treaty to conserve genetic resources of wild species was proposed at the IUCN World National Parks Congress in 1982. A similar recommendation by the World Resources Institute was proposed for the U.S. Government to develop an international convention. However, one key question that needs to be addressed before implementation is whether a new global treaty could be adopted and enforced in time to address the problem. In addition, consideration must be given to financial and technical resources still needed for treaties that currently play a role in resource conservation.

Existing treaties have been difficult to implement because of a lack of administrative machinery (e.g., well-funded and staffed secretariats); lack of financial support for on-the-ground programs (e.g., equipment, training, and staff); and lack of reciprocal obligations that serve as incentives to comply (21). A possible exception is CITES, which has mechanisms to facilitate reciprocal trade controls and a technical secretariat, although inadequately funded.

Creating protected areas is the conventional approach in most international conservation programs. The modern interpretation of protected areas includes the full range of conservation uses, from strict protection to multiple use (44). The question of alternatives and supportive measures outside protected areas has also become a growing concern. The 1984 State of the Environment Report of the Organisation for Economic Cooperation and Development (OECD), for example, urges that protected areas are not enough. Environmentally sensitive policies for nondesignated lands are also needed (60). This conclusion is reinforced by IUCN's Commission on Ecology:

> The idea of basing conservation on the fate of particular species or even on the maintenance of a natural diversity of species will become even less tenable as the number of threatened species increases and their refuges disappear. Natural areas will have to be designed in conjuction with the goals of regional development and justified on the basis of ecological processes operating within the entire developed region and not just within natural areas.

Land-use planning may help integrate environmentally sensitive policies in nondesignated areas. Control options to safeguard genetic diversity outside protected areas could also be explored (21,22). Where private land is involved, general controls could be enforced by imposing restrictions on land use or by instituting a permit system. These practices are commonly used for nature conservation and environmental protection in many western countries, particularly Europe. Permits could be required for all activities likely to harm certain natural habitats or ecosystems. This approach requires legislation to authorize the requirement, procedures, decisions on the conditions to be imposed, and activities excluded from the permit requirement.

Nonstatutory protection of specific sites could be achieved through voluntary agreements between the landowner and conservation authorities. Such agreements are more attractive when the landowner is offered certain incentives, such as tax subsidies or deductions, for preserving sites. In the United States, such "conservation easements" are valuable mechanisms for conserving private lands (77).

Zoning ordinances could become a powerful conservation tool if extended not only to construction but to all changes in land use, including agriculture. Programs to preserve areas where only small natural or seminatural sites remain within cultivated fields, for example, are also important. Such efforts can help maintain at least a minimum amount of natural vegetation in hedgerows, tree groves, riparian, and other areas. Giving conservation advice to farmers about the value of protected lands would be an important component of such controls.

In many countries, however, land management agencies have little or no authority to oversee activities of other agencies or to veto actions that would be detrimental to maintaining the land's natural condition. Although a variety of land-use planning tools are being considered, two prerequisites exist for using them:

1. to strengthen the technical capacity to identify, inventory, and monitor valuable natural areas; and
2. to provide the legal authority to protect such areas.

Offsite Activities

Offsite maintenance of biological diversity is assuming increased prominence due to concern over loss of genetic resources. Its prominence is also the result of a greater appreciation of the important role that offsite maintenance of wild species can play in conserving species diversity, especially when linked to onsite programs. However, a number of major resources remain unprotected in the existing framework. These include medicinal plants; some industrial plants, such as rubber; a number of animals, including wild and domesticated varieties and possibly some marine species, for which commercial breeding techniques are evolving (58).

To cover existing gaps in maintaining plant genetic resources, efforts could be made to extend IBPGR's mandate to assume responsibilities for medicinal plants, industrial plants, and minor crops. IBPGR has already expressed reluctance to assume principal responsibility for these areas, noting that in many cases, such efforts should be relegated to national programs (79). Another option, however, is creation of a new group to cover these particular interests. Such an effort should try to establish some organizational affiliation capitalizing on the expertise already acquired by IBPGR.

Perhaps the most blatant gap, however, is in the area of animal genetic resources. Although FAO and UNEP have initiated investigations in this area, no national, regional, or international programs have yet emerged. An international board on animal resources could be established, with a mandate and approach similar to IBPGR's. But instead of establishing a network of national programs, a more reasonable approach might include creating a network of regional programs, promoting conservation of animal germplasm and monitoring endangered livestock breeds.

Additional international exchange of information is also needed, particularly with respect to what is conserved in smaller collections, such as those maintained by university faculty or private breeders. This exchange often occurs informally through working networks of researchers. In some cases, however, improved data management systems may be appropriate.

Integration

Diversity maintenance programs require complementary efforts between onsite and offsite conservation, and finding the balance of emphasis is key. The first session of the FAO Commission on Plant Genetic Resources discussed building this integration by establishing national plant genetic resource centers that would be closely linked to offsite genebanks and protected area management (41). Such efforts will require improved cooperation at international and national levels, along with creative use of existing laws and programs to meet emerging management and scientific needs.

CHAPTER 10 REFERENCES

1. American Phytopathological Society, *National Work Conference on Microbial Collections of Major Importance to Agriculture* (St. Paul, MN: APS, 1981). *In*: Halliday and Baker, 1985.
2. American Seed Trade Association, "Position Paper of the American Seed Trade Association on FAO International Undertaking on Plant Genetic Resources," May 1984.
3. Anonymous, "European Parliament Votes To Adopt FAO Undertaking," *Diversity* (9):24-25, 1986a.
4. Anonymous, "New Patent Policy Stirs Seed Industry," *Diversity (9):6*, winter 1986b.
5. Barton, J.H., "The International Breeder's Rights System and Crop Plant Innovation," *Science* 216:1070-1075, June 4, 1982.
6. Barton, J.H., "International/Legal Framework for *Ex-Situ* Conservation of Biological Diversity," OTA commissioned paper, 1985.
7. Batisse, M., "Developing and Focusing the Biosphere Concept," *Nature and Resources* (in press).
8. Berlan, J.P., and Lewontin, R., "Breeders' Rights and Patenting Life Forms," *Nature* 322:785-788, Aug. 28, 1986.
9. Bliss, F.A., "Market and Institutional Factors Affecting the Plant Breeding System," OTA commissioned paper, 1985.
10. Boardman, R., *International Organization and the Conservation of Nature* (Bloomington, IN: Indiana University Press, 1981).
11. Burhenne, W.E., *International Environmental Law—Multilateral Treaties Series*, vols. I-IV (Berlin, Federal Republic of Germany: Erich Schmidt Verlag, 1984). *In*: Lausche, 1985.
12. Butler, L.J., and Marion, B.W., "The Plant Variety Protection Act and the Importance of Public Plant Breeders and Competition in the Seed Industry," *Journal of Agronomic Education* 14(1), 1985.
13. Chiarappa, L., and Karpati, J.F., "Plant Quarantine and Genetic Resources," In: 1984, per 11-23 J.H.W. Holden, and J.T. Williams (eds.) London.
14. Cleary, S., Bureau of Oceans and International

Environmental and Scientific Affairs, U.S. Department of State, Washington, DC, personal communication, 1986.
15. Collins, N., "International Inventory and Monitoring for the Maintenance of Biological Diversity," OTA commissioned paper, 1985.
16. Colwell, R.R., "A World Network of Environmental, Applied, and Biotechnological Research," *ASM News* 49(2):72-73, 1983.
17. Colwell, R.R., notes pertaining to MIRCEN Director's Council, personal communication, July 1, 1986.
18. Consultative Group on International Agricultural Research (CGIAR), *1984 Report on the Consultative Group and the International Agricultural Research It Supports* (Washington, DC: October 1984).
19. Convention for the Protection and Development of the Marine Environment of the Wider Caribbean Region, "Message From the President of the United States to the 98th Congress," 2d sess., Treaty Doc. No. 98-13 (Washington, DC: U.S. Government Printing Office, 1984). *In*: Lausche, 1985.
20. Curtis, C., president, The Oceanic Society, Washington, DC, personal communication, September 1985. *In*: Lausche, 1985.
21. de Klemm, C., "Conservation of Wetlands, Legal and Planning Mechanisms," *Proceedings of the Conference of the Contracting Parties to the Convention on Wetlands of International Importance Especially as Waterfowl Habitat* (Gland, Switzerland: International Union for the Conservation of Nature and Natural Resources, 1984). *In*: Lausche, 1985.
22. de Klemm, C., "Protecting Wild Genetic Resources for the Future: The Need for a World Treaty," *National Parks, Conservation, and Development—The Role of Protected Areas in Sustaining Society, Proceedings of the World Congress on National Parks, Bali, Indonesia, October 11-22, 1982* (Washington, DC: Smithsonian Institution Press, 1984). *In*: Lausche, 1985.
23. Dooge, J.C.I., and Baker, F.W.G., "International Governmental and Non-Governmental Scientific Cooperation," *Ecology in Practice, Part II: The Social Response*, F.D.Castri, F.W.G. Baker, and M. Hadley (eds.) (Dublin: Tycooly International Publishing, 1984).
24. Dresser, B., and Liebo, S., "Technologies To Maintain Animal Germplasm," OTA commissioned paper, 1986.
25. Flesness, N., "Status and Trends of Wild Animal Diversity," OTA commissioned paper, 1985.
26. Food and Agriculture Organization (FAO) of the United Nations, Report of the Conference of FAO, 22d sess., Rome, Nov. 5-23, 1983, C83/REP (Rome: 1983).
27. Food and Agriculture Organization (FAO) of the United Nations, Commission on Plant Genetic Resources, *Base Collections of Plant Genetic Resources*, CPGR: 85/4 (Rome: 1984).
28. Food and Agriculture Organization (FAO) of the United Nations, First Session of the Commission on Plant Genetic Resources, Rome, Mar. 11-15, 1985 (Rome: 1985).
29. Foose, T.S., "The Relevance of Captive Populations to the Conservation of Biotic Diversity." *In*: C.M. Schonewald-Cox, et al. *Genetics and Conservation* (Menlo Park, CA: The Benjamin/Cummings Publishing Co., Inc., 1983).
30. Gregg, W.P., MAB coordinator, U.S. National Park Service, personal communication, 1986.
31. Halle, M., deputy director of Conservation for Development Center, IUCN, personal communication, Gland, Switzerland, 1986.
32. Halliday, J., and Baker, D., "Technologies To Maintain Microbial Diversity," OTA commissioned paper, 1985.
33. Hawkes, J.G., *Plant Genetic Resources: The Impact of International Agriculture Research Centers*, Consultative Group on International Agricultural Research (CGIAR), Study paper No. 3 (Washington, DC: The World Bank, 1985).
34. Henson, E.L., "An Assessment of the Conservation of Animal Genetic Diversity at the Grassroots Level," OTA commissioned paper, 1985.
35. Hewitt, W.B., and Chiarappa, L., *Plant Health and Quarantine in International Transfer of Genetic Resources* (Cleveland, OH: Chemical Rubber Co. Press, 1977).
36. Hobbilink, H., "Home-Grown Solutions," *South*, January 1986, p. 112.
37. Hodges, J., "Animal Genetic Resources in the Developing World: Goals, Strategies, Management and Current Status," *Proceedings of Third World Congress on Genetics Applied to Livestock Production*, Lincoln, NE, vol. 10, July 16-22, 1986.
38. International Board for Plant Genetic Resources, *A Global Network of Genebanks* (Rome: 1983).
39. International Union for Conservation of Nature and Natural Resources, and World Wildlife Fund, *Plant Advisory Group: Minutes of Third Meeting*, Quetame, Colombia July 6-7, 1986.

40. International Union for Conservation of Nature and Natural Resources, *1985 United Nations List of National Parks and Protected Areas* (Gland, Switzerland and Cambridge, England: 1985).
41. International Union for Conservation of Nature and Natural Resources, "Report of the First Meeting of the ECG Working Group on *In Situ* Conservation of Plant Genetic Resources," Apr. 11-12, 1985 (Gland, Switzerland: 1985). *In*: Lausche, 1985.
42. International Union for Conservation of Nature and Natural Resources, "Report on IUCN's Participation in the First Session of the FAO Commission on Plant Genetic Resources," Rome, Mar. 11-15, 1985 (Gland, Switzerland: IUCN, 1985). *In*: Lausche, 1985.
43. International Union for Conservation of Nature and Natural Resources, *World Conservation Strategy* (Gland, Switzerland: 1980).
44. International Union for Conservation of Nature and Natural Resources, *Categories, Objectives, and Criteria for Protected Areas* (Gland, Switzerland: 1978).
45. Kahn, R., "Assessment of Plant Quarantine Practices," OTA commissioned paper, 1985.
46. Keya, S.O., Freire, J., and DaSilva, E.J., "MIRCENs: Catalytic Tools in Agricultural Training and Development," *Impact of Science on Society* 142:142-151, 1986.
47. Lausche, B., "International Laws and Associated Programs for *In-Situ* Conservation of Wild Species," OTA commissioned paper, 1985.
48. Lausche, B., World Wildlife Fund, personal communication, June 1986.
49. Lesser, W.H., and Masson, R.T., *An Economic Analysis of the Plant Variety Protection Act* (Washington, DC: American Seed Trade Association, 1983).
50. Loyns, R.M.A., and Begleiter, A.J., "An Examination of Potential Economic Effects of Plant Breeder Rights on Canada," Working Paper, Policy Coordination Bureau, Consumer and Corporate Affairs Canada, no date.
51. Lucas, G., and Oldfield, S., "The Role of Zoos, Botanical Gardens, and Similar Institutions in the Maintenance of Biological Diversity," OTA commissioned paper, 1985.
52. Lyster, S., *International Wildlife Law* (Cambridge, England: Grotius Publications, Ltd., 1985).
53. MacFadyen, J.T., "United Nations: A Battle Over Seeds," *The Atlantic* 256(5):36-44, November 1985.
54. Malone, T.F., and Roederer, J.G. (eds.), *Global Change*, The Proceeding of a Symposium sponsored by the International Council of Scientific Unions (ICSU) during its 20th General Assembly in Ottawa, Canada on Dec. 25, 1984 (New York: Cambridge University Press, 1985).
55. McGowan, V.F., and Skerman, V.B.D. (eds.), *World Directory of Collections of Micro-organisms* (Brisbane, Australia: World Data Center, 1982).
56. Mooney, P.R., "The Law of the Seed," *Development Dialogue* 1(2), 1983.
57. Moulton, W., "Constraints to International Exchange of Animal Germplasm," OTA commissioned paper, 1985.
58. National Council on Gene Resources, *Anadromous Salmonid Genetic Resources: An Assessment and Plan for California*, prepared for California Gene Resources Program, 1982.
59. National Research Council, *UNESCO Science Program: Impacts of U.S. Withdrawal and Suggestions for Alternative Interim Arrangements, A Preliminary Assessment* (Washington DC: National Academy Press, 1984).
60. Organisation for Economic Cooperation and Development (OECD), *1984 State of the Environment Report* (Paris: 1985).
61. Perrin, R.K., Kunnings, K.A., and Ihnen, L.A., *Some Effects of the U.S. Plant Variety Protection Act of 1970*, Economics Research Report No. 46 (Raleigh, NC: North Carolina State University, Department of Economics and Business, 1983). *In*: Bliss, 1985.
62. Perry, J.S., "Managing the World Environment," *Environment* 28(1):10-15, 37-40, January/February 1986.
63. Plucknett, et al., "Crop Germplasm Conservation and Developing Countries," *Science* 220: 163-169, Apr. 8, 1983.
64. Porter, J.R., "The World View of Culture Collections," *The Role of Culture Collections in the Era of Molecular Biology*, R.R. Colwell (ed.) (Washington, DC: American Society for Microbiology, 1976). *In*: Halliday and Baker, 1985.
65. Rubinoff, I., "The Preservation of Tropical Moist Forests: A Plan for the South," unpublished, 1984.
66. Simon, J., "Disappearing Species, Deforestation, and Data," *New Scientist*, May 15, 1986, pp. 60-63.
67. Stiles, S., "Biotechnology: Revolution or Evolution for Agriculture?" *Diversity* 8:29-30, winter 1986.
68. Sun, M., "The Global Fight Over Plant Genes," *Science* 231:445-447, Jan. 31, 1986.
69. Sweatman, I.M., "The World Conservation

Bank," Project of the International Wilderness Leadership Foundation, 1986.
70. Tooze, W., "Seeds of Discord," *Barron's*, July 30, 1984.
71. United Nations Educational, Scientific, and Cultural Organization, "Action Plan for Biosphere Reserves," *Nature and Resources* XX(4):11-22, 1984.
72. United Nations Educational Scientific, and Cultural Organization, International Co-ordinating Council for the Programme on Man and the Biosphere, Ninth Session, MAB/ICC-9/CONF.4, Paris, Oct. 23, 1986.
73. United Nations Environment Programme, *Register of International Treaties and Other Agreements in the Field of the Environment* (Nairobi, Kenya: 1984). *In*: Lausche, 1985.
74. United Nations Environment Programme, *Global Resources Information Database* (Nairobi: UNEP/GEMS, 1985).
75. United Nations Environment Programme, "UNEP and Protected Areas—Review of Joint Activities Carried Out by UNEP and FAO, IUCN and the UNESCO MAB Programme 1973-1982," paper presented at the World National Parks Congress, Bali, Indonesia, Oct. 11-22, 1982 (Nairobi: 1982).
76. United Nations Conference on the Human Environment, "Declaration on the Human Environment," *Keesing's Contemporary Archives*, Sept. 16-23, 1972, pp. 25476-25479. *In*: Lausche, 1985.
77. U.S. Congress, Office of Technology Assessment, *Grassroots Conservation of Biological Diversity in the United States*, OTA-BP-F-38 (Washington, DC: U.S. Government Printing Office, February 1986).
78. Walter, S.M., and Birks, H., "Botanic Gardens and Conservation," WWF Plants Campaign—Plant pac No. 5 (Gland, Switzerland: World Wildlife Fund, 1984).
79. Williams, J.T., "A Decade of Crop Genetic Resources Research." *In*: J.H.W. Holden and J.T. Williams, *Crop Genetic Resources: Conservation and Evaluation* (Boston: George Allen & Unwin, 1984).
80. Williams, S.B., "Protection of Plant Varieties and Parts as Intellectual Property," *Science* 225:18-23, July 6, 1984.
81. Witt, S.C., *Biotechnology and Genetic Diversity* (San Francisco: California Agricultural Lands Project, 1985).
82. Witt, S.C., "FAO Still Debating Germplasm Issues," *Diversity* (8):24w25, winter 1986.
83. Wood, H.W., Jr., "The United Nations World Charter for Nature: The Developing Nations' Initiative To Establish Protections for the Environment," *Environmental Law Quarterly* 12(4): 977-998, 1985.
84. World Resources Institute, *Tropical Forests: A Call for Action*, Report of an International Task Force convened by the World Resources Institute, The World Bank, and the United Nations Development Programme (Washington DC: 1985).

Chapter 11
Biological Diversity and Development Assistance

CONTENTS

Table

Boxes

Chapter 11
Biological Diversity and Development Assistance

HIGHLIGHTS

- The United States has a stake in maintaining biological diversity in developing countries. Many of these nations are in regions where biological systems are highly diverse, pressures that degrade diversity are most pronounced, and the ability to forestall a reduction in diversity is least well developed.
- With recent amendments to the Foreign Assistance Act and earmarking of funds, the United States has defined maintaining biological diversity as an important objective in U.S. development assistance. It is unclear, however, whether the Agency for International Development (the principle U.S. development assistance agency) can effectively promote conservation of biological diversity.
- Development assistance can help improve the capacity of developing countries to maintain diversity by 1) building public support; 2) establishing an information base; 3) building institutional support; 4) promoting planning and management; 5) increasing technical capacity; and 6) increasing the direct economic benefits from sustainable use of biological resources.
- Multilateral development banks strongly influence the nature of resource development in developing countries. Recent congressional pressures to encourage these banks to place greater emphasis on environmental implications of their activities, including threats to biological diversity, have met with some success. Continued monitoring of progress in this area is necessary to enhance progress made to date.

INTRODUCTION

Concern about the loss of biological diversity is acute for developing countries for several reasons. First, the level of diversity is greater in developing countries particularly in tropical locations, than it is in industrial countries. Second, biological diversity is less well-documented in developing countries. Third, conversions of natural ecosystems to human-modified landscapes are more pronounced and likely to accelerate in developing countries due to the combined pressures of population growth and poverty. Finally, developing countries characteristically lack both the technical and financial resources to address these issues.

The United States has a stake in maintaining biological diversity, particularly in developing countries. The rationale for assisting developing countries rests on the following:

1. recognition of the substantial benefits of a diversity of plants, animals, and microorganisms;
2. evidence that degradation of ecosystems can undermine U.S. support of economic development efforts; and
3. esthetic and ethical motivations to avoid irreversible loss of unique life forms (see box 11-A).

Box 11-A.—U.S. Stake in Maintaining Biological Diversity in Developing Countries

Political Interests

- The United States has strong commitments to world peace, economic and social stability, and maintenance of the Earth's basic life-support systems—commitments that require concern about the integrity and long-term productivity of the world's natural resource base.
- U.S. public institutions and private firms conduct activities that directly and indirectly affect biological resources of other nations and, therefore, are in positions to influence the attitudes and actions of host governments and local citizens on biological diversity maintenance.
- Political stability can be compromised as a result of a breakdown of ecological systems. Civil unrest in countries such as Haiti and El Salvador has been attributed, in part, to degradation of natural resources.

Economic Interests

- The non-oil-exporting developing nations purchase one-third of all U.S. exports. Adverse domestic resource conditions seriously affect the ability of these countries to buy U.S. goods and services.
- Many of the natural reservoirs of crop genetic diversity are located in developing countries. Without a diverse base for crop breeding, the development of high-yielding tree and crop varieties characteristic of the Green Revolution cannot be sustained.
- Over the years, the United States has invested billions of dollars in international development assistance programs that could be undercut by loss of biological diversity associated with resource degradation.
- The United States has lent billions of dollars to developing countries. Continued declines in natural resources will reduce the ability of these nations to pay their debts.

Humanitarian Interests

- The United States is committed to meeting basic needs and supporting developing countries' economic and social development, which in turn is linked inextricably to the quality and integrity of the world's natural resource base.
- The United States increasingly is being requested by governments and international development organizations to provide technical assistance and financial support for conservation-related activities in developing countries.

Environmental Interests

- The United States shares with South and Central America and the Caribbean area hundreds of species of migratory animals—birds, insects, marine turtles, mammals—whose survival depends on maintaining suitable habitats.
- The United States is committed to help preserve the world's flora, fauna, and vulnerable ecosystems by virtue of domestic legislation and national policies, and by being party to a large number of international conventions and agreements. Principal among these measures are the Endangered Species Act of 1973, the Convention on International Trade in Endangered Species of Wild Flora and Fauna, and the Convention on Nature Protection and Wildlife Preservation in the Western Hemisphere.

Educational and Scientific Interests

- Advances in medicine depend heavily on research animals found in developing countries. The United States accounted for more than half the estimated 30,000 primates traded internationally in 1982 for use in medical research.
- Rich arrays of living systems demonstrate the many ways organisms can cope with variable and often unfavorable physical and biotic environments. Hence, areas such as tropical forests provide unparalleled opportunities to understand complex processes of evolutionary interaction.

SOURCE: Adapted from U.S. Interagency Task Force, *The World's Tropical Forests: A Policy, Strategy, and Program for the United States* (Washington, DC: U.S. Department of State, 1980).

INTEGRATION OF ECONOMIC DEVELOPMENT AND BIOLOGICAL DIVERSITY MAINTENANCE

Interests and activities of development agencies and conservation organizations have merged in recent years, in light of the changing perspectives of these two groups. Historically, conservation organizations and development agencies planned their efforts independently in developing countries (64). Conservation groups focused almost exclusively on natural areas, promoting protection from human exploitation and preservation of particular wild species and their habitats. In contrast, development organizations focused on raising the standard of living in both rural and urban areas and concentrated on the major agricultural species.

Increasingly, development assistance agencies *and* developing country governments are establishing policies that recognize the importance of environmental factors in development strategies. These policies stem from a growing awareness in development planning of the costs of ignoring environmental factors. The greater reliance of developing-country economies on their natural resource base—soils, fisheries, and forests—underlies this growing appreciation for sustainability in development initiatives.

Planning began to include environmental considerations in cost-benefit and similar analyses during the 1970s. The emphasis was on mitigating side effects, such as pollution and salinization. By the late 1970s, development agencies began to include components to sustain the resource base that affected a project. Watershed protection above irrigation systems received funding, for instance. Development assistance in the early 1980s supported projects to deal directly with the problems associated with natural resource degradation, such as fuelwood shortages in arid regions.

Although maintaining biological diversity has not become an objective of assistance projects, these steps led toward development that generally caused less resource degradation and thus generally benefited diversity maintenance. In the 1983 Amendment to the Foreign Assistance Act (described in the next section), Congress directed the Agency for International Development (AID) to support projects that have maintenance of biological diversity as a specific objective, such as establishing protected areas and controlling poaching.

Conservation organizations, in turn, realized that their traditional emphasis on establishing parks and protected areas would be insufficient to protect biological diversity and began to broaden their approach. These groups have increasingly realized that failure to account for the needs of rural people jeopardizes the long-term success of conservation projects.

A clear manifestation of conservationists' efforts to reorient their activities is the development of the World Conservation Strategy (WCS). This document links conservation with development and provides policy guidelines for determining development priorities that secure sustainable use of resources (20). The WCS has three principal objectives: 1) the maintenance of essential ecological processes, 2) the preservation of genetic diversity, and 3) the sustainable use of species and ecosystems. The document is used to increase dialog on the interests and approaches of the development and conservation communities. It has been only partially successful, however. The WCS has been effective in narrowing the gap of conservation and development interests in policy documents, but on a practical basis this gap remains.

Part of the problem with linking development and conservation lies in the failure to identify common criteria and benefits. Conservation activities generally justify projects by biological and esthetic criteria. For example, conservation organizations would draw attention to the Tuatara (*Sphenodon punctatus*) of New Zealand because it is the last remaining species of an entire order of reptiles (32). Unique or spectacular habitats are also given special attention. Conservation organizations also focus on spectacular species of birds or mammals, largely in response to the esthetic interests of contributors.

Development initiatives, on the other hand, are directed by economic criteria. Internal rates of return and similar economic analyses, for example, are important steps in justifying particular projects. This emphasis can be detrimental for biological diversity because many values associated with maintaining diversity are difficult to measure (see ch. 2) and thus are undervalued in development project decisions (28). The standard economic approach may be unable to account for the loss of biological diversity, where time horizons are long, benefits are diffuse, and losses are irreversible (37). The problem is particularly acute for weakened economies where overexploiting renewable resources to meet immediate needs often undermines the chances for long-term sustainability of resources.

Lack of institutional overlap also presents problems in defining common ground among development and conservation interests. Responsibilities for natural resources are generally split among agencies (e.g., agriculture, forestry, and wildlife). Despite efforts by developing countries to establish offices responsible for broader environmental issues, the agencies are frequently unable to add conservation components to development activities, let alone to compete with other agencies for financial or administrative support.

Another management problem that can hinder efforts to protect a particular habitat or species is the imbalance between the means devoted to conservation enforcement and the market value of the protected resource. The salaries of officials assigned to enforce conservation measures can be extremely low compared to the worth of the resources they are guarding. Perhaps a more difficult dilemma is trying to dissuade local populations from exploiting or degrading protected areas when subsistence requirements and lack of alternatives compel them to do so.

This problem raises a central question in defining the role of development assistance in maintaining biological diversity. Should development assistance support diversity maintenance if such initiatives have adverse impacts on the people it is intended to help? Current legislation (discussed later in this chapter) stresses the beneficial aspects of maintaining diversity in overall development. But some diversity maintenance projects can conflict with local development interests. For instance, conflict can arise by denying access to resources on protected lands. Wildlife conservation efforts in proximity to agricultural lands may also threaten crops, domestic livestock, and even humans (9).

In examining the issue of possible conflicts between development and diversity maintenance, it is perhaps useful to define two approaches to maintaining diversity. First is the symptomatic approach. This is the approach typically undertaken by environmental groups and is often directed at protecting a particular species and its habitat. Because of the focused nature of this approach, needed interventions, usually involving strict protective measures, are often easy to define. However, such a program can be costly and difficult to implement, especially if initiated only after threats reach a critical point. Problematic from a development perspective is the case where strict protective measures impinge on the interests of local populations.

Alternatively, there is a curative approach to threats to diversity. This approach attempts to address the root causes of the threats to diversity. It generally involves a much broader array of initiatives and is less focused on diversity per se. It emphasizes the human element of the conservation equation.

The greatest threats to diversity in developing countries stems less from the impacts of development than from a lack of development. Addressing the root causes of threats to diversity will therefore need to emphasize the availability of opportunities for individuals in developing countries to enhance their quality of life. This is the approach generally taken by development assistance agencies in their efforts to elevate standards of living by creating employment opportunities and increasing access to education, health care, and family planning.

Both approaches will be necessary in meeting the challenges of diversity maintenance. Within the context of U.S. interests to promote diversity maintenance through the channels of development assistance, it is important to stress areas of overlap between these two approaches. That is, emphasis should be placed on promoting the type of projects that, on the one hand, promote opportunities for local populations and, on the other hand, maintain the diversity within biological systems.

This approach is based on the proposition that the best way to maintain diversity within a development initiative is to use that diversity. Examples abound of efforts to capitalize on diversity maintenance in areas ranging from tourism to biological resource development (9). This utilitarian approach should be approached with caution, however. It is important to ensure that initiatives will be environmentally, economically, and institutionally sustainable over the long term. Identifying possibilities for multiple uses of an area or biological resource should be stressed. Further, it is important to ensure that the benefits of such interventions actually accrue to the people affected.

A consensus exists that long-term conservation must have a base of support at the national level and account for the interests and participation of local populations. It seems reasonable, therefore, to stress these criteria in development assistance projects supporting biological diversity. These criteria provide consistency in U.S. interests in conservation and development and promote projects most likely to succeed. Cases will arise in which the particular focus of protection inevitably conflicts with local demands. Resolving such conflicts is the responsibility of local or national governments, although foreign assistance can be useful, especially in providing resources to facilitate or compensate for a particular intervention. Whether such support should be considered under particular development assistance or through other channels is not clear.

The greatest opportunities, however, lie in taking a more forward-looking and anticipatory approach by helping countries define strategies and policies to preempt such conflicts. Support for planning, management, and inventory of diversity, promoting in-country expertise, and constituencies to support diversity maintenance initiatives help reduce the incidence of conflict between development and diversity maintenance. In the final analysis, the success of U.S. support for maintaining diversity in developing countries will depend on success in promoting the capacity in the developing countries themselves.

U.S. RESPONSE

After nearly a decade of legislative and administrative concern about the role of U.S. foreign assistance in environmental protection (see box 11-B), the case for U.S. action to conserve diversity in developing countries was recognized in Section 119 of the Foreign Assistance Act (FAA), added by Congress as part of the International Environment Protection Act of 1983 (Public Law 180-64). This amendment includes the following:

- authorizes the President to furnish assistance to countries in protecting and maintaining wildlife habitats and in developing sound wildlife management and plant conservation programs (Sec. 119(b));
- directs the Administrator of AID, in consultation with the heads of other appropriate government agencies, to develop a U.S. strategy including specific policies and programs to protect and conserve biological diversity in developing countries (Sec. 119(c)); and
- requires the President to report annually to Congress on the implementation of Section 119 (Sec. 119(d)).

Section 119 signals Congress' belief that U.S. development assistance should specifically initiate projects traditionally undertaken by conservation organizations. In effect, AID has been directed to deal not only with the foundations

Box 11-B.—Amendments to Foreign Assistance Act Concerning International Environmental Protection

Congressional concern with international environmental protection has increased markedly over the last decade. U.S. foreign assistance programs began incorporating environmental concerns in the late 1970s when a series of amendments to the Foreign Assistance Act defined the Agency for International Development's (AID) mandate in the area of environment and natural resources. These amendments gave specific emphasis to promoting efforts to halt tropical deforestation, a major threat to conserving biological diversity.

- **1977:** Amended Section 102 to add environment and natural resources to areas AID should address.
- **1977:** Added new Section 118 on "Environment and Natural Resources," authorizing AID to fortify "the capacity of less developed countries to protect and manage their environment and natural resources" and to "maintain and where possible restore the land, vegetation, water, wildlife, and other resources upon which depend economic growth and well-being, especially that of the poor."
- **1978:** Amended Section 118, requiring AID to carry out country studies in the developing world to identify natural resource problems and institutional mechanisms to solve them.
- **1978/79:** Amended Section 103 to emphasize forestry assistance, acknowledging that deforestation, with its attendant species loss, constituted an impediment to meeting basic human needs in developing countries.
- **1981:** Amended Section 118, making AID's environmental review regulations part of the act, and added a subsection (d), expressing that "Congress is particularly concerned about the continuing and accelerating alteration, destruction, and loss of tropical forests in developing countries." Instructs the President to take these concerns into account in formulating policies and programs relating to bilateral and multilateral assistance and to private sector activities in the developing world.
- **1983:** Added Section 119, directing AID in consultation with other Federal agencies to develop a U.S. strategy on conserving biological diversity in developing countries.
- **1986:** Redesignated Section 118 as Section 117 with the new Section 118 addressing tropical forest issues.
  Amended Section 119, which among other things earmarked money for biological diversity projects.

SOURCE: Adapted from B. Rich and S. Schwartzman, "The Role of Development Assistance in Maintaining Biological Diversity *In-Situ* in Developing Countries," OTA commissioned paper, 1985.

of the threats but also with some of the consequences.

The *U.S. Strategy on the Conservation of Biological Diversity: An Interagency Task Force Report to Congress* was delivered to Congress in February 1985, in response to Section 119. This report was followed by an annual report, *Progress in Conserving Biological Diversity in Developing Countries FY1985*, which outlines implementation of Section 119 a year later.

The strategy has been criticized for lack of commitment to action, even though it contains 67 recommendations. Its most concrete aspect is allocation of responsibilities among agencies, but this is done without any indication of funding mechanisms. Some critics have questioned whether the strategy advances a cohesive plan and whether U.S. Government agencies are significantly increasing their allocation of resources to address this issue (54,58). Severe budget constraints undoubtedly limit the degree to which new programs can be put forward. It is therefore critical for agencies to establish clear priorities and to indicate which actions need to be taken and how much they will cost.

AID drafted an Action Plan on Conserving Biological Diversity in Developing Countries, to apply the general recommendations to specific agency programs and policies (51). It proposes specific actions based on strategy recommendations and assigns them a priority of near-term (within the next two fiscal years) or long-term (requiring additional or redirected resources). However, it is clear that initiatives are determined by funding restrictions rather than by critical needs.

Another difficulty with the draft action plan is reflected in responses from various AID missions. Reviews of the draft express skepticism that specific initiatives can be implemented at the mission level, based solely on the broad, generalized directions it contains. Recent congressional earmarkings of the AID budget to support diversity projects further emphasizes the need to develop a more refined strategy for identifying priority projects.

Despite the criticisms of AID's draft action plan, it represents the agency's effort to identify its responsibilities for about half of the 67 recommendations contained in the strategy. Other Interagency Task Force members have yet to identify how their resources and expertise could be applied to the strategy. Development of action plans by other Federal agencies may be a useful way to identify strengths and opportunities within each agency, to identify areas for cooperation, and to provide a way to examine agency commitments more effectively.

IMPLEMENTATION OF U.S. INITIATIVES: THE AGENCY FOR INTERNATIONAL DEVELOPMENT

Overall, AID has developed an extensive set of guidelines and procedures for programs to incorporate concerns for the environment. To this extent, it deserves high marks compared with other development assistance agencies, both bilateral and multilateral. Less evident, however, are indications that these procedures are being consistently implemented. Critics question AID's incorporation of environmental assessments of project development at a stage when modifications can be easily made (67).

Several factors limit AID's implementation of biological diversity initiatives in developing countries, including a belief by the agency that it is adequately addressing biological diversity, declining budgets and staff to initiate projects, and an inadequate number of trained personnel to address conservation issues.

Defining the maintenance of biological diversity as a priority is viewed with some trepidation at the highest levels of AID (27). The issue is seen as one among many priorities (e.g., women in development, child welfare, and so on) identified in the Foreign Assistance Act. Although such mandates have been partially effective, their numbers, the frequency of changes, and the lack of priority among them may hamper efficient management of agency resources (16,53,60).

AID has been forced to allocate declining resources in response to various congressional mandates. It is unlikely that programs to safeguard diversity can compete successfully for an increased share of the AID budget. Reviews of AID's implementation of environmental projects provide reason to be skeptical (16,41).

Because diversity conservation is related to many factors (e.g., poverty, population pressure, pollution, and agricultural policies), AID believes its obligations are largely addressed by conventional assistance projects (41). For instance, the February 1985 task force report to Congress identified 253 projects as having a conservation component (62). Few of these, however, are the types of projects identified in Section 119. Most involve more indirect contributions, such as reducing destructive pressures on habitats.

These indirect initiatives are critical, of course. Without them, the long-term prospects for biological diversity would be dismal. Perhaps projects identified in Section 119 should be viewed as supplemental measures or as attempts to designate important conservation areas while they can still be easily protected. One concern, however, is that Section 119, as the central piece of legislation addressing concerns for maintaining diversity in developing countries, may define biological diversity, and the initiatives to conserve it too narrowly.

Congress has expressed dissatisfaction with the level of funding AID has directed to meeting the provisions of Section 119 by earmarking $2.5 million for diversity projects in fiscal year 1987. This amount represents the only specified funding for environmental projects contained in the FAA. That this appropriation is intended to account for diversity on three continents, however, stresses the need to allocate this funding judiciously. Also of concern is the impact of this earmarking on support for other conservation initiatives, such as those in Sections 117 and 118 of FAA that lack any specific funding provisions.

Yet simply allocating new funds for diversity projects may not be an adequate response. If projects are proposed to meet a spending target without allocations based on an established set of priorities, efforts may be inefficient or even counterproductive.

The agency's commitments to biological diversity projects and to acting on environmental concerns have been eroded by the Gramm-Rudman-Hollings Act (70). Overall, 4.3 percent of AID's 1986 budget was sequestered, but the Office of Forestry, Natural Resources, and the Environment (FNR) had its budget cut 25 percent (26). Such reductions indicate where agency priorities lie and add credence to claims that despite a commitment to environmental concerns, commitment in the form of resource allocation lags.

It should also be noted, however, that the two major funding sources (the Agricultural, Rural Development, and Nutrition account and the Selected Development Activities account) that support most environmental projects also suffered disproportionate cuts—15.5 and 20.6 percent (50). These reductions reflect congressional, not AID, appropriations.

One proposed way to increase the emphasis and visibility of environmentally related issues is to elevate FNR to a bureau (10). Because many of the funding allocation decisions are made at the bureau level, this change in status may increase the share of resources devoted to diversity projects. Such an action, on the other hand, could isolate a newly established bureau.

An alternative is to establish a separate funding source, such as a Forestry, Natural Resources, and Environment account, for various bureaus and offices as well as overseas missions to draw on. Several functional accounts (e.g., Agriculture, Rural Development, and Nutrition; and Population and Health) already exist. Establishing an additional account will likely be seen as further constraining AID's flexibility. It would, however, place resources behind congressional concerns for biological diversity and the environment and natural resources generally, as outlined in Section 119 as well as Sections 117 and 118.

Another approach would seek to incorporate biological diversity concerns into AID development activities at different levels of the agency ranging from general policy documents at the agency level to more strategic efforts at the regional bureau and missions levels. AID could prepare a policy determination (PD) document on biological diversity that would serve as a general statement that maintaining diversity is an explicit objective of the agency.

Existence of a PD could mean that consideration of diversity concerns would, where appropriate, become an integral part of sectoral programming and project design. Further, it would require that projects be reviewed and evaluated by the Bureau of Program and Policy Coordination for consistency with the objectives of the PD. Because of the increase in bureaucratic provisions this would create, the formulation of a PD on diversity would probably not be well received within AID.

The three regional bureaus (i.e., Africa, Asia and Near East, and Latin America and the Caribbean) could also prepare documents that identify important biological diversity initiatives in their regions. The Asia and Near East Bureau, in fact, has already prepared such a document. But the lack of agency commitment and the hesitancy of the bureau to redirect scarce funds have reduced the document's utility thus far. The Africa Bureau is currently completing a natural resources management plan that includes an assessment of regional priorities for biological diversity maintenance.

The development of such reports for each regional bureau is considered an effective way to identify priorities for projects, especially given the earmarking of funds. A network of specialists and information sources already exists to help identify priority areas. For example, committees of the International Union for the Conservation of Nature and Natural Resources (IUCN), and especially its Conservation Monitoring Center in Cambridge, England, are major sources of such information.

AID country-level environmental profiles can also identify priorities for diversity projects. The agency has completed 50 preliminary Phase I profiles and 17 in-depth Phase II profiles (see table 11-1). AID has also supported "state of the environment" reports in five countries, which are similar to environmental profiles but generally prepared within the country by a local group (18).

The most important focus of biological diversity strategies is at the mission level, where projects are implemented. Congress has already mandated that Country Development Strategy Statements and other country-level documents prepared by AID address diversity concerns. Most missions, however, lack the expertise or adequate access to expertise needed to address this provision of Section 119 as amended.

Table 11-1.—Country Environmental Profiles Undertaken or Supported by the Agency for International Development

Areas with profiles	Phase I profile	Phase II profile	State of the environment report
Asia/Near East/North Africa	15	1	2
Latin America/Caribbean	14	14	2
Sub-Saharan Africa	21	2	1
Total	50	17	5

SOURCE: International Institute for Environment and Development, Environmental Planning and Management Project, "Country Environmental Profiles, Natural Resource Assessments and Other Reports on the State of the Environment," Washington, DC, May 1986.

AID has recently developed a concept paper to explore the desirability of establishing a diversity project within AID's Bureau of Science and Technology. Benefits of such a project include centralizing access to funding and perhaps expertise on biological diversity. The preliminary nature of the concept paper, however, makes more critical assessment premature.

In response to AID funding cuts, staff cuts, and a move to cut management units, conservation groups have proposed several ways to loosen up money for biological diversity projects (2,6). Of particular interest are calls for greater use of Public Law 480 funds for conservation projects. This option has both precedence (52) and the potential to increase activities in this area. It would enable a relatively small dollar amount to be supplemented with larger amounts of foreign currency. The use of excess foreign currencies by the U.S. Fish and Wildlife Service (discussed later in this chapter) provides further opportunities.

Matching grants provided to conservation organizations offers another cost-effective way to promote projects. AID matching grants to World Wildlife Fund-U.S. for its Wildlands and Human Needs Projects and to The Nature Conservancy International for its network of Conservation Data Centers are good examples of such public/private cost-sharing initiatives.

Another constraint to implementing Section 119 is the lack of adequately trained personnel in environmental sciences within AID (6,10,67). Although AID designates an environmental officer at each mission, the person may have little background in environmentally related issues. The duties of an environmental officer are included with numerous other duties; few AID personnel are full-time environmental officers.

The agency could recruit personnel with environmental science backgrounds and provide further training to officers to address this problem. Developing-country professionals could also be enlisted as environmental officers within the missions. This action would be consistent with recent agency emphasis on reducing the U.S. presence in AID missions for economic as well as security reasons.

Taking advantage of expertise that exists within other U.S. agencies (e.g., National Park Service, Fish and Wildlife Service, the National Oceanic and Atmospheric Administration, Smithsonian Institution, and Peace Corps) could also significantly enhance the effectiveness of development assistance. The agency already has a Resource Services Support Agreement with the U.S. Department of Agriculture to provide forestry expertise and services. Such mechanisms can be used to establish a formal agreement with agencies such as the Department of the Interior to provide AID missions with access to conservation expertise. In addition, other agencies such as the Peace Corps are already supporting some projects in the field that focus on biological diversity. Increased collaboration between AID and the Peace Corps can be mutually beneficial.

Section 119 states the following:

> . . . whenever feasible, the objectives of this section shall be accomplished through projects managed by appropriate private and voluntary organizations, or international, regional, or national nongovernmental organizations [NGOs] that are active in the region or country where the project is located.

A number of NGOs are already working with AID in developing capacity to maintain diversity in developing countries. These include important initiatives in the areas of conservation data centers, of supporting development of national conservation strategies, and of implementing field projects. AID is also using a private NGO to maintain a listing of environmental management experts. Such partnership could continue to be encouraged by Congress through oversight hearings, for instance.

THE ROLE OF MULTILATERAL DEVELOPMENT BANKS

Multilateral development banks (MDBs) are the largest providers of development assistance and have considerable influence on development policy and financing. In this capacity, they are uniquely situated to influence environmental aspects of development (40). In 1983, the World Bank, the Inter-American Development Bank, the African Development Bank, and the Asian Development Bank in 1983 loaned at least $20 billion to fund projects in developing countries—nearly three times the amount committed by the U.S. Agency for International Development, the largest bilateral agency. Funds loaned by MDBs are supplemented by larger amounts from governments of recipient countries, and many projects receive cofinancing from other development agencies and private banks. For every dollar loaned by the World Bank, for example, more than 2 additional dollars are raised from other sources (41).

Many countries modify their development policies in response to MDB suggestions and pressures. An important element is the developing-country sector work of the MDBs-policy documents produced as background material to help identify priorities in lending.

MDB's influence on policy can be the single most important influence in many countries on the development model adopted (41). Because agricultural, rural development, and energy policies can have profound effects on habitats, diversity in developing countries can be significantly affected by MDB policies.

The most immediate effect of MDBs on maintaining biological diversity may be support for creating protected areas. The World Bank has been the leader among development banks in this area—the bank has financed the protection of 59,000 square kilometers in 17 countries. It has funded entire conservation projects—for instance, a wildlife reserve and tourism project in Kenya. More often, it has included conservation components in larger projects—for instance, a protected area in conjunction with an irrigation project in Indonesia. In this case, the designated area protects tropical forest and wildlife while providing key watershed management services. Even conservation components that represent a small fraction of a project's total cost can play a substantial role in preserving diversity.

The performance of MDBs in preserving diversity depends on their more general environmental policies and the degree to which these policies are implemented. In this regard, the banks have issued statements emphasizing the need for sound environmental management projects.

The World Bank, the Inter-American Development Bank, the Asian Development Bank, and six other multilaterals in 1980 signed a "Declaration of Environmental Policies and Procedures Relating to Economic Development." As a result, these organizations formed the Committee of International Development Institutions on the Environment (CIDIE), under the auspices of the United Nations Environment Programme (UNEP). CIDIE has met five times since 1980 to exchange information on progress and plans of MDBs for improving their environmental performance. Under the terms of agreement, the agencies will perform systematic environmental analyses of activities, fund programs and projects designed to solve environmental problems, manage resources sustainably, and provide support for improving environmental policymaking institutions and their capacity to implement environmental controls in developing countries.

A study prepared by the International Institute for Environment and Development for the fourth CIDIE meeting found, however, that the commitment of MDBs to sound environmental management in development projects was not effectively translated into action. The study came to the following conclusions:

> The fact that we found so little evidence of the application of existing guidelines suggests that either they have been tried and found useless, or that agencies have not made sufficient resources and incentives available to sustain their use. We suggest that some agencies never put some guidelines into operation because their function is to improve public relations. . . . In many cases, staff do not use guidelines because agencies do not require their use, nor provide appropriate training and resources, nor establish any institutional penalties for failing to use them (16).

A number of congressional hearings have brought to light evidence of serious ecological problems resulting from projects supported by MDBs (54,55,56,57). Through testimony presented at these hearings, several categories of projects were identified that may directly contribute to large-scale environmental destruction. Categories cited as problematic included large-scale cattle ranching (especially in the tropics), hydroelectric power projects and irrigation systems, and resettlement projects (41). Evidence of low economic returns and high environmental costs associated with a number of these projects suggest that greater scrutiny of environmental impacts should be applied before MDBs provide financing.

Following these hearings, the House Subcommittee on International Development Institutions and Finance issued a series of recommendations to the U.S. Treasury Department, in effect proposing a U.S. environmental policy for MDBs (41). These recommendations were largely supported by the Treasury Department, the lead Federal agency for U.S. participation in these organizations. Included were calls for increased environmental staffing and mandatory procedures for project review, and for the U.S. executive directors of the MDBs to try to modify or oppose projects that would erode the natural resource base. Recommendations also emphasized the needs for institution-building and training in conservation, improved management of protected areas, involvement of indigenous peoples in development planning, and withdrawal of support from projects that cause extensive damage to habitats in species-rich areas.

Because U.S. influence in MDBs has traditionally been strong, a concerted effort from the U.S. executive directors no doubt could improve MDB environmental performance and make significant contributions to maintaining biological diversity. Emergence of a Wildlands Management Policy at the World Bank may, in part, reflect congressional and public attention on the subject. The recently approved policy sets guidelines for the management of natural areas in bank projects. These include avoiding conversion of wildlands of special concern, giving preference to using already converted lands, compensating for the loss of wildlands by setting aside similar areas, and preserving relevant wildland areas.

To maintain momentum, however, continued congressional oversight and input from U.S. executive directors is likely to be needed, such as in efforts to enlist greater environmental expertise within the banks. Language contained in the fiscal year 1986 appropriations bill clearly reflects congressional interest on this subject (21).

Consideration could also be given to promoting the approach to diversity maintenance embodied in the recent World Bank policy. To this end, U.S. representatives could be encouraged to establish a similar approach within CIDIE.

PROMOTION OF CAPACITY AND INITIATIVES IN DEVELOPING COUNTRIES

A large number of initiatives at the international level have addressed various aspects of diversity maintenance in developing countries (see ch. 10). These range from international meetings to treaties and conventions such as the Convention on International Trade in Endangered Species of Wild Flora and Fauna. Such initiatives can be important in raising awareness of the issue and of national responsibilities. They can effectively set standards, monitor progress, serve as promotional work, and establish legal norms (8). An international perspective also enables interested parties to define global priorities. However, translating these initiatives into concrete activities requires that they be implemented and supported at the national and local levels, underlining the importance of developing national capacities and constituencies to address loss of diversity.

The responsibility for maintaining biological diversity within a country's borders ultimately falls on national governments. Yet it can be argued that national governments have responsibilities to the international community. Avoiding loss of genetic resources that may meet the needs of future generations and maintaining diversity because it represents the biological heritage of the planet are commonly heard arguments in this regard.

These arguments may be insufficient or unconvincing for many developing countries, especially when national resources would have to be devoted to maintaining diversity, yet the benefits would accrue outside their borders. In other cases, a country may acknowledge its national interests in maintaining diversity but lack the resources—both financial and technical—to stem the loss.

Six priority areas where U.S. bilateral assistance could promote abilities and initiatives in developing countries have been identified: building public support, establishing an information base, building institutional support, promoting planning and management, increasing technical capacity, and increasing economic benefits derived from wild species. Although described separately, these areas are mutually reinforcing.

Building Public Support

Creating a favorable climate of public opinion is critical to the success of conservation programs. Developing countries commonly lack an organized base of citizen support; in the few cases where support has existed, in Ecuador for example, it has been a key element in efforts to launch programs.

A study of poor farmers in Costa Rica found that:

> . . . farmers could not comprehend the concept of "untouchable" forest reserves. The values of outdoor recreation, wildlife, and biological diversity may be seen by wealthy policymakers . . . but they are generally alien to poor farmers struggling for survival (45).

Consequently, efforts to protect habitats may depend on demonstrating to rural populations that they will benefit from such activities and on soliciting their support in project design and implementation (36).

Benefits to those in rural areas can be in the form of actual financial compensation, as in the Amboseli game reserve in Kenya. Here, Masai pastoralists participated in designing a conservation program, and they now benefit financially from the arrangement through tourist revenues and through employment opportunities (63). Alternatively, local support can be solicited by convincing people of the importance of maintaining diversity. In Malaysia, for example, public support was marshaled to protect the Batu caves from quarrying by pointing out that the durian, a highly valued fruit crop, depends on cave-nesting bats for pollination (36).

The opening of the Kuna Indian Udirbi Tropical Forest Reserve, a 5,000-acre park on Panama's Atlantic coast, resulted from integrating local peoples' desire to protect a forested area of cultural and religious importance with the establishment of income-generating facilities for visiting scientists and naturalists. The project is unusual because it was initiated by the Kuna themselves and had unanimous support. A number of organizations (including the Centro Agronomico Tropical de Investigacion y Ensenanze, the Smithsonian Tropical Research Institute, AID, the Inter-American Foundation, and the World Wildlife Fund-U.S.) have provided technical and financial support, although both the benefits and management responsibilities are being directed toward the Kuna (41,69).

Emphasis on environmental education is another strategy for building public support (36). A major constraint at all school levels is the shortage of appropriate teaching materials in local languages (67). Furthermore, most textbooks use examples drawn from temperate zone ecosystems, which can be difficult for students in the tropics to understand. Development of teaching materials could help remedy this.

In Costa Rica, the World Wildlife Fund's conservation and education program, working with the Ministry of Education and educators and conservationists from local universities, developed educational material in Spanish for elementary school ecology courses. The material was tested by 70 teachers in 11 schools, reaching 2,000 students in 1982. The success of the program led to its adoption by the Ministry of Education and to the distribution of materials to all public elementary schools in the country in 1984. World Wildlife Fund expanded the program into Colombia and Honduras in 1984 and to Brazil and Guatemala in 1985 (4).

Mobilizing public support through mass information campaigns has also been successful in developing countries. In Malaysia, for example, numerous private voluntary and nongovernmental organizations, such as the

Photo credit: G. Lieberman

An education project in Costa Rica funded by the World Wildlife Fund allows elementary school students to study ecology with textbooks in their native language. Above, sixth graders study relationships between different plant species. The program, begun in 1982, has been expanded to Colombia, Honduras, Guatemala, and Brazil.

Malayan Nature Society, the Friends of the Earth, and the Consumers' Association of Penang, conduct information programs to develop public understanding (1).

In a number of campaigns, flagship species are identified. These are species with high esthetic appeal that are often endemic to a country, and consequently capable of generating public interest and pride in the nation's biota. For instance, the yellow-tailed woolly monkey —Peru's largest and most endangered primate— is the centerpiece of a campaign to protect its cloud forest habitat in a project begun in 1984 by the World Wildlife Fund-U.S. in conjunction with the Natural History Museum of Lima and the Peruvian Conservation Foundation (69). Although this approach has been criticized for focusing inordinate attention on large mammals at the expense of other endangered taxa, it has been effective in rallying public support around certain species, promoting public awareness and in the process protecting other endangered species through habitat preservation.

Support for indigenous private and voluntary organizations has also been identified as an important component of building public support. Bolstering such organizations can create reliable recipients and managers of conservation funding with the potential of becoming self-supporting, a national constituency for exerting pressure on decisionmakers, public awareness for biological diversity, and a grassroots capacity to respond quickly and flexibly where governments cannot or will not (13,59). Monitoring development projects for undesirable environmental impacts is another important role for these groups.

Experience has shown, however, that this approach has certain constraints (18,59,60). These include saturating particular groups with funding and distorting the natural growth of these small organizations. AID, as a large agency usually dealing with large amounts of money, may be reluctant to initiate contact with many small organizations to promote small-scale projects. These concerns can be addressed by working more closely with umbrella nongovernmental organizations (e.g., the Environmental Liaison Centre in Nairobi) or through American groups that have local counterparts or affiliates in developing countries. Another option is to have agencies with more experience working at the grassroots level (e.g., the Peace Corps or the Inter-American Foundation) take a lead in this area.

Establishing an Information Base

Conducting an inventory and monitoring the biota are two key steps that facilitate corrective action in situations where human activity threatens diversity (5). An inventory can combine a traditional biological survey with the most modern technology such as remote sensing. It might also simply involve pulling together information on the status, distribution, and threats to major ecosystems and species to determine conservation priorities and affect land-use decisions.

Monitoring biological diversity refers to surveillance of the distribution and abundance of flora and fauna. The purpose is to detect adverse impacts on species or habitats, assess the extent to which human activities are responsible, and then promote corrective measures wherever possible (5).

Although nationally instituted programs to conduct inventories and monitor biological diversity are rare, a few examples do serve as models. The Mexican National Research Institute for Biological Resources (INIREB, from the full title in Spanish), for instance, prepares an inventory of plant and animal resources, studies threatened and endangered species, establishes reserves and protects habitats of ecological importance, develops alternative land-use strategies, and trains professionals in conservation-oriented fields. The range of activities undertaken by INIREB indicates the balanced approach of this organization.

Promoting national or regional databases to monitor biological diversity is an effective way to synthesize information and help define research and conservation priorities. A number of international organizations have developed databases of use to governments, assistance agencies, and conservation organizations. Still, promoting in-country capacity for such activities is an important goal. First, these databases can provide a finer evaluation (i.e., of higher resolution), defining local priorities within a regional context, than is possible with information covering larger areas. Second, the process can foster in-country expertise and bolster environmental effectiveness.

A major initiative to develop country-level Conservation Data Centers (CDCs) in Latin America and the Caribbean is currently being undertaken by The Nature Conservancy International (TNCI). CDCs are modeled on the State Heritage Programs begun 15 years ago in the United States. To date, six CDCs have been established in partnership with local institutions, with plans to expand this to 35 programs by the end of the decade. In terms of bolstering national capacity, the strengths of CDC programs lie in their employment of scientists (a zoologist, a botanist, an ecologist, and a data handler); their emphasis on institutionalizing the system; and their pressure to have local collaborating agencies adopt operational funding after 3 to 5 years (13).

The CDC programs devote little attention, however, to public education components. Furthermore, although the programs assemble existing information difficult for foreign institutions (e.g., from world museums and herbaria), they do little to provide new information in a region where at least five-sixths of the organisms are unknown (38). Overall these programs are very useful in identifying areas of conservation interest. Accordingly, the U.S. Fish and Wildlife Service has contracted with TNCI to develop databases on distribution of natural plant communities and to identify areas of high endemism and diversity in Latin America (25).

In lieu of formal CDCs, which could take considerable time, resources, and effort to disseminate broadly, some developing countries could benefit from more modest systems (35). A simple computer in the office in a ministry or university could record existing studies and represent a major improvement in national capacity.

Inventorying and monitoring biological resources are also important in maintaining genetic diversity among domesticated species. The rate at which farmers are replacing traditional, genetically diverse crop varieties with more uniform, high-yielding varieties is the subject of much concern in industrial and developing countries. Considerable effort to collect and store germplasm has already been made for major crop varieties, with less done for minor crops and wild relatives.

Efforts have been made to collect data, including prototypes for national databases, on threatened breeds of livestock in developing countries (12). But, information on genotype loss is inadequate to focus initiatives. USDA could provide assistance in this area through increased support to the FAO and the International Board for Plant Genetic Resources, for example, to help develop abilities to monitor losses of livestock and crop genetic resources.

Building Institutional Support

The greatest obstacles to addressing the loss of diversity are less technical than economic and political. Consequently, building institutional capacity—in both the public and private sectors—is of paramount importance. However, institution-building through development assistance is a difficult process that requires both long-term commitment and a strong appreciation of national sovereignty.

Concern about the environment is a relatively new addition to the political agendas of developing countries—for many, it dates to the 1972 U.N. Conference on the Human Environment held in Stockholm, Sweden. At that time, much of the attention on environmental problems in developing countries was generated from outside, notably from industrial countries. Most lacked a national constituency among government agencies, scientists, environmental groups, or the general public that perceived a threat stemming from degradation of the environment (17).

A great deal has changed since then. The Stockholm Conference accentuated pollution problems and the need for industrial standard setting—concerns most developing country governments felt were industrial country problems (23). Since then, environmental concerns have broadened to emphasize conservation of natural resources. Developing countries are on average six times more dependent on a productive resource base—soils, fisheries, and forests—which provides rationale for greater developing country concerns in this area (43).

Discussions on environmental issues are now being initiated by developing countries. The number of environmental agencies has increased since 1972 from about one dozen to 110 (43). However, most agencies have been ineffective in addressing environmental concerns. This ineffectiveness is due to the constraints discussed earlier, including a lack of personnel, training, and resources; an inability to compete with established interests; and a lack of legal authority.

Encouraging the development of institutional capacity is not easy, but U.S. development assistance agencies have the experience and the legal mandate to help in the process. Initiatives to enhance the stature, effectiveness, and resources of agencies responsible for conservation have been identified (10). These initiatives include requiring developing country officials to submit comments on environmental and natural resource aspects of U.S. development assistance projects and soliciting greater input from ministries in AID's development of country environmental profiles and natural resource assessments (10).

The process of infusing an awareness of biological resources in overall development planning was an objective in an AID-supported natural resources profile undertaken by the Thai Development Research Institute—a national policy analysis group (22). The process is important because it involves identification of needs and responsibilities of the 24 agencies in Thailand responsible for natural resources. Ultimately, the profile should be incorporated into the country's 5-year development plan.

An environmental profile of Paraguay illustrates the importance of the process, as much as the product, for infusing awareness of biological diversity throughout a country's institutions (66). This AID-supported project, carried out by the National Planning Secretariat of the Presidency, involved some two dozen Paraguayan scientists, technicians, and other specialists. The emphases on increasing reliance on national scientists and policymakers, on a broad intersectoral approach, and on support from the highest levels of government are keys to meeting the objectives of building institutions.

Promoting Planning and Management

As pressures on natural resources in developing countries increase, the need to integrate conservation and development interests will become more critical. Planning and management strategies should be included in resource development initiatives—from habitat protection onsite to germplasm storage offsite—and these initiatives should consider wild species as well as domesticates.

Developing a national strategy to conserve biological diversity should account for the mixed objectives for maintaining the array of species, and the mixed status of these groups (29). A biological continuum of ecosystems, species, populations, and varieties fills various needs, and various management programs and techniques are appropriate. Consequently, management objectives and technologies and the links between them should be taken into account, as well as the most urgent problems to address (29).

One activity that addresses this problem is the development of national conservation strategies (NCSs), which are general policy statements on the role of conservation in development planning (19). AID began support of an NCS for Nepal in fiscal year 1985 through the International Union for the Conservation of Nature and Natural Resources (IUCN), and it is continuing to assist in the preparation or implementation of similar strategies for Sri Lanka, the Philippines, and Zimbabwe (52). Although the general nature of these documents may limit their usefulness in implementing specific projects, they can be important vehicles for presenting the case for maintaining biological diversity (evidenced by the NCS for Zambia) (44).

The lack of management plans for specific protected areas has been identified as a major problem in almost all developing countries. Without them, most areas suffer from inappropriate development, sporadic and inconsistent management, and lack of clearly defined management objectives. Ways to develop such plans have been proposed and are being applied to six major protected areas: Amboseli, Kenya; Simen Mountains, Ethiopia; Sapo, Liberia; Khao, Thailand; Sinharaja, Sri Lanka; and Amboro, Bolivia (44). A country may also analyze its existing parks and protected areas to develop plans for an orderly allocation of natural areas (44). Although few examples of such plans exist, methods for doing this analysis have also been developed. Systems are currently in place in Brazil, Indonesia, and Dominica (44).

In situ genebanks have received some attention as a way to conserve gene pools of wild economic plants (see ch. 5). The strategy has particular relevance for developing countries, where most of the ancestral stock of current economic species occurs. General guidelines for managing such units have been developed (34). Sri Lanka (for wild medicinal plants), India (for citrus and sugarcane), and Mexico (for teosinte) have either prepared or are developing plans for *in situ* genebanks. Efforts are under way to expand this strategy to tropical South America (35).

Maintaining diversity through traditional parks and protected areas is becoming difficult for some nations for economic and political reasons, and it is likely to become less common in the future. Setting aside land for a single use can often be an economic impossibility. Some nations, particularly small countries and islands, do not have the large, undisturbed tracts of land. The trend is toward integrating reserves as part of overall development plans, rather than adding them later as areas separate from development.

Few approaches, however, have considered the role of human activities in ecological processes affecting protected areas (see ch. 5 for further discussion). Strategies for conserving diversity are starting to consider this. Conservationists are beginning to promote strategies that surround protected areas with zones of compatible land use (such as the UNESCO biosphere reserve program) and to encourage the use of regional plans to manage resources (such as the Organization of American States' integrated regional development planning).

The potential of botanic gardens and zoological gardens as a management tool in developing countries is unclear, but it could be enhanced through links with other institutions and with existing international networks (see ch. 10) (24). These institutions occupy a unique position because of their links between onsite and offsite efforts. One example proving successful is the Rio de Janeiro Primate Center that is involved with the captive breeding and reintroduction of the golden lion tamarin (69).

Concern over loss of agriculturally important resources suggests a need to devote more attention to better management of germplasm collection, storage, and use in developing countries. Preliminary studies have been conducted on the feasibility of enhancing national programs in animal germplasm maintenance. A number of obstacles have been identified: technical constraints, problems of isolation of breeds, disease control, funding sources for long-term facilities, and political concerns, such as where to locate genebanks and who owns them (15).

As mentioned earlier, several regional institutions have already identified threatened breeds of livestock and maintained data on them. This work is also a starting point for enhancing regional capacities to develop offsite storage facilities. These institutions, which could benefit from financial or technical support, include the Inter-African Bureau for Animal Resources in Nairobi, Kenya; International Livestock Centre for Africa in Addis Ababa, Ethiopia; Asociacion Latinoamericana de Produccion Animal in Maracay, Venezuela; and the Society for the Advancement of Breeding Research in Asia and Oceania in Kuala Lampur, Malaysia (39).

The number of crop genetic resource programs in developing countries has increased dramatically over the last decade. In part, this increase reflects an awareness of the importance of collecting, maintaining, and evaluating plant germplasm as a prerequisite to meeting future food requirements. Much of the change is also credited to the International Board for Plant Genetic Resources (IBPGR), which has played a catalytic role in encouraging and supporting national genebanks.

Ten years ago, only a handful of genebank collections existed, primarily in industrial countries. As of 1985, 72 countries—45 of them in the developing world—had long- or medium-term germplasm storage facilities in operation or under construction (33). IBPGR currently has agreements with 31 countries (25 of them developing ones) to serve as international base collections for long-term storage of plant germplasm. As the network of long-term collections approaches its goal of 50, covering 40 major crops before the end of the century, greater attention will be focused on bolstering medium-term collections, 100 of which have already been identified. Facilitating medium-term collections is particularly important for those developing countries where the costs and technical requirements make the establishment of long-term facilities impractical.

The operation and effectiveness of various national plant germplasm programs is uneven. Particularly disconcerting has been the failure of some national programs to respond to an IBPGR Seed Storage Advisory Committee recommendation to rectify inadequacies and improve scientific standards at existing facilities (65).

Increasing Technical Capacity

The availability of trained personnel is another constraint to conservation. The problem has been studied intensively in the Latin American region and in Africa since the mid-1970s (11,31,46,47,48,68). However, neither governments nor international or bilateral development assistance agencies have come forward with sufficient funding to meet the needs outlined in these studies.

For a total of 50 developing countries, there are only six technical colleges established to meet regional training needs for protected area managers: at Bariloche in Argentina, the Centro Agronomico Tropical de Investigacion y Ensenanze in Costa Rica, the Ecole de Fauna in Cameroon, the College of African Wildlife Management in Tanzania, the Wildlife Institute of India in Dehra Dun, and the School of Conservation Management in Indonesia at Bogor (44). Most of these colleges need external support, and all could be encouraged to augment biological diversity concerns in their curricula.

The efforts of several U.S. Federal agencies to provide training, technical assistance, and distribution of technical information hold potential for increasing technical capacity in developing countries. Those involved include the U.S. Fish and Wildlife Service (FWS), National Park Service (NPS), U.S. Forest Service, the Smithsonian Institution, and National Oceanic and Atmospheric Administration. Activities have been outlined in several documents (e.g., ref. 61). For example, congressional legislation to implement the Western Hemisphere Convention directs FWS to devote attention to personnel development in Latin America. This development has been accomplished through several initiatives, with special emphasis on training wildlife biologists, where possible, through in-country workshops. The Foreign Service Currency Program allows FWS to provide training in Egypt, India, and Pakistan. Authorized in Section 8(a) of the Endangered Species Act, this program allows excess foreign currencies to be used toward conserving threatened or endangered species in those countries (25).

AID and other government agencies have developed cooperative arrangements with several U.S. universities, other scientific institutions (e.g., botanic and zoological institutions), and private conservation organizations. These arrangements provide avenues to direct assistance funding toward increasing technical capacity and training of country personnel.

The University of Michigan, through funding from Federal agencies (e.g., NPS), has international seminars that provide training in areas such as park management, forest management, and coastal-marine management. FWS has undertaken several projects with World Wildlife Fund-U.S. to promote expertise in species and habitat conservation. The University of Florida, in conjunction with a program offered by the National Zoo's Conservation and Research Center in Front Royal, VA, provides hands-on research and training to developing-country students (4).

U.S. development assistance could promote technical training through national and regional germplasm conservation and storage programs. Although most of the support for training currently comes from international organizations, principally the IBPGR and the Food and Agriculture Organization of the United Nations (FAO), USDA could enhance its activities in this area through the National Plant Germplasm System and the Forest Service. The thrust of these U.S. agency efforts, however, may be better directed at identifying areas where assistance could be channeled through existing training programs.

Specific training on conserving animal resources has been organized through FAO and UNEP. A 2-week course (taught in English) is offered through the University of Veterinary Science in Budapest, Hungary. The primary goal of this course is to provide developing-country participants with an overview of the present state of theory and practice (3). Although this type of training usefully draws attention to the importance of animal genetic resources, conservation strategies will depend on a commitment by national governments to avoid haphazard crossing of indigenous breeds and to monitor the most endangered ones (15).

Training and management are also critical for operating plant germplasm storage facilities. A 1-year graduate program in conservation and use of plant resources at the University of Birmingham in England has provided training to more than 100 developing-country scientists (14). Some graduates now direct genetic resources programs in their home countries. IBPGR has also established a training program (taught in French) at Gembloux, Belgium, and a training program to be taught in Spanish is under consideration (14). Some 500 developing-

Photo credit: International Board for Plant Genetic Resources

Genetic resources conservation requires a cadre of qualified personnel. The University of Birmingham in England has an international postgraduate program in genetic conservation.

country scientists have benefited from IBPGR-supported courses on plant genetic resource management and from internship programs at international agricultural research centers. In addition, IBPGR has helped incorporate relevant courses in universities in several developing countries (65). Despite these advances, training in genetic resource conservation and use still needs increased attention.

Increasing Direct Economic Benefits of Wild Species

One of the most forceful arguments for the need to maintain biological diversity has been the potential that wild species hold to improve the quality of human life. The examples of perennial corn and rosy periwinkle (an anti-leukemia drug) are commonly cited in the literature on this subject. For the most part, however, this rationale has been expounded by scientific, conservation, and political groups in industrial countries, where motivations as well as technologies to exploit genetic resources are comparatively well-developed.

The point has been less forcefully argued or acted on in developing countries. The reason may be because these countries have been unable to capitalize on their biological resources; the products and profits from them—for many reasons, including differences in levels of technology, research facilities, and interest—accrue elsewhere. Given that the greatest diversity of potentially important organisms is located in developing countries (e.g., centers of diversity of crop species and moist tropical forests as sources of medicinal products), enhancing the incentives for developing countries is critically important.

Various mechanisms exist to promote identification and development of biological resources in developing countries. Supporting research by developing-country scientists, such as through the AID Program in Science and Technology Cooperation (49), offers opportunities not only to promote development of indigenous biological resources but also to cultivate scientific expertise and supporting

Photo credit: United Nations/photo by S. Stokes

Crocodile farm in Papua New Guinea has potential to provide direct economic benefits and encourages protection of biological resources.

Photo credit: UN/UNDP photo 154002, S. Maines

To reduce dependence on tea, rubber, and coconut exports, Sri Lanka is promoting the cultivation of minor export crops such as citronella.

institutions as well. Ethnobotanical surveys and research represent another promising avenue for encouraging greater recognition of the importance and opportunites of maintaining biological diversity. Wildlife-based tourism and other wildlife utilization enterprises offer further possibilities. However, these should be approached with some caution to ensure that benefits actually accrue to the country and account for the interests of local populations (9).

Loss of agricultural genetic resources in developing countries is a pronounced concern. Addressing it will depend on enhancing capacity in national agricultural programs and increasing awareness of the potential of germplasm to contribute to development needs. Continued U.S. support for International Agricultural Research Centers, especially the International Board for Plant Genetic Resources, serves an important role in this regard. Bilateral programs through the U.S. Department of Agriculture, such as the one that currently exists with Mexico, could also be promoted. Accounting for the unique contributions of traditional agricultural systems will also need special attention. Ongoing research provides strong evidence on the importance and potential of these high diversity, low input systems in addressing the particular needs and limitations of most developing-country agriculturalists (42). Greater support for research in investigating and improving indigenous agricultural systems is seen as a high priority for development assistance. Increasing attention is also being addressed at incorporating traditional agroecosystems within biosphere reserves programs (30).

The prospects and promises of biotechnology have prompted a few developing countries to place a premium on developing their capacities in this field. Although biotechnology's contributions to biological diversity maintenance is mixed, the incentives it may provide developing countries to protect and develop their genetic resources argues for supporting developing-country expertise. Access to technical procedures, however, is generally restricted to countries with well-developed capabilities. A large number of developing countries could apply these technologies to exploit genetic resources if access to information, training, and technology were improved. Microbiological Research Centers, otherwise known as MIRCENS (see ch. 10), place a strong emphasis on training developing-country scientists. The recently created International Center for Genetic Engineering and Biotechnology, established by the United Nations Industrial Development Organization, also has as its main function the dissemination of these technologies to developing countries.

CHAPTER 11 REFERENCES

1. Aiken, S.R., and Leigh, C.H., "On the Declining Fauna of Peninsular Malaysia in the Post-Colonial Period," *Ambio* 14(1):15-22, 1985.
2. Baldi, P., Spivy-Weber, F., and Chapnick, B., *Foreign Assistance Funding Alternatives* (Washington, DC: National Audubon Society, 1986).
3. Bodo, I., "Report on the FAO/UNEP Training Courses on Animal Genetic Resources Conservation and Management," *Animal Genetic Resources Conservation by Management, Data Banks, and Training* (Rome: Food and Agriculture Organization of the United Nations, 1984).
4. Cohn, J., "Creating a Conservation Ethic," *Americas*, November-December 1985.
5. Collins, M., "International Inventory and Monitoring for the Maintenance of Biological Diversity," OTA commissioned paper, 1985.
6. Conservation Foundation, "Third World Progress Is Painfully Slow," *Conservation Foundation Letter* (Washington, DC: 1986).
7. Dasmann, R.F., "Ecological Principles for Economic Development: Ten Years Later," paper presented to IUCN 16th technical meeting, Madrid, Spain, Nov. 5-14. 1984.
8. Dias, C., president, International Center for Law in Development, New York, personal communications, May 1986.
9. Eltringham, S.K., *Wildlife Resources and Economic Development* (New York: John Wiley & Sons, 1984).

10. Environment and Energy Study Institute Task Force, "A Congressional Agenda for Improved Resource and Environmental Management in the Third World: Helping Developing Countries Help Themselves," Washington, DC, 1985.
11. Fahrenkrog, E., *Final Report: Study for the Establishment of an Inter-American Training Center for Management and Operations Personnel of National Parks and Similar Areas* (Washington, DC: World Wildlife Fund, 1978). *In*: Saunier and Meganck, 1985.
12. Food and Agriculture Organization of the United Nations, *Animal Genetic Resources Conservation by Management, Data Banks, and Training* (Rome: FAO, 1984).
13. Goebel, M., The Nature Conservancy International, Arlington, VA, personal communications, June 1986.
14. Hawkes, J.G., *Plant Genetic Resources*, CGIAR Study Paper No. 3 (Washington, DC: World Bank, 1985).
15. Hodges, J., "Annual Genetics Resources in the Developing World: Goals, Strategies, Management and Current Status," *Proceedings of Third World Congress on Genetics Applied to Livestock Production*, Lincoln, NE, vol. 10, July 16-22, 1986.
16. Horberry, J., "Accountability of Development Assistance Agencies: The Case of Environmental Policy," *Ecology Law Quarterly* 23:817-869, 1986.
17. Howard-Clinton, E.G., "The Emerging Concepts of Environmental Issues in Africa," *Environmental Management* 8(3):187-190, 1984.
18. International Institute for Environment and Development, "Environmental Planning and Management Project, Country Environmental Profiles, Natural Resource Assessment and Other Reports on the State of the Environment," Washington, DC, May 1986.
19. International Union for Conservation of Nature and Natural Resources, Conservation for Development Center, *National Conservation Strategies: A Framework for Sustainable Development* (Gland, Switzerland: 1984).
20. International Union for Conservation of Nature and Natural Resources/United Nations Environment Programme/World Wildlife Fund, *World Conservation Strategy* (Gland, Switzerland: IUCN, 1980).
21. Kasten, Sen. R.W., "Development Banks: Subsidizing Third World Pollution," *The Washington Quarterly*, summer 1986, pp. 109-114.
22. Kux, M., U.S. Agency for International Development, Rosslyn, VA, personal communications, June 1986.
23. Lausche, B., World Wildlife Fund-U.S., Washington, DC, personal communications, June 1986.
24. Lucas, G., and Oldfield, S., "The Role of Zoos, Botanical Gardens, and Similar Institutions in the Maintenance of Biological Diversity," OTA commissioned paper, 1985.
25. Mason, L., "International Assistance Programs of the Department of the Interior, Forest Service, and Advisory Council on Historic Preservation," U.S. Congress, House Committee on Interior and Insular Affairs, Subcommittee on Public Lands, Hearings, Oct. 8, 1985.
26. McPherson, M.P., administrator of Agency for International Development, letter to Congressman Gus Yatron, chairman, U.S. Congress, House Committee on Foreign Affairs, Subcommittee on Human Rights and International Organizations, Mar. 21, 1986a.
27. McPherson, M.P., Agency for International Development Oversight Hearing, Apr. 21, 1986, Senate Committee on Foreign Relations, Washington, DC, 1986b.
28. Mueller-Dombois, D., Karawinata, K., and Handley, L.L., "Conservation of Species and Habitats: A Major Responsibility in Development Planning." *In*: Carpenter, R. (ed.), *Natural Systems for Development: What Planners Need to Know* (New York: Macmillan Publishing Co., 1983).
29. Namkoong, G., "Conservation of Biological Diversity by *In-situ* and *Ex-situ* Methods," OTA commissioned paper, 1986.
30. Oldfield, M.L., and Alcorn, J.B., "Conservation of Traditional Agroecosystems: A Reevaluation," *Bioscience*, March 1987.
31. Organization of American States, *Final Report: Technical Meeting on Education and Training for the Administration of National Parks, Wildlife Reserves and Other Protected Areas*, Regional Scientific Technological Development Program, Sept. 25-29, 1978, Mérida, Venezuela, 1978. *In:* Saunier and Meganck, 1985.
32. Orians, G., and Kunin, W., "An Ecological Perspective on the Valuation of Biological Diversity," OTA commissioned paper, 1985.
33. Plucknett, D., Smith, N., Williams, J.T., and Anishetty, N.M., *Gene Banks and The World's Food* (Princeton, NJ: Princeton University Press, forthcoming).
34. Prescott-Allen, R., "Management of *In Situ*

Gene Banks," *Proceedings of the 24th Working Session of CNPRA*, Madrid, Spain, 1984. *In*: Thorsell, 1985.

35. Prescott-Allen, R., consultant, British Columbia, Canada, personal communications, May 1986.
36. Rambo, T., "Socioeconomic Considerations for the *In-Situ* Maintenance of Biological Diversity in Developing Countries," OTA commissioned paper, 1985.
37. Randall, A., "Human Preferences, Economics, and Preservation of Species," *The Preservation of Species: The Value of Biological Diversity*, B.G. Norton (ed.) (Princeton, NJ: Princeton University Press, 1986).
38. Raven, P., Missouri Botanical Gardens, St. Louis, MO, personal communications, June 1986.
39. Rendel, J., *Animal Genetic Resources Data Banks* (Rome: Food and Agriculture Organization of the United Nations, 1984).
40. Rich, B., "The Multilateral Development Banks, Environmental Policy and the United States," *Ecology Law Quarterly*, spring 1985.
41. Rich, B., and Schwartzman, S., "The Role of Development Assistance in Maintaining Biological Diversity *In-Situ* in Developing Countries," OTA commissioned paper, 1985.
42. Richards, P., *Indigenous Agricultural Revolution* (Boulder, CO: Westview Press, 1985).

42a. Saunier, R., and Meganck, R., "Compatibility of Development and the *In-Situ* Maintenance of Biological Diversity in Developing Countries," OTA commissioned paper, 1985.

43. Talbot, L., "Helping Developing Countries Help Themselves: Toward a Congressional Agenda for Improved Resource and Environmental Management in the Third World," working paper (Washington, DC: World Resources Institute, 1985).
44. Thorsell, J., "The Role of Protected Areas in Maintaining Biological Diversity in Tropical Developing Countries," OTA commissioned paper, 1985.
45. Thrupp, A., "The Peasant View of Conservation," *Ceres* 14(4):31-34, 1981.
46. United Nations Environment Programme, Projet de Programme d'Action pour l'Education et la Formation en Matiere d'Environnement en Afrique, UNEP/W.G. 87/2., 1983. *In*: Saunier and Meganck, 1985.
47. United Nations Environment Programme, Red Regional de Instituciones para la Formación Ambiental, Informe de Actividades de 1982-1983 y Sugerencias de Programación para 1984, UNEP/I.G. 47/4, 1984. *In*: Saunier and Meganck, 1985.
48. United Nations Environment Programme, Final Report, Fourth Inter-Governmental Regional Meeting on the Environment in Latin America and the Caribbean, Cancún, Mexico, UNEP/I.G. 57/8, 1985. *In*: Saunier and Meganck, 1985.
49. U.S. Agency for International Development, The Office of the Science Advisor, *Development Through Innovative Research* (Washington, DC: 1985).
50. U.S. Agency for International Development, *Congressional Presentation Fiscal Year 1987: Main Volume* (Washington, DC: 1986).
51. U.S. Agency for International Development, "Draft AID Action Plan on Conserving Biological Diversity in Developing Countries," prepared by Bureau of Science and Technology, Office of Forestry, Environment, and Natural Resources, Washington, DC, January 1986.
52. U.S. Agency for International Development, "Progress in Conserving Biological Diversity in Developing Countries FY1985," Washington, DC, February 1986.
53. U.S. Congress, Congressional Research Service, "U.S. Foreign Assistance in an Era of Declining Resources: Issues for Congress in 1986," prepared by L.Q. Nowels, CRS Rpt. No. 86-95F, Washington, DC, 1986.
54. U.S. Congress, House of Representatives, Committee on Foreign Affairs, Subcommittee on Human Rights and International Organizations, *U.S. Policy on Biological Diversity* (hearings on June 6, 1985) (Washington, DC: U.S. Government Printing Office, 1985).
55. U.S. Congress, House of Representatives, *Environmental Impact of Multilateral Development Bank-Funded Projects*: Hearing before the Subcommittee on International Development Institutions and Finance of the House Committee on Banking, Finance, and Urban Affairs, 98th Cong., 1st sess., 1983.
56. U.S. Congress, House of Representatives, *Draft Recommendations on the Multilateral Development Banks and the Environment*: Hearing before the Subcommittee on International Development Institutions and Finance of the House Committee on Banking, Finance, and Urban Affairs, 98th Cong., 2d sess., 1984.
57. U.S. Congress, House of Representatives, *Tropical Forest Development Projects—Status of Environmental and Agricultural Research*:

Hearing before the Subcommittee on Natural Resources, Agriculture Research and Environment of the House Committee on Science and Technology, 98th Cong., 2d sess., 1984.

58. U.S. Congress, Senate Committee on Foreign Relations, *Hearings: Issues on Biological Diversity and Tropical Deforestation,* Mar. 19, 1986 (Washington, DC: U.S. Government Printing Office, 1986).
59. U.S. Congress, Office of Technology Assessment, *Grassroots Conservation of Biological Diversity in the United States—Background Paper #1,* OTA-BP-F-38 (Washington, DC: U.S. Government Printing Office, February 1986).
60. U.S. Congress, Office of Technology Assessment, *Continuing the Commitment: A Special Report on the Sahel*, OTA-F-308 (Washington, DC: U.S. Government Printing Office, August 1986).
61. U.S. Department of State, *Conserving International Wildlife Resources: The United States Response* (Washington, DC: U.S. Government Printing Office, 1984).
62. U.S. Interagency Task Force, *U.S. Strategy on the Conservation of Biological Diversity: An Interagency Task Force Report to Congress* (Washington, DC: Agency for International Development, 1985).
63. Western, D., "Amboseli National Park: Human Values and the Conservation of a Savanna Ecosystem," *National Parks, Conservation, and Development*, proceedings of the World Congress on National Parks, Bali, Indonesia, Oct. 11-22, 1982, J.A. McNeely and K.R. Miller (eds.) (Washington, DC: Smithsonian Institution Press, 1984).
64. Western, D., "Conservation-Based Rural Development," *Sustaining Tomorrow: A Strategy for World Conservation and Development*, F.R. Thibodeau and H.H. Field (eds.) (Hanover, NH: University Press of New England, 1984).
65. Williams, J.T., "A Decade of Crop Genetic Resources Research," *Crop Genetic Resources: Conservation and Evaluation*, J.H.W. Holden and J.T. Williams (eds.) (London: George Allen & Unwin, 1984).
66. Winterbottom, R., International Institute for Environment and Development, Washington, DC, personal communications, June 1986.
67. World Resources Institute, *Recommendations for a United States Strategy To Conserve Biological Diversity in Developing Countries* (Washington, DC: 1984).
68. World Wildlife Fund, *Strategy for Training in Natural Resources and Environment* (Washington, DC: 1980).
69. World Wildlife Fund-U.S., *Annual Report 1984* (Washington, DC: WWF, 1984).
70. Yatron, G., chairman, U.S. Congress, House Committee on Foreign Affairs, Subcommittee on Human Rights and International Organizations, letter to M. Peter McPherson, administrator, Agency for International Development, Jan. 29, 1986.

Appendixes

Appendix A
Glossary of Acronyms

AAZPA —American Association of Zoological Parks and Aquariums
AID —U.S. Agency for International Development
AIDS —acquired immune deficiency syndrome
AMBC —American Minor Breeds Conservancy
APHIS —Animal and Plant Health Inspection Service (USDA)
ARS —Agricultural Research Service (USDA)
ATCC —American Type Culture Collection
BGCCB —Botanical Gardens Conservation Coordinating Body
CACs —Crop Advisory Committees (USDA)
CAMCORE—The Central America and Mexico Coniferous Resources Cooperative
CAST —Council for Agricultural Science and Technology
CDCs —Conservation Data Centers (TNCI)
CDSS —Country Development Strategy Statements
CEQ —Council for Environmental Quality
CGIAR —Consultative Group on International Agricultural Research
CIAT —Centro Internacional de Agricultura Tropical (CGIAR, Colombia)
CIDIE —Committee of International Development Institutions on the Environment
CIMMYT —Centro Internacional de Mejoramiento de Maiz y Trigo (CGIAR, Mexico)
CIP —Centro Internacional de la Papa (CGIAR, Peru)
CITES —Convention on International Trade in Endangered Species of Wild Flora and Fauna
CMC —Conservation Monitoring Center (IUCN)
CPC —Center for Plant Conservation
CRGO —Competitive Research Grants Office (USDA)
CSRS —Cooperative State Research Service (USDA)
DNA —deoxyribonucleic acid
DOE —U.S. Department of Energy
ECG —Ecosystem Conservation Group (FAO, UNEP, UNESCO, and IUCN)
ELISA —enzyme-linked immunosorbent assay
EPA —U.S. Environmental Protection Agency
ESF —Economic Support Funds (AID)
FAA —Foreign Assistance Act
FAO —Food and Agriculture Organisation (U.N.)
FNR —Forestry, Natural Resources, and the Environment Office (AID)
FWS —Fish and Wildlife Service (Department of the Interior)
FWS/ESO —Fish and Wildlife Service, Endangered Species Office (Department of the Interior)
GAO —General Accounting Office (U.S. Congress)
GEMS —Global Environment Monitoring System (UNEP)
GIS —Geographic Information Systems
GRID —Global Resources Information Database (GEMS,UNEP)
GRIN —Germplasm Resources Information Network (NPGS,USDA)
HPLC —high-performance liquid chromatography
IARCs —International Agricultural Research Centers
IBPGR —International Board for Plant Genetic Resources
ICBP —International Council for Bird Preservation
IITA —International Institute of Tropical Agriculture (CGIAR, Nigeria)
IMS —Institute of Museum Services
IPPC —International Plant Protection Convention
IRDP —Integrated Regional Development Planning (OAS)
IRRI —International Rice Research Institute (CGIAR, Philippines)
ISIS —International Species Inventory System
IUCN —International Union for the Conservation of Nature and Natural Resources
IUPOV —International Union for the Protection of New Varieties of Plants
MAB —Man in the Biosphere Program (UNESCO)
MARC —Meat Animal Research Center (USDA)
MDBs —Multilateral Development Banks
MIRCENs —Microbiological Resource Centers
NAS —National Academy of Sciences
NCS —National Conservation Strategy

NGO —nongovernmental organization
NMFS —National Marine Fisheries Service (Department of Commerce)
NMR —nuclear magnetic resonance
NOAA —National Oceanic and Atmospheric Administration (Department of Commerce)
NPGS —National Plant Germplasm System (USDA)
NPS —National Park Service (Department of the Interior)
NRRL —Northern Regional Research Laboratory (USDA)
NSF —National Science Foundation
NSSL —National Seed Storage Laboratory (NPGS, USDA)
OAS —Organization of American States
OECD —Organisation for Economic Cooperation and Development
OTA —Office of Technology Assessment (U.S. Congress)
PASA —Participating Agency Service Agreement
PBR —plant breeders' rights
PD —Policy Determination (USAID)
PNV —potential natural vegetation
PVPA —Plant Variety Protection Act
R&D —Research and Development
RFLP —restriction fragment length polymorphisms
RNA —Research Natural Area
RNA —ribonucleic acid
RPIS —Regional Plant Introduction Station (USDA)
RSSA —Resource Services Support Agreements
SAF —Society of American Foresters
SCMU —Species Conservation Monitoring Unit (IUCN)
SCS —Soil Conservation Service (USDA)
SMCRA —Surface Mining Control and Reclamation Act
SSPs —Species Survival Plans (AAZPA)
TNC —The Nature Conservancy
TNCI —The Nature Conservancy International
UN —United Nations
UNEP —United Nations Environment Programme
UNESCO —United Nations Educational, Scientific, and Cultural Organization
USDA —U.S. Department of Agriculture
WCS —World Conservation Strategy

Appendix B
Glossary of Terms

Artificial insemination: A breeding technique, commonly used in domestic animals, in which semen is introduced into the female reproductive tract by artificial means.

Biochemical analysis: The analysis of proteins or DNA using various techniques, including electrophoretic testing and restriction fragment length polymorphism (RFLP) analysis. These techniques are useful methods for assessing plant diversity and have also been used to identify many strains of micro-organisms.

Biogeography: A branch of geography that deals with the geographical distribution of animals and plants.

Biological diversity: The variety and variability among living organisms and the ecological complexes in which they occur.

Biologically unique species: A species that is the only representative of an entire genus or family.

Biosphere reserves: Established under UNESCO's Man in the Biosphere (MAB) Program, biosphere reserves are a series of protected areas linked through a global network, intended to demonstrate the relationship between conservation and development.

Biota: The living organisms of a region.

Biotechnology: Techniques that use living organisms or substances from organisms to make or modify a product. The most recent advances in biotechnology involve the use of recombinant DNA techniques and other sophisticated tools to harness and manipulate genetic materials.

Breed: A group of animals or plants related by descent from common ancestors and visibly similar in most characteristics. Taxonomically, a species can have numerous breeds.

Breeding line: Genetic lines of particular significance to plant or animal breeders that provide the basis for modern varieties.

Buffer zones: Areas on the edge of protected areas that have land use controls and allow only activities compatible with protection of the core area, such as research, environmental education, recreation, and tourism.

Captive breeding: The propagation or preservation of animals outside their natural habitat, involving control by humans of the animals chosen to constitute a population and of mating choices within that population.

Centers of diversity: The regions where most of the major crop species were originally domesticated and developed. These regions may coincide with centers of origin.

Chromatography: A chemical analysis technique whereby an extract of compounds is separated by allowing it to migrate over or through an adsorbent (such as clay or paper) so that the compounds are distinguished as separate layers.

Clonal propagation: The multiplication of an organism by asexual means such that all progeny are genetically identical. In plants, it is commonly achieved through use of cuttings or *in vitro* culture. For animals, embryo splitting is a method of clonal propagation.

Community: A group of ecologically related populations of various species of organisms occurring in a particular place and time.

Critical habitats: A technical classification of areas in the United States that refers to habitats essential for the conservation of endangered or threatened species. The term may be used to designate portions of habitat areas, the entire area, or even areas outside the current range of the species.

Cryogenic storage: The preservation of seeds, semen, embryos, or micro-organisms at extremely low temperatures, below $-130°$ C. At these temperatures, water is absent, molecular kinetic energy is low, diffusion is virtually nil, and storage potential is expected to be extremely long.

Cryopreservation: *See* cryogenic storage.

Cultivar: International term denoting certain cultivated plants that are clearly distinguishable from others by one or more characteristics and that when reproduced retain their distinguishing characteristics. In the United States, "variety" is considered to be synonymous with cultivar (derived from "cultivated variety").

Cutting: A plant piece (stem, leaf, or root) removed from a parent plant that is capable of developing into a new plant.

Cycad: Any of an order of gymnosperms of the family cycadaceae. Cycads are tropical plants that resemble palms but reproduce by means of spermatozoids.

Database: An organized collection of data that can be used for analysis.

DNA: Deoxyribonucleic acid. The nucleic acid in chromosomes that codes for genetic information.

The molecule is double stranded, with an external "backbone" formed by a chain of alternating phosphate and sugar (deoxyribose) units and an internal ladder-like structure formed by nucleotide base-pairs held together by hydrogen bonds.

Domestication: The adaptation of an animal or plant to life in intimate association with and to the advantage of man.

Ecology: A branch of science concerned with the interrelationship of organisms and their environment.

Ecosystem: An ecological community together with its physical environment, considered as a unit.

Ecosystem diversity: The variety of ecosystems that occurs within a larger landscape, ranging from biome (the largest ecological unit) to microhabitat.

Electrophoresis: Application of an electric field to a mixture of charged particles in a solution for the purpose of separating (e.g., mixture of proteins) as they migrate through a porous supporting medium of filter paper, cellulose acetate, or gel.

Embryo transfer: An animal breeding technique in which viable and healthy embryos are artificially transferred to recipient animals for normal gestation and delivery.

Endangered species: A technical definition used for classification in the United States referring to a species that is in danger of extinction throughout all or a significant portion of its range. The International Union for the Conservation of Nature and Natural Resources (IUCN) definition, used outside the United States, defines species as endangered if the factors causing their vulnerability or decline continue to operate.

Endemism: The occurrence of a species in a particular locality or region.

Equilibrium theory: A theory of island biogeography maintaining that greater numbers of species are found on larger islands because the populations on smaller islands are more vulnerable to extinction. This theory can also be applied to terrestrial analogs such as forest patches in agricultural or suburban areas or nature reserves where it has become known as "insular ecology."

Exotic species: An organism that exists in the free state in an area but is not native to that area. Also refers to animals from outside the country in which they are held in captive or free-ranging populations.

***Ex-situ*:** Pertaining to study or maintenance of an organism or groups of organisms away from the place where they naturally occur. Commonly associated with collections of plants and animals in storage facilities, botanic gardens, or zoos.

Extinct species: As defined by the IUCN, extinct taxa are species or other taxa that are no longer known to exist in the wild after repeated search of their type of locality and other locations where they were known or likely to have occurred.

Extinction: Disappearance of a taxonomic group of organisms from existence in all regions.

Fauna: Organisms of the animal kingdom.

Feral: A domesticated species that has adapted to existence in the wild state but remains distinct from other wild species. Examples are the wild horses and burros of the West and the wild goats and pigs of Hawaii.

Flora: Organisms of the plant kingdom.

Gamete: The sperm or unfertilized egg of animals that transmit the parental genetic information to offspring. In plants, functionally equivalent structures are found in pollen and ovules.

Gene: A chemical unit of hereditary information that can be passed from one generation to another.

Gene-pool: The collection of genes in an interbreeding population.

Genetic diversity: The variety of genes within a particular species, variety, or breed.

Genetic drift: A cumulative process involving the chance loss of some genes and the disproportionate replication of others over successive generations in a small population, so that the frequencies of genes in the population is altered. The process can lead to a population that differs genetically and in appearance from the original population.

Genotype: The genetic constitution of an organism, as distinguished from its physical appearance.

Genus: A category of biological classification ranking between the family and the species, comprising structurally or phylogenetically related species or an isolated species exhibiting unusual differentiation.

Germplasm: Imprecise term generally used to refer to the genetic information of an organism or group of organisms.

Grow-out (growing-out): The process of growing a plant for the purpose of producing fresh viable seed to evaluate its varietal characteristics.

Habitat: The place or type of site where an organism naturally occurs.

Hybrid: An offspring of a cross between two genetically unlike individuals.

Inbreeding: Mating of close relatives resulting in increased genetic uniformity in the offspring.

***In-situ*:** Maintenance or study of organisms within an organism's native environment.

***In-situ* gene banks:** Protected areas designated specifically to protect genetic variability of particular species.

Interspecies: Between different species.

Intrinsic value: The value of creatures and plants independent of human recognition and estimation of their worth.

Inventory: Onsite collection of data on natural resources and their properties.

***In vitro*:** (Literally "in glass"). The growing of cells, tissues, or organs in plastic vessels under sterile conditions on an artificially prepared medium.

Isoenzyme (Isozyne): The protein product of an individual gene and one of a group of such products with differing chemical structures but similar enzymatic function.

Landrace: Primitive or antique varieties usually associated with traditional agriculture. Often highly adapted to local conditions.

Living collections: A management system involving the use of offsite methods such as zoological parks, botanic gardens, arboretums, and captive breeding programs to protect and maintain biological diversity in plants, animals, and microorganisms.

Micro-organisms: In practice, a diverse classification of all those organisms not classed as plants or animals, usually minute microscopic or submicroscopic and found in nearly all environments. Examples are bacteria, cyanobacteria (blue-green algae), mycoplasma, protozoa, fungi (including yeasts), and viruses.

Minor breed: A livestock breed not generally found in commercial production.

Modeling: The use of mathematical and computer-based simulations as a planning technique in the development of protected areas.

Morphology: A branch of biology that deals with form and structure of organisms.

Multiple use: An onsite management strategy that encourages an optimum mix of several uses on a parcel of land or water or by creating a mosaic of land or water parcels, each with a designated use within a larger geographic area.

Native: A plant or animal indigenous to a particular locality.

Offsite: Propagation and preservation of plant, animal, and micro-organism species outside their natural habitat.

Onsite: Preservation of species in their natural environment.

Open-pollinated: Plants that are pollinated by physical or biological agents (e.g., wind, insects) and without human intervention or control.

Orthodox seeds: Seeds that are able to withstand the reductions in moisture and temperature necessary for long-term storage and remain viable.

Pathogen: A specific causative agent of disease.

Phenotype: The observable appearance of an organism, as determined by environmental and genetic influences (in contrast to genotype).

Phytochemical: Chemicals found naturally in plants.

Population: A group consisting of individuals of one species that are found in a distinct portion of the species range and that interbreed with some regularity and therefore have a common set of genetic characteristics.

Predator: An animal that obtains its food primarily by killing and consuming other animals.

Protected areas: Areas usually established by official acts designating that the uses of these particular sites will be restricted to those compatible with natural ecological conditions, in order to conserve ecosystem diversity and to protect and study species or areas of special cultural or biological significance.

Provinciality effect: Increased diversity of species because of geographical isolation.

Recalcitrant seeds: Seeds that cannot survive the reductions in moisture content or lowering of temperature necessary for long-term storage.

Recombinant DNA technology: Techniques involving modifications of an organism by incorporation of DNA fragments from other organisms using molecular biology techniques.

Restoration: The re-creation of entire communities of organisms closely modeled on communities that occur naturally. It is closely linked to reclamation.

Riparian: Related to, living, or located on the bank of a natural watercourse, usually a river, sometimes a lake or tidewater.

Serological testing: Immunologic testing of blood serum for the presence of infectious foreign disease agents.

Somaclonal variations: Structural, physiological, or biochemical changes in a tissue, organ, or plant that arise during the process of *in vitro* culture.

Species: A taxonomic category ranking immediately below genus and including closely related, morphologically similar individuals that actually or potentially interbreed.

Species diversity: The number and variety of species found in a given area in a region.

Species richness: Areas with many species, especially the equatorial regions.

Spectroscopy: Any of several methods of chemical analysis that identify or classify compounds based on examination of their spectral properties.

Stochastic: Models, processes, or procedures that are based on elements of chance or probability.

Subspecies: A distinct form or race of a species.

Taxon: A taxonomic group or entity (pl. taxa).

Taxonomy: A hierarchical system of classification of organisms that reflects the totality of similarities and differences.

Threatened species: A U.S. technical classification referring to a species that is likely to become endangered within the foreseeable future, throughout all or a significant portion of its range. These species are defined as vulnerable taxa outside the United States by the IUCN.

Tissue culture: A technique in which portions of a plant or animal are grown on an artificial culture medium in an organized (e.g., as plantlets) or unorganized (e.g., as callus) state. (*See* also *in vitro* culture.)

Variety: *See* cultivar.

Wild relative: Plant species that are taxonomically related to crop species and serve as potential sources for genes in breeding of new varieties of those crops.

Wild species: Organisms captive or living in the wild that have not been subject to breeding to alter them from their native state.

Wildlife: Living, nondomesticated animals.

Appendix C

Participants of Technical Workgroups

Five technical workgroups comprised of preeminent biological, physical, and social scientists from the public and private sectors provided input and reviewed much of the technical information for this study. OTA greatly appreciates the time and effort of each member of these workgroups.

VALUES WORKGROUP

Gardner M. Brown, Jr.
Department of Economics
University of Washington
Seattle, WA

Malcolm McPherson
Harvard Institute for International Development
Cambridge, MA

Margery L. Oldfield
Department of Zoology
University of Texas
Austin, TX

Holmes Rolston, III
Department of Philosophy
Colorado State University
Fort Collins, CO

PLANTS WORKGROUP

Robert Hanneman
Department of Horticulture
University of Wisconsin
Madison, WI

Thomas Orton
DNA Plant Technology Corp.
Watsonville, CA

Robert Stevenson
American Type Culture Collection
Rockvillc, MD

Jack Kloppenburg
Department of Rural Sociology
University of Wisconsin
Madison, WI

Jake Halliday
Battelle-Kettering Research Lab
Yellow Springs, OH

Gary Nabhan
Desert Botanical Garden
Phoenix, AZ

ANIMAL WORKGROUP

Betsy Dresser
Director of Research
Cincinnati Wildlife Research Federation
Cincinnati, OH

Keith Gregory
USDA/ARS
U.S. Meat Animal Research Center
Clay Center, NE

David Notter
Department of Animal Science
Virginia Polytechnic Institute
Blacksburg, VA

Warren Foote
International Sheep and Goat Institute
Utah State University
Logan, UT

Ulysses Seal
VA Medical Center
Minneapolis, MN

Dale Schwindaman
USDA/APHIS
Veterinary Services
Hyattsville, MD

NATURAL ECOSYSTEMS—U.S. WORKGROUP

Robert Jenkins
The Nature Conservancy
Arlington, VA

Bruce Jones
USDI/FWS
Office of Endangered Species
Arlington, VA

Orie Loucks
Holcomb Research Institute
Butler University
Indianapolis, IN

Hal Salwasser
US Forest Service
Washington, DC

Dan Tarlock
Chicago Kent College of Law
Chicago, IL

Nancy Foster
NOAA/Marine Sanctuaries
Washington, DC

NATURAL ECOSYSTEMS—INTERNATIONAL WORKGROUP

Raymond Dasmann
Department of Environmental Studies
University of California
Santa Cruz, CA

Barbara Lausche
World Wildlife Fund/Conservation Foundation
Washington, DC

Gene Namkoong
Department of Genetics
North Carolina State University
Raleigh, NC

Kathy Parker
Office of Technology Assessment
Washington, DC

Daniel Simberloff
Department of Zoology
Florida State University
Tallahassee, FL

Appendix D

Grassroots Workshop Participants

George Fell
Natural Lands Institute
Rockford, IL

Elizabeth Henson
American Minor Breeds Conservancy
Pittsboro, NC

Hans Neuhauser
The Georgia Conservancy, Inc.
Savannah, GA

Edward Schmitt
Brookfield Zoo
Brookfield, IL

Jonathan A. Shaw
Bok Tower Gardens
Lake Wales, FL

Kent Wheally
Seed Savers Exchange
Decorah, IA

Appendix E
Commissioned Papers and Authors

This report was possible because of the valuable information and analyses contained in the background reports commissioned by OTA. These papers were reviewed and critiqued by the advisory panel, workgroups, and outside reviewers. The papers are available in a separate six-part volume through the National Technical Information Service.[1]

VALUES

Papers	*Author(s)*
An Economic Perspective of the Valuation of Biological Diversity	Alan Randall University of Kentucky
Values and Biological Diversity	Bryan G. Norton New College of the University of South Florida
An Ecological Perspective on the Valuation of Biological Diversity	Gordon Orians and William Kunin University of Washington
Impact of Changing Technologies on the Valuation of Genetic Resources (1st draft only)	Lawrence A. Riggs Genrec
Critical Assessment of the Value of and Concern for the Maintenance of Biological Diversity	Malcolm F. McPherson Harvard Institute for International Development

MANAGED SYSTEMS—PLANTS

Status and Trends of Wild Plants	Hugh Synge Conservation Monitoring Center
Status and Trends of Agricultural Crop Species	Christine Prescott-Allen PADATA, Inc.
Diversity and Distribution of Wild Plants	Marshall Crosby and Peter H. Raven Missouri Botanical Garden
New Technologies and the Enhancement of Plant Germ Plasm Diversity	Tom J. Orton DNA Plant Technology Corp.
Plant Germplasm Storage Technologies	Leigh Towill, Eric E. Roos, and Phillip C. Stanwood National Seed Storage Laboratory
Technologies To Evaluate and Characterize Plant Germplasm	Norm F. Weeden and David A. Young Cornell University
Assessment of Plant Quarantine Practices	Robert P. Kahn USDA-APHIS-PPQ
Assessment of Market Institutional Factors Affecting the Plant Breeding System	Frederick A. Bliss University of Wisconsin
Historical Analysis of Genetic Resources Valuation	Jean Pierre Berlan-Universite Daix Marsaille Richard Lowentin-Harvard University
Technologies To Maintain Microbial Diversity	Jake Halliday and Dwight D. Baker Battelle-Kettering

[1]The National Technical Information Service, 5285 Port Royal Rd., ATTN: Sales Department, Springfield VA 22161 (703) 487-4650.

MANAGED SYSTEMS—ANIMALS

Status and Trends of Wild Animal Diversity	Nate Flesness ISIS—Minnesota Zoo
Status and Trends of Domesticated Animals	Hank A. Fitzhugh, Will Getz, and F.H. Baker Winrock International
Technologies To Maintain Animal Germplasm (draft form)	Betsy Dresser—Cincinatti Wildlife Federation Stan Liebo—Real Vista International
Concepts and Strategies To Maintain Domestic and Wild Animal Germplasm	David R. Notter—Virginia Polytechnic Institute Thom J. Foose—Minneapolis Zoo
Management of Animal Disease Agents and Vectors Potentially Hazardous for Animal Germplasm Resources	Werner P. Heuschele San Diego Zoo
Constraints to International Exchange of Animal Germplasm	William M. Moulton Consultant

NATURAL ECOSYSTEMS—U.S.

Status and Trends of U.S. Natural Ecosystems	David Crumpacker University of Colorado
Geographic Information Systems As A Tool To Assist in the Preservation of Biological Diversity	Duane Marble SUNY—Buffalo
Planning and Management Techniques for Natural Systems	Herman H. Shugart University of Virginia
A Science of Refuge Design and Management	Daniel Simberloff Florida State University
Assessment of Application of Species Viability Theories	Mark L. Shaffer U.S. Agency for International Development
Diversity of Marine/Coastal Ecosystems	Maurice P. Lynch and Carleton Ray Coastal Environment Associates, Inc.
Federal Laws and Policies Pertaining to the Maintenance of Biological Diversity on Federal and Private Lands	Michael Bean Environmental Defense Fund
Mandates to Federal Agencies To Conduct Biological Inventories	Lynn Corn Congressional Research Service
Ecological Restoration As A Strategy for Conserving Biological Diversity	Bill Jordan—University of Wisconsin Arboretum Edith Allen—Utah State University Rob Peters—WWF/Conservation Foundation

NATURAL ECOSYSTEMS—INTERNATIONAL

Status and Trends of Natural Ecosystems Worldwide	Jeremy Harrison Conservation Monitoring Center
International Inventory and Monitoring for the Maintenance of Biological Diversity	N. Mark Collins Conservation Monitoring Center
The Role of Protected Areas in Maintaining Biological Diversity in Tropical Developing Countries	James W. Thorsell IUCN

Role of Traditional Agriculture in Preserving Biological Diversity	Gene Wilken Colorado State University
Compatibility of Development and the *In-Situ* Maintenance of Biological Diversity in Developing Countries	Richard Saunier and Richard Meganck Organization of American States
The Role of Development Assistance in Maintaining Biological Diversity *In-Situ* in Developing Countries	Bruce Rich and Stephen Schwartzman Environmental Defense Fund
Socioeconomic Considerations for the *In-Situ* Maintenance of Biological Diversity in Developing Countries	Terry Rambo East-West Center
Conservation of Biological Diversity by *In-Situ* and *Ex-Situ* Methods	Gene Namkoong North Carolina State University
International Institutional/Legal Frameworks for *Ex-Situ* Conservation of Biological Diversity	John Barton International Technology Development
International Laws and Associated Programs for *In-Situ* Conservation of Wild Species	Barbara Lausche World Wildlife Fund/Conservation Foundation
International Laws and Associated Programs for Conservation of Biological Diversity—Joint Options for Congress	Barton/Lausche
Living Collections (Plants and Animals)	Grenville Lucas and S. Oldfield Kew Gardens
Technologies to Maintain Tree Germplasm Diversity	Frank T. Bonner US Forest Service
Causes of Loss of Biological Diversity	Norman Myers Consultant

GRASSROOTS

Role of Grassroots Activities in the Maintenance of Biological Diversity: Living Plants Collection of North American Genetic Resources	Gary Nabhan and Kevin Dahl Native Seeds Search
Report on Grass Roots Genetic Conservation Efforts	Cary Fowler Rural Advancement Fund
An Assessment of the Conservation of Animal Genetic Diversity at the Grassroots level	Elizabeth Henson American Minor Breeds Conservancy
Grassroots Groups Concerned with *In-Situ* Preservation of Biological Diversity in the U.S.	Elliott A. Norse Ecological Society of America

Index

Index

NOTE: Page numbers appearing in italics are referring to information mentioned in the boxes.

NOTE: Page numbers appearing in italics are referring to information mentioned in the boxes.

NOTE: Page numbers appearing in italics are referring to information mentioned in the boxes.

NOTE: Page numbers appearing in italics are referring to information mentioned in the boxes.

NOTE: Page numbers appearing in italics are referring to information mentioned in the boxes.